READING AND UNDERSTANDING MULTIVARIATE STATISTICS

EDITED BY

LAURENCE G. GRIMM

AND PAUL R. YARNOLD

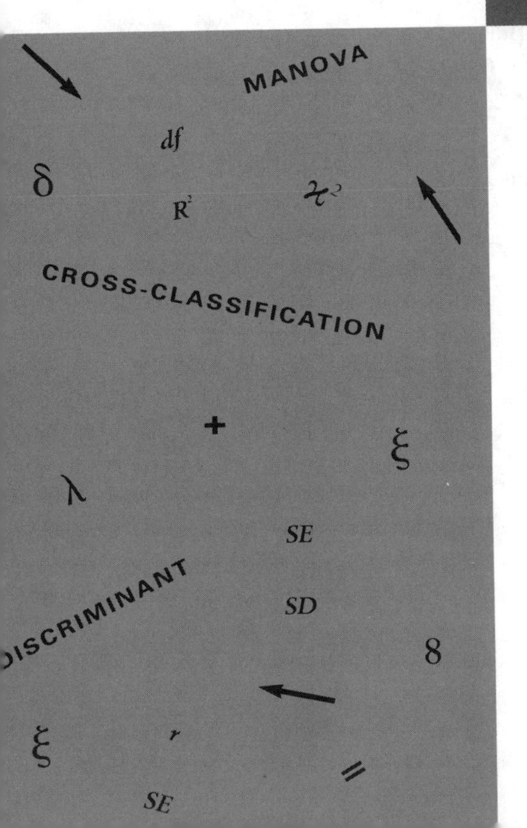

AMERICAN
PSYCHOLOGICAL
ASSOCIATION
WASHINGTON, DC

Fifth printing September 1998

Published by the
American Psychological Association
750 First Street, NE
Washington, DC 20002

Copies may be ordered from
APA Order Department
P.O. Box 92984
Washington, DC 20002

In the UK and Europe, copies may be ordered from
American Psychological Association
3 Henrietta Street
Covent Garden, London
WC2E 8LU England

This book was typeset in Futura and New Baskerville by Easton Publishing Services, Inc., Easton, MD

Printer: Data Reproductions Corporation, Auburn Hills, MI
Cover designer: Grafik Communications, Ltd., Alexandria, VA
Technical/production editor: Kathryn Lynch

Library of Congress Cataloging-in-Publication Data
Reading and understanding multivariate statistics / edited by Laurence G. Grimm and Paul R. Yarnold.
p. cm.
Includes bibliographical references and index.
ISBN 1-55798-273-2 (acid-free paper)
1. Multivariate analysis. 2. Psychometrics. I. Grimm, Laurence G. II. Yarnold, Paul R.
QA278.R43 1994
001.4′225—dc20
94-29873
CIP

British Library Cataloguing-in-Publication Data
A CIP record is available from the British Library.

Printed in the United States of America

Contents

Contributors

Fred B. Bryant, Loyola University of Chicago
Joseph A. Durlak, Loyola University of Chicago
Laurence G. Grimm, University of Illinois at Chicago
Laura Klem, University of Michigan
Mark H. Licht, Florida State University
Willard Rodgers, University of Michigan
A. Pedro Duarte Silva, Universidade Católica Portuguesa, Porta, Portugal
Loretta J. Stalans, Loyola University of Chicago
Antonie Stam, University of Georgia
Kevin P. Weinfurt, Georgetown University
Raymond E. Wright, Marianjoy Rehabilitation Hospital and Clinics, Wheaton, Illinois
Paul R. Yarnold, Northwestern University Medical School and University of Illinois at Chicago

218-234
283-289
296-305

Preface

In the last 20 years, the use of multivariate statistics in research has become commonplace. Indeed, it is difficult to find empirically based articles that do not use one or another multivariate analysis. Although the increased used of multivariate analyses has allowed researchers to answer increasingly complex research questions, many consumers of research results have been left behind.

Before 1975, most graduate programs in the behavioral sciences (e.g., psychology and education) and business did not require a course in multivariate statistics. As a consequence, behavioral scientists who attended graduate school before 1975 had to learn multivariate statistics on their own.

Even if one has been exposed to formal training in an array of multivariate techniques, it is not unusual to have forgotten all but those analyses that are used in one's immediate area of research. For example, those researchers who routinely use multiple regression may have little knowledge of discriminant analysis. It is not only active researchers who may have a narrow acquaintance with multivariate statistics. Instructors not engaged in research, practicing clinical psychologists, psychiatrists, physicians, program consultants, and graduate students in the behavioral sciences often do not possess the requisite knowledge to understand the diversity of multivariate applications that frequently appear in the literature. Consequently, behavioral scientists unfamiliar with multivariate techniques tend to skip over the Results section of a research article that uses multivariate techniques.

One solution to this problem is for everyone who is weak in multivariate analyses to take a course in multivariate statistics. However, it is extremely unlikely that a person who does not intend to use these analyses in his or her own research will devote so much time and effort to learn advanced statistics. Another solution would be to seek out a multivariate statistics text. However, most texts are not written for the consumer of research or the novice to a particular statistical method. They are much more focused on the fine details of algorithms and formulas for analyzing data.

So, another book on how to analyze multivariate data is not needed; there are already several good texts for this purpose on the market. What is needed, however, is a book that presents the fundamental conceptual aspects of multivariate techniques and acquaints the reader with the assumptions, statistical notations, and research contexts for various procedures. It is our purpose in this book to do just that.

To accomplish this, each chapter is written in a manner that allows the reader to transfer what has been learned from the chapter to an understanding of published work that uses the analysis under discussion. Consequently, when encountering an article that uses one of the multivariate analyses included in this book, the reader will be able to understand why the author used the particular analysis, the meaning of the statistical symbols associated with the analysis, the assumptions of the analysis, how to interpret summary tables or diagrams, and how substantive conclusions logically flow from the analysis. Graduate students should find the book a particularly useful supplement to their standard texts in multivariate statistics.

The chapters are written for an audience that has never had any formal exposure to multivariate statistics; a grounding in univariate statistics is necessary, however. Concepts and symbols are presented with a minimum reliance on formulas. The authors provide example applications of each statistical analysis, underlying assumptions and mechanics of the analysis, and an extensive discussion of an interesting working example. Some analyses are more complex than others and, thus, require greater persistence on the part of the reader. With careful attention, however, the reader should be able to study the chapters and achieve an understanding of many multivariate procedures without getting lost in the mathematical aspects of statistical analyses, which, mercifully, are not presented.

Although the authors have attempted to write their chapter as a self-contained treatment of a topic, it may be helpful to read some chapters before others, depending on the reader's level of expertise in multivariate statistics. Cross-reference to other chapters in the text are provided within each chapter. For example, the reader who is unfamiliar with multiple regression should read chapter 2 before tackling the chapter on path analysis.

The chapters in this book do not exhaust the entire range of multivariate analyses. On the basis of our familiarity with research in the behavioral sciences, we have selected *some* of the most common analyses. However, the topics addressed in this text apply to numerous fields of study, including psychology, business, education, medicine, biology, and criminal justice.

Among the APA staff who have been most helpful to us in bringing our efforts to fruition, we would like to thank Julia Frank-McNeil and Mary Lynn Skutley in Acquisitions and Development, who shared our vision regarding the need for a nonmathematical, conceptual treatment of multivariate statistics. Peggy Schlegel, our Development Editor, ushered the book through all stages of substantive revision. Not only did Peggy provide encouragement throughout the project, she served as an in-house reviewer and offered many helpful suggestions for improving the organization and content of each chapter. Kathryn Lynch, who copyedited the manuscript and coordinated the production stages of the book, contributed much to the final product through her editorial expertise and careful attention to detail. Victoria Boyle, our indexer, did a superb job in organizing all of the information we provided in this book.

We were also fortunate to have the input of several external reviewers, who dedicated substantial time and effort in helping us determine the kind of revisions that would be beneficial. They include Steven Breckler, Robert Cudek, Bert Green, Jr., Kevin Murphy, Deborah Schnipke, Dawn Snipes, James Stevens, and Stanley Wasserman.

Finally, we are most indebted to the contributing authors. Their expertise, patience, and responsiveness made our editorial work enjoyable and enhanced our knowledge of multivariate statistics.

Introduction to Multivariate Statistics

Laurence G. Grimm and Paul R. Yarnold

How do gender and years of education relate to income?

What dimensions underlie subjects' responses to a list of 60 mood states, or what dimensions underlie subjects' preferences for one type of car versus another?

How well can whether students will drop out of high school be predicted on the basis of information concerning gender, IQ, and number of siblings?

Within the psychotherapy outcome literature, what is the relative effectiveness of cognitive–behavioral therapy versus medication in the treatment of depression? Also, are these findings consistent across studies, and if not, what differences among the studies could explain any inconsistent findings?

Multivariate analyses, which may be used to address complex research questions such as those above, have become increasingly popular over the last 20 years. Perhaps the greatest impetus for the use of multivariate techniques has been the development of mainframe and microcomputer software packages as well as textbooks that assist researchers in the use of such software (J. P. Stevens, 1986; Tabachnick & Fidell, 1989). Complex statistical analyses that were available 20 years ago were prohibitively time-consuming if performed with only the help of a hand calculator. Now large data sets can be analyzed in seconds.

With the availability of software packages for analyzing multivariable studies, researchers are designing studies in ways that allow for deeper understanding of the complex relations among sets of variables. The days of measuring a single dependent variable in a simple, between-groups design are quickly passing. It is understandable that some researchers might yearn for the old days. The correlation matrix has given way to causal modeling or confirmatory factor analysis. The univariate *F* test has graduated to multivariate analysis of variance (MANOVA), in which appear strange new statistics like Wilks's lambda. Log-linear analyses of multiway contingency tables allow the researcher to move beyond the simple chi-square analysis of one- and two-way contingency tables. And regression analysis has become increasingly complex, as we now must make sense of logistic, hierarchical, and stepwise regression, with new options such as backward and forward elimination strategies. Not all of these complex analyses are new, but few people had the motivation to learn multivariate techniques before there were "canned" software packages. The terrain, however, is shifting.

We compared the 1976 and 1992 volumes of two well-known psychology journals: *Journal of Consulting and Clinical Psychology* (*JCCP*) and *Journal of Personality and Social Psychology* (*JPSP*). We counted the number of articles that used at least one multivariate statistical analysis. For *JCCP*, the percentage of articles that used multivariate analyses rose from 9% to 67% in that time period. In 1976, 16% of the articles in *JPSP* used a multivariate statistical analysis, whereas 57% of the articles in 1992 used multivariate analyses. One would think that such an increase in the use of multivariate analyses would be accompanied by an increased emphasis on this topic in required graduate statistics courses. However, the results of a study by Aiken, West, Sechrest, and Reno (1990) suggest otherwise, at least in the field of psychology.

On the basis of a survey of 186 psychology departments, Aiken et al. (1990) concluded that "the statistical and methodological curriculum has advanced little in 20 years" (p. 721). When respondents were asked to judge the competencies of their graduate students to apply various techniques of statistics in their own research (Aiken et al., 1990, Table 5, p. 726), only a minority of students were deemed competent to perform an array of multivariate techniques. For example, the percentage of students judged competent to perform MANOVA was 18%, path analysis 2%, confirmatory and exploratory factor analysis 3% and 12%, respectively, multidimensional scaling 2%, meta-analysis 5%, and "other multivariate procedures" 11%.

Like it or not, multivariate statistics will grow in popularity, and graduate programs will have to prepare students for this new reality. In addition, consumers of research articles must attain at least a cursory knowledge of multivariate statistics or risk becoming alienated from the research process. Furthermore, active researchers, even those who use multivariate techniques, tend to stay with a narrow set of design and data-

pendent variables. The significant *F* value, however, may be due to a combination of the dependent variables. Thus, it should not be surprising if, after a significant multivariate *F* statistic is obtained, none of the subsequent univariate *F* tests are significant.

The concept of a linear composite is ubiquitous in multivariate analyses. Imagine that a researcher wants to use several variables to predict
arital satisfaction in a multiple regression analysis. For instance, the
iables used to predict marital satisfaction might be the couples' years
*Rea*marriage, number of children, and annual income. Instead of using
arrah predictor variable separately to predict marital satisfaction, the in-
resedual predictors are combined to form a single composite variable. The
withtion between this new variable and marital satisfaction is reflected in
ana multiple correlation, or *R*. Each single variable contributes to the
ceptuall correlation by means of its strength of association with marital
in sinaction. The multiple correlation reflects the best way to combine, or
have at, the predictors so that the relation between the composite variable
deeperarital satisfaction is maximized.
symbolst another example of establishing composite variables is found in
only assualysis. For instance, a relatively large number of individual items
univariate re grouped to form a small number of composite variables,
For grai, that represent underlying dimensions of the test.
standard texts now to a discussion of various scales of measurement that
courses as prepne the variables of a study.
may already use
multiple regressio
criminant analysis). surement

The book is equ ehavioral sciences collect data that are in the form
multivariate statistics: is not the actual numbers that are of interest, but
istrators, and a range sent. Measurement is the assignment of numbers
science research results ding to predetermined rules. Because there are
tempted to skip over the g numbers, the same number can have a dif-
multivariate statistics, there on the rules used to assign the number. S. S.
may have in their area of int levels of measurement that differ in the
Our approach in selecting rding the underlying characteristic di-
yses that many of our readers a pply.
literature or to use in their own re
meant to be comprehensive. Also, w
multivariate. For instance, multiple re
a multivariable strategy rather than nominal scale is a labeling activity.
there are several predictor variables bu designation of females as 1 and

males as 2 in a data set, and social security numbers are examples of the use of a nominal scale. When using a nominal scale, one cannot interpret the numbers as anything other than the names of things. Moreover, if males are assigned the number 2 and females are assigned the number 1—a procedure known as *dummy coding*—it does not mean that males are twice as much as females.

When two categories are used, the nominal measure is called *dichotomous*. Gender is a dichotomous variable. If a variable is defined by the presence or absence of an attribute, it is a dichotomous variable. Variables involving categorization into two or more categories are variously referred to as *qualitative, categorical*, or *nonmetric*. In addition, tabulation of the number, or percentage, of occurrences in a category is a *frequency count*. The chi-square and log-linear analyses are based on data in the form of frequency counts. It may seem odd to include categorization as a form or level of measurement, but as Coombs (1953) stated, "this level of measurement is so primitive that it is not always characterized as measurement, *but it is a necessary condition for all higher levels of measurement*" (p. 473).

An essential requirement of nominal scaling is that subjects be classified into *mutually exclusive* and *exhaustive* categories. In other words, each subject or observation is assigned to one and only one category, and all observations or subjects are classified into the specified categories. For example, suppose a researcher wants to identify which set of variables predicts response to treatment. All subjects are assigned to one of two categories: responder or nonresponder. According to the rule of mutual exclusivity and exhaustiveness, subjects can only be assigned to one or the other category, not both, and all subjects are assigned to one or the other category. The use of a nominal scale requires a consistent application of an assignment rule. Therefore, in our example, the researcher must clearly define what is meant by responder versus nonresponder. In this instance, a responder might be defined as someone who falls within the "normal" range of anxiety and a nonresponder as someone who falls outside of this range.

Note that the difference between categories is one of kind rather than degree; this is a fundamental characteristic of the nominal scale of measurement. In addition, the members of a category are viewed as similar with respect to the criterion or criteria used in the categorization schema. Whether the investigator has used a good method of categorization can only be judged within a practical or theoretical context (or both).

In later chapters are many instances in which categorical variables are used in an analysis. For example, logistic regression and discriminant analysis are used when the dependent variable is categorical. In chapter 6, Rodgers discusses cross-classification tables and the concept of *odds* in a research design in which variables are defined categorically. In addition, MANOVA involves the use of categorical, independent variables.

Ordinal Scale

An ordinal scale shares a feature of the nominal scale in that observations are categorized; however, ordinal numbers have a particular relation to one another. Larger numbers represent a greater quantity of something than do smaller numbers. The ordinal scale represents a rank ordering of some attribute. Even though ordinal scales are quantitative, there is only a limited sense in which quantity is implied. Rankings reflect more or less of something, but not how much more or less. The difference between the winner and runner-up may not be the same amount of difference as that between the 4th- and 5th-place designates. In other words, the intervals between adjacent ranks are not constant over the entire range of ranks. In chapter 5, Stalans discusses multidimensional scaling applied to variables measured on an ordinal scale.

Interval Scale

A third level of measurement is the interval scale. An interval scale possesses the qualities of the nominal scale in that different numbers represent different things. The interval scale is also like the ordinal scale in that different numbers reflect more or less of something. In addition, however, the interval scale has the property that numerically equal distances on the scale represent equal distances on the dimension underlying the scale. The Fahrenheit scale is an interval scale. The difference between the temperature of 80° and 85° is the same amount of heat (measured in units of mercury) as the difference between 90° and 95°.

The distinction between an ordinal scale and an interval scale is not always easily made, especially in the behavioral sciences. For instance, is the interval between an IQ of 100 and one of 105 the same as the interval between an IQ of 45 and one of 50? The numerical distance between the scores is the same, but it is not the numerical distance between numbers that defines the difference between ordinal and interval scales. It is the underlying dimension that the scale is tapping that is important. Is the difference in *amount of intelligence* between an IQ score of 100 and one of 105 the same as the difference in the amount of intelligence between an IQ score of 45 and one of 50? As another example, if the highest score possible on a measure of depression is 30, is the amount of depression between scores of 20 and 30 the same as the amount of depression between scores of 5 and 15? Many statistical tests require that the data represent an underlying dimension of equal intervals. Although it is open to question whether or not a variable can be scaled as equal intervals, behavioral scientists are willing to assume that most of their measures are interval scales, even in the absence of an empirical demonstration that this is the case (cf. Cliff, 1993).

Many of the analyses discussed in this text are applied to continuous scales. When multiple regression is used to predict scores on a self-report inventory, it is assumed that the criterion variable is measured using a continuous scale of measurement. MANOVA assumes that dependent variables are continuous, and discriminant analysis uses independent variables that are continuous.

Ratio Scales

A ratio scale possesses all of the properties of an interval scale, with the addition of a meaningful absolute zero point. Data that are collected using an interval or ratio scale are referred to as *metric data*. The Fahrenheit scale is not a ratio scale because zero is not the complete absence of heat. A measure of length is a ratio scale because there is an absolute zero point (i.e., a point of no length). One characteristic of a ratio scale is that a number that is mathematically twice as large as another number represents twice as much of whatever is being measured. IQ, for example, does not possess this characteristic since someone with an IQ of 100 is not twice as intelligent as someone with an IQ of 50. On the other hand, because the underlying dimension of height has an absolute zero point, a person whose height is 80 in. is twice as tall as someone whose height is 40 in. Time is another variable that has an absolute zero point. Therefore, one subject's reaction time of .50 s is twice as fast as another subject's reaction time of .25 s.

In the behavioral sciences, the kinds of variables that lend themselves to ratio scales are rare, and some measures may only appear to be measured on a ratio scale. No measures of achievement, aptitude, personality traits, or psychopathology have a meaningful absolute zero point.

Without an absolute zero point to serve as an anchor point, two different scores can be interpreted relatively only; a score that is twice the size of another score does not represent twice the amount of the thing being measured. Generally speaking, when selecting an appropriate statistical analysis, the most important distinctions are between nominal scales (categorical data), ordinal scales (ranked data), and continuous measures (interval and ratio scales).

Relation Between Scales of Measurement and Statistical Analyses

As Pedhazur and Schmelkin (1991) succinctly stated, "measurement supplies the numbers used in statistical analyses" (p. 25). Statistical analyses are directly linked to measurement. The scale of measurement "allows for" various statistical manipulations of the data. Clearly it is meaningless to compute the arithmetic mean and standard deviation of ordinal numbers. Accordingly, it is inappropriate to use parametric statistics when analyzing ordinal data. Yet there are many instances in which one has to wonder if behavioral scientists have quietly agreed to assume that certain kinds of data are interval, rather than ordinal, so as to analyze the data using the more powerful parametric statistical analyses. Researchers would most likely use a parametric analysis when comparing the difference between mean IQ of two groups, despite the unsubstantiated assumption of equal intervals of intelligence among adjacent IQ scores. In addition, can a psychologist safely assume that measures of attitudes assign numbers that represent equal spacing of an underlying dimension? Yet parametric statistical analyses are routinely applied to measures of attitude. Some statisticians (Borgatta, 1968; Gardner, 1975; Nunnally, 1978) argue that the distinction between ordinal scales and interval scales cannot be clearly defined. Gardner (1975) suggested that some variables lie in a "gray" area between an interval and ordinal scale, for example, mental ability and attitudes. Nunnally (1978) and Labovitz (1972) are of the opinion that the harm done by treating variables that fall within the intermediate region of ordinal and interval scales is offset by the greater sensitivity and power inherent in the use of parametric statistics applied to these variables. Nevertheless, it is safe to say that the results of an analysis that assumes a certain scale of measurement is suspect when this assumption is violated.

Statisticians will continue to debate the appropriate correspondence among levels of measurement and methods of data analysis. All would agree, however, that more attention needs to be paid by behavioral scientists to increasing the precision of measurements.

Chapters in This Book

The following chapters in this text discuss a wide range of frequently used statistical procedures. Why are there different types of multivariate analyses? Because each multivariate procedure is appropriate for a different type of research question. That is, each analysis has its own objective or purpose. In addition, however, in deciding which multivariate procedure is appropriate for the analysis of one's data, it is important also to consider the measurement scale of each of the variables in the research design.

The remainder of this chapter provides an admittedly oversimplified summary of the chapters in the textbook. Our primary purpose here is to briefly introduce each analysis in the book, with special attention to the objective or purpose of the statistical procedure and the nature of the data used in the analysis.

Multiple Regression Analysis

Multiple regression is typically used when one attempts to predict a single continuous variable (often called a *dependent* or *criterion variable*) using two or more continuous or nominal variables (often called *independent* or *predictor variables*). For example, multiple regression could be used to predict income, using gender and years of education as predictors. Income and education are continuous measures, and gender is a nominal measure. One common use of multiple regression is in theory testing, in which the relations among criterion and predictor variables are examined. Another popular use of multiple regression is in forecasting, in which a regression equation is developed using one sample of data and then the equation is used to predict the value of the criterion variable in instances for which the predicted value is unknown. For example, a personnel manager might be interested in predicting the performance of a prospective employee, on the basis of scores from inventories that measure need for achievement and sociability. A multiple regression equation that was derived from a sample of employees would be used, in which the variables performance, need for achievement, and sociability were known.

In chapter 2, Licht provides a detailed discussion of many of the most important aspects of multiple regression.

Path Analysis

Path analysis begins with a hypothesized causal model in which the investigator clearly states the predicted relations among a set of variables. The hypothesized manner in which the variables are related are often illustrated in a schematic diagram. It is necessary to measure each variable in the path model; the type of data used in the analysis is the same as is appropriate for multiple regression. Indeed, in chapter 3, Klem points out that "path analysis can be viewed as an extension of *multiple regression*. . . . '*X* causes *Y*' is a regression model, whereas '*X* causes *Y* and *Y* causes *Z*' is a path analysis" (p. 65).

Consider the following hypothetical causal model of the etiology of psychiatric symptoms among victims of a crime. There are four measured variables: premorbid psychological problems (i.e., the degree to which the person experienced psychiatric symptoms before the crime), the amount of violence during the crime, the degree to which the person experienced a sense of helplessness during the crime, and the severity of symptoms subsequent to the crime. In the path model, the following is hypothesized:

1. The most important (immediately prior) determinant of the development of posttraumatic psychological symptoms is the degree to which the person felt helpless during the crime.

2. Two variables are important (immediately prior) determinants of the experience of helplessness: a history of psychological problems and the amount of violence during the crime.

3. A history of psychological problems and the amount of violence during the crime will independently and directly determine the severity of posttraumatic symptoms. However, the direct link between premorbid status and violence on posttraumatic symptoms will be weaker than the direct link between helplessness and posttraumatic symptoms.

A series of multiple regression equations is used to establish a pattern of relations among the variables and to evaluate the degree to which the obtained data conform to the hypothesized relations among variables. It is also possible to compare which of two models best fits the observed data. For example, one might compare the model described above with an alternative model that does not specify helplessness as a mediating variable but, rather, hypothesizes direct links between posttraumatic symptoms and premorbid psychological problems, amount of violence, and helplessness. In chapter 3, Klem provides a comprehensive introduction to this important methodological approach, which is becoming increasingly popular in the social sciences and medicine.

Factor Analysis

In chapter 4, Bryant and Yarnold discuss principal components analysis (PCA) and both exploratory (EFA) and confirmatory factor analysis (CFA). PCA and EFA are dimension-reduction techniques. Given a collection of continuous variables—for example, items of a questionnaire—PCA (and EFA) can be used to identify a smaller set of variables, called *eigenvectors* (factors). These new variables explain the majority of variation among the original set of variables. For example, imagine that a questionnaire is administered to a sample of subjects. The questionnaire consists of 60 items that a panel of experts developed to assess mood states. The results of a PCA might indicate that three eigenvectors—Anger, Happiness, and Sadness—explained 85% of the total variation in subjects' responses to the set of 60 items. Thus, rather than viewing the questionnaire as measuring 60 different variables, one could explain 85% of the total variation in responses to these original variables in terms of three new variables.

CFA allows the researcher to test more precise hypotheses than are possible to test using PCA or EFA. In CFA, for example, one can specify which items belong on which factor and can also test hypotheses about the way in which the factors relate to each other. For example, when analyzing the mood state questionnaire using CFA, the investigator will specify which items belong on the Anger factor, which should go with the Happiness factor, and which items would constitute the Sadness factor. In addition, hypotheses can be tested, such as Happiness and Sadness are negatively correlated, Anger and Happiness are uncorrelated, and Anger and Sadness are positively correlated. As in path analysis, CFA requires a prior theoretical model, which is compared with the obtained data. Moreover, CFA can be used to identify which of two or more models provides the best "explanation," or fit, of the data.

Multidimensional Scaling (MDS)

MDS is appropriate when one wants to understand patterns underlying the similarity of members of a category. For example, a market researcher may want to know why people like certain cars more than others. The researcher selects 11 cars, including Mercedes, Jaguar, Honda Accord,

Buick Century, Lincoln Town Car, and Corvette. The researcher asks adults who are visiting car lots to rate how much they like each of the cars. With MDS, the researcher will use the correlations among their liking ratings for the 11 cars to discover why some cars are preferred over other cars. MDS represents similarity between preferences for two cars as the distance between two cars in a space. That is, the cars are arranged so that cars that have similar preference ratings are close together and cars that have different preference ratings are far apart. Thus, as the represented distance between any two cars increases, it means that they are less similar on preference ratings. MDS first arranges the cars along a line, and this arrangement represents *Dimension 1*.

Generally, one dimension does not adequately capture the pattern of similarity judgments (in this case, correlations of preferences). MDS also identifies a second dimension, called *Dimension 2*, which explains more of the pattern of similarity judgments that could not be explained by Dimension 1. As the MDS solution moves from one dimension to two dimensions, the cars are no longer located on a line, but rather in a plane. The location of these stimuli in the plane reveals the pattern of the cars that represents the subjects' preference ratings. The cars that appear close together on the plane are seen as more similar than the cars that are located relatively farther away from one another. MDS also provides a measure of "stress," which tells how well this created pattern fits with the pattern among the correlations. When MDS identifies a third dimension, cars are now located in three-dimensional space, or in a cube. Cars that are close to one another in this space are deemed more similar than cars that are relatively farther from one another. In this manner, MDS proceeds to determine the smallest number of dimensions necessary to establish an acceptable level of fit between implied and observed similarity ratings. The location of the stimuli in this *smallest space* is a graphic representation of the system of preference ratings.

How does one determine the nature of these MDS dimensions? This is accomplished by measuring variables that, on the basis of theory, would explain the basis for the similarity judgments. In the automobile example, ratings can be obtained from a sample of subjects regarding cost, size, foreign or American made, and so on. One can then use these data in a regression analysis to predict the scores of the stimuli on the MDS dimension(s). Each stimulus (type of car) has a score, which is its location on each MDS dimension. If the MDS scores of the stimuli can be adequately predicted, then one obtains insight into the variables that influence the judgments of similarity among these stimuli. In chapter 5, Stalans provides an introduction to one type of MDS: nonmetric two-way multidimensional scaling. Her chapter is intended to acquaint the reader with the types of problems that are best suited for MDS procedures and with how to interpret the statistical results of a study that uses MDS.

Cross-Classification Analysis

In chapter 6, Rodgers discusses the analysis of cross-classified data. A cross-classification table, also called a *contingency table*, consists of frequency counts for each cell of a table that is created when one *crosses* two or more qualitative factors. For example, a person might be interested in tabulating the number of males and females in each of five different job classifications in a corporation. The purpose of a cross-classification analysis is to determine if two or more variables are associated. In this example, one could determine if gender was unequally distributed across job classifications.

A chi-square analysis is typically used to test for an association in contingency tables involving two cross-classified variables. However, it is easy to imagine adding a third variable to this design example, perhaps race or marital status. In this design, traditional chi-square analysis is no longer appropriate. Rather, a family of multivariate procedures, collectively known as *linear probability models*, have been devised for the analysis of multiway contingency tables. These analyses include, for example, log-linear, logistic regression, and probit models. Central to the understanding of these approaches is the concept of odds ratios. Rodgers's chapter illustrates how odds ratios are computed and interpreted in the context of multiway contingency tables. In our example, results might indicate that the odds of being in the upper echelons of the corporation differ depending on the gender of the employee. Perhaps the most popular type of linear probability model used for analysis of such data is logistic regression analysis.

Logistic Regression

Logistic regression can be used when predictor variables are qualitative (e.g., gender) or quantitative (e.g., test scores) and the criterion variable is dichotomous. (If the criterion variable is continuous, multiple regression would quite likely be used.) Logistic regression can be extended to the case in which the criterion variable has more than two levels, called *multinomial logistic regression*.

Imagine that a study is conducted, comparing high school students who graduate with students who drop out. Graduation status is the dependent measure. Predictor variables include gender, IQ, and number of siblings. With these data, logistic regression can be used to determine the increase in the probability of dropping out of school that is associated with an increase in the value of a predictor variable while controlling for the other variables. Here, the analysis identifies the relative importance of variables that are predictive of graduation status. In addition, logistic regression may be used to develop a classification model that can be used to identify students at risk of dropping out of school, solely on the basis of their scores on the independent variables. In chapter 7, Wright discusses logistic regression—a procedure that is becoming increasingly popular in a variety of disciplines.

Multivariate Analysis of Variance (MANOVA)

MANOVA is appropriate when one's design involves one or more categorical independent variables ("treatments") and two or more continuous dependent variables. As in analysis of variance (ANOVA), one can examine the effects of each independent variable separately as well as the effects of combinations, or interactions among independent variables, provided the design is factorial. In this design context, why would one prefer the use of MANOVA instead of multiple ANOVAs? First, conducting separate univariate F tests on each dependent variable can inflate the probability of a Type I error—rejecting a true null hypothesis. Second, and more important, it is possible to obtain a significant multivariate effect even though separate ANOVAs might indicate that the groups do not differ with respect to any one dependent variable. These problems are circumvented when using MANOVA, which allows a simultaneous test across all dependent variables. That is, MANOVA finds a linear combination of the dependent measures that maximizes separation among groups. MANOVA also provides a test statistic for which a p value for the linear composite may be determined.

In chapter 8, Weinfurt discusses a variety of procedures for interpreting a multivariate effect that has been identified using MANOVA, as well as numerous statistics typically associated with a MANOVA. If a multivariate effect is found, a common practice is to follow the MANOVA with a series of univariate F tests. However, this approach will not reveal if the groups differ with respect to a linear composite of two or more dependent variables. Discriminant analysis is often used to specify a combination of dependent variables that maximally separates the groups.

Discriminant Analysis (DA)

In chapter 9, Silva and Stam discuss descriptive and predictive DA. In both forms of DA, the criterion or dependent variable is categorical, and the independent variables are continuous. The primary purpose of descriptive DA is to understand the way in which the independent variables serve to differentiate among the different categories of the criterion variable. In contrast, the purpose of predictive DA is to use information about a subjects' scores on the independent variables to predict their actual category membership status. As an example, imagine that for a random sample of college basketball players, one is interested in studying the relationship between whether or not players are drafted into the NBA (dependent variable) and various quantitative aspects of their college performance, such as their average points per game, height, and vertical leap (independent variables). Descriptive DA would be used to understand how well these variables serve to discriminate between the two classes of players (i.e., those who make it to the NBA vs. those who do not). Note that in this application, descriptive DA is similar to MANOVA, with the designations of independent and dependent variables reversed. In using MANOVA, one uses NBA draft status as the independent variable and the performance measures as the dependent variables to test for a significant difference among the means of the dependent variables (or, more specifically, a linear composite of the dependent variables).

Coaches, recruiters, and players might be more interested in predictive DA in this application. Here, the objective of predictive DA would be to use some combination of the independent variables to predict whether a given player will or will not make it to the NBA.

If this research also involved qualitative measures to predict players' college performance, such as type of injury, if any, then discriminant analysis would be inappropriate and logistic regression analysis would be the procedure of choice.

Meta-Analysis

Each of the procedures discussed thus far was applied in the context of a single study. Sometimes specific research questions receive a great deal of attention by the scientific community, and many studies are conducted on the same phenomenon. In such cases, it is possible to use meta-analysis to quantitatively summarize the cumulative findings of a body of literature. As discussed by Durlak in chapter 10, meta-analysis quantifies the data from each study in terms of descriptive features of the study and transforms the results of each study into a common metric, called the *effect size* (ES). ES can be indexed by either the strength of association among variables or the magnitude of between-groups differences. For the sample of studies, it is first determined whether the corresponding effect sizes are similar, or homogeneous. If the ESs are homogeneous, then one may use meta-analytic procedures to estimate the ES for the literature considered in its entirety. On the other hand, if the ESs are not homogeneous, then one searches for groups of studies that do have homogeneous ESs.

Imagine that for a body of literature, there exists a group of studies that show similarly large ESs and another group of studies that show similarly weak ESs. What differentiates the two groups of studies? To answer this question, attributes of each individual study—such as demographic characteristics of the sample, differences in measurement instruments, or other methodological differences—are analyzed in an effort to identify the variables that account for the discrepant findings. For instance, a meta-analysis of the psychotherapy literature might reveal that one type of treatment is more effective for one behavior disorder and another treatment approach is more effective for a different disorder. Other salient factors might emerge, such as the experience level of the therapist or whether behavioral observations or self-report measures were used to define treatment outcome. Through the use of meta-analyses, researchers can identify variables that are important in understanding the phenomenon under investigation and highlight issues that need further study.

Conclusion

The use of multivariate statistics across diverse fields of study is becoming common. Although many multivariate procedures are extensions of univariate/bivariate analyses, the person having only experience in univariate/bivariate statistics may experience the realm of multivariate statistics as not only daunting, but completely foreign. Certainly the researcher who uses multivariate analyses must take the responsibility for the appropriate selection of techniques and have an in-depth understanding of the myriad aspects of the analysis. It is unnecessary to master multivariate analyses to understand the Results section of an article that uses such analyses. What is essential, however, is that the consumer of research findings be able to follow the statistical reasoning used by the investigators. We hope that this textbook reduces some of the intimidation that accompanies contact with multivariate statistics. Reading in the area of statistics, by nature, involves effort. With careful attention, however, the reader should be able to study the following chapters and achieve an understanding of many multivariate procedures without getting lost in the mathematical aspects of statistical analyses, which mercifully have not been presented.

References

Aiken, L. S., West, S. G., Sechrest, L., & Reno, R. R. (1990). Graduate training in statistics, methodology, and measurement in psychology: A survey of PhD programs in North America. *American Psychologist, 45*, 721–734.

Borgatta, E. F. (1968). My student, the purist: A lament. *Sociological Quarterly, 8*, 29–34.

Cliff, N. (1993). What is and isn't measurement. In G. Keren & C. Lewis (Eds.), *A handbook for data analysis in the behavioral sciences: Methodological issues* (pp. 59–94). Hillsdale, NJ: Erlbaum.

Coombs, C. H. (1953). Theory and methods of social measurement. In L. Festinger & D. Katz (Eds.), *Research methods in the behavioral sciences* (pp. 471–535). New York: Dryden.

Gardner, P. L. (1975). Scales and statistics. *Review of Educational Research, 45*, 43–57.

Labovitz, S. (1972). Statistical usage in sociology: Sacred cows in ritual. *Sociological Methods & Research, 1*, 13–37.

Nunnally, J. (1978). *Psychometric theory* (2nd ed.). New York: McGraw-Hill.

Pedhazur, E. J., & Schmelkin, L. P. (1991). *Measurement, design, and analysis: An integrated approach*. Hillsdale, NJ: Erlbaum.

Stevens, J. P. (1986). *Applied multivariate statistics for the social sciences*. Hillsdale, NJ: Erlbaum.

Stevens, S. S. (1951). Mathematics, measurement, and psychophysics. In S. S. Stevens (Ed.), *Handbook of experimental psychology* (pp. 1–49). New York: Wiley.

Tabachnick, B. G., & Fidell, L. S. (1989). *Using multivariate statistics* (2nd ed.). New York: Harper Collins.

2 Multiple Regression and Correlation

Mark H. Licht

It is very difficult to understand the research, and thus keep up-to-date on developments, in any area of the behavioral and social sciences without the ability to understand the results of multiple regression and correlational analyses (MRC). Virtually every issue of most popular journals publishing applied or basic research in these areas contains reports of studies that use these procedures. This chapter is written for individuals who have limited knowledge of the detailed statistical aspects of MRC. It is intended to help the readers attain a conceptual understanding of the results of the most common and direct uses of these analyses.

My greatest concern in undertaking this task is that "a little knowledge can be a dangerous thing" with regard to interpreting statistics. As is discussed, there are certain complexities and disagreements among "experts" concerning some interpretations of MRC analyses. Thus, simple guidelines for interpretation can often be misleading or misused. However, it is my hope that this chapter will aid in the comprehension of research that uses MRC by enhancing the conceptual understanding of this procedure while also increasing the awareness of the complexities involved in interpreting multiple regression.

As described by Wiggins (1973), the historical roots of regression/correlational analyses date back to the work of Francis Galton and Karl

---

I would like to thank the following people for their thoughtful comments on various drafts of this chapter: Barbara G. Licht, Richard Wagner, Ellen Crawford, Anita D. McQuillen, and Elaine Goff. I also want to thank the editors of this book, Larry Grimm and Paul R. Yarnold, and two anonymous reviewers for their helpful suggestions. I am certain that the input provided by these colleagues, students, and friends helped to make this a better chapter than it otherwise would have been. Of course, any shortcomings are solely my responsibility.

Pearson in the late nineteenth century. These procedures became closely associated with the "individual-differences" branch of psychology, which attempted to clarify differences among individuals by studying naturally occurring relationships among variables. In contrast, the "experimental" branch of psychology attempted to manipulate and control variables in laboratory settings, in efforts to discover general laws that would apply to all individuals. This branch favored the use of data-analytic procedures such as the analysis of variance (ANOVA), developed by Ronald Fisher.

One of the many unfortunate consequences of the schism that developed between the individual-difference and experimental branches of psychology has been an inappropriate stereotyping of these data-analytic procedures. Many experimental psychologists appeared to view correlational data analyses as "less scientifically respectable" (Cohen & Cohen, 1983, p. 5) and appropriate only when one was using a correlational research design (i.e., no random assignment or manipulation of independent variables). Fisherian ANOVA and related techniques were considered most appropriate for experimental research designs (i.e., random assignment and manipulation of independent variables). However, there is no justification for generalizing the differences between correlational and experimental research designs to procedures for analyzing data.

Ironically, Fisher's initial approach to testing the significance of group differences was not the now famous means squares ANOVA procedures, but rather MRC (see Tatsuoka, 1975). However, in that precomputer era, the computational complexity of MRC precluded its practical application. Thus, Fisher turned to the computationally feasible ANOVA procedures. However, one can, and many statistical computer programs do, accomplish the tasks of ANOVA and analysis of covariance (ANCOVA) by means of MRC. In fact, the Fisherian analysis of variance procedures simply represent a special, more limited case of MRC (Cohen & Cohen, 1983; Pedhazur, 1982; Tatsuoka, 1975).

Conceptually, MRC determines the statistical significance of differences among groups of subjects (i.e., the basic task of ANOVA) by determining whether there is significant prediction of subjects' scores on the dependent variable from knowledge of their group membership. Explanation of how this is accomplished is beyond the scope of this chapter. However, this chapter should provide the basic knowledge of MRC concepts necessary for understanding the excellent presentations of these uses of MRC found elsewhere (see, e.g., Cohen & Cohen, 1983, or Pedhazur, 1982).

With the development and widespread accessibility of high-speed computers to carry out the computations, the popularity of MRC has grown dramatically for a broad range of data-analytic problems. MRC is now widely recognized as a flexible and general approach to analyzing data for a variety of research designs and questions. For example, the independent variables can be continuous or categorical, naturally occurring or experimentally manipulated, and correlated or uncorrelated, and the relationship between independent and dependent variables can be linear or curvilinear (Cohen & Cohen, 1983; Pedhazur, 1982).

Multiple regression/correlation is closely related to its less powerful relative, *bivariate* (also called *simple* or *zero-order*) regression/correlation, in which there are only two variables. In fact, some view MRC as the multivariate extension of bivariate regression/correlation. Although often categorized with other multivariate procedures, MRC is probably more accurately classified as a univariate procedure because it involves only one dependent variable. In many ways, MRC is to bivariate regression/correlation as multifactor ANOVA is to one-factor ANOVA. That is, bivariate regression/correlation and one-factor ANOVA both involve one independent variable and one dependent variable, whereas MRC and multifactor ANOVA involve multiple independent variables but still one dependent variable. For discussions of true multivariate extensions of MRC, such as canonical analysis, which use multiple dependent variables, see, for example, Cohen and Cohen (1983) or Pedhazur (1982).

It is common to divide studies that use bivariate or multiple regression/correlation into two types: (a) those that attempt to *predict* events or behavior for practical decision-making purposes in applied settings and (b) those that attempt to *understand* or *explain* the nature of a phenomenon for purposes of testing or developing theories. As discussed in more detail elsewhere (e.g., Pedhazur, 1982), prediction and explanation are basic goals in science and are inextricable concepts. Thus, studies directed primarily at prediction provide information for explanation and vice versa. Despite the somewhat artificial nature of the distinction between prediction and explanation, it is useful to examine interpretations of MRC separately for these two purposes.

The next two sections of this chapter, Applied Prediction and Theoretical Explanation, each begin with a brief overview of the use of MRC for that purpose. The abstract descriptions of the procedures and concepts presented in these overviews are then illustrated in actual examples from the literature. These examples are then further used to help explain the results and interpretations of MRC analyses. Each of these sections ends with a brief statement of important conclusions about the use of

MRC for those purposes. Discussions of several important methodological and conceptual issues are avoided in these sections to minimize the complexity of the initial descriptions of results and interpretations. However, these issues are discussed later in the chapter, under General Methodological Considerations and Assumptions. The chapter ends with General Conclusions concerning the use of MRC and Suggestions for Further Reading.

Commonly encountered MRC terms, symbols, and their definitions are presented in the Glossary at the end of the chapter. It may be helpful to refer to the Glossary while reading the chapter or when reading MRC studies. Whereas some of the entries are self-explanatory, others require further clarification, which is provided in text.

Applied Prediction

Overview

The use of MRC for practical prediction purposes is most commonly found in educational, vocational, forensic, and clinical settings. It is used to determine the utility of a set of *predictor variables* (e.g., demographics, test scores, or behavioral observations) for predicting another important event or behavior, called the *criterion variable* (e.g., school performance, job performance, violence, attempted suicide, or test scores).[1]

The greater potential predictive power of multiple over bivariate regression/correlation is easily seen in that the absolute level of prediction must be at least as good, and most likely better, with multiple predictors than with any one of these predictors taken by itself. For example, when attempting to predict the criterion of success in graduate school, one must do at least as well, if not better, by using both undergraduate grades and Graduate Record Examination (GRE) scores as predictors as one would by using undergraduate grades alone. That is, even if GRE scores were to add nothing to the prediction above and beyond undergraduate grades,

1. It is not uncommon to find the predictor variables referred to as *independent variables* and the criterion referred to as the *dependent variable*. Although this is more commonly found when MRC is used for theoretical purposes, it is also found in prediction studies. Technically, these alternative terms are only accurately used when the predictors have been manipulated, as in a true experimental research design. However, because these terms are frequently used interchangeably in the MRC literature, one should not depend on the terminology used to indicate the nature of the research design used to acquire the data. In this chapter, the more inclusive terms, *predictors* and *criterion*, are used throughout.

including GRE in the MRC analysis would not reduce the absolute accuracy of prediction because it would simply be ignored. Furthermore, any predictive utility of GREs that is not also provided by grades would operate to enhance the predictive accuracy. However, as is discussed later, one must guard against including unnecessary predictors (i.e., predictors that do not make meaningful contributions above and beyond the other predictors) because the potential for sample specific findings increases with each additional predictor.

As is illustrated in the example that follows, for purposes of prediction, a *derivation study* is conducted in which MRC is used to derive a linear formula or equation, referred to as the *multiple regression equation* and consisting of a weighted sum of two or more variables. This equation specifies how the scores on the predictors should be combined to produce the most accurate predicted scores on the criterion. Also, MRC provides estimates of the accuracy of the prediction and of the degree of relationship between this linear combination of predictors and the criterion.

Because the linear equation from the derivation study is "custom made" for the sample used in that study, it is not expected to predict as well for another sample that differs in any way (Wiggins, 1973). That is, the predictive accuracy and degree of relationship are expected to shrink when the equation is used with other samples. Two procedures are available to determine the amount of shrinkage that might occur. One procedure, also illustrated in the following example, involves conducting a second study, called the *cross-validation study*. Here, data collected from a new sample are used to evaluate how well the formula from the derivation study actually predicts for other people from the same population. Note that this is not a replication study, in that the results of a second multiple regression analysis are not simply being compared with the original results. Rather, the results of the first analysis are being evaluated by using them with data from a new sample.

The second procedure for determining the amount of shrinkage is to obtain an estimate by means of one of several formulas (e.g., Cohen & Cohen, 1983; Pedhazur, 1982). These so-called *shrinkage formulas* correct for the number of predictors relative to the number of subjects (Wiggins, 1973). The use of empirical cross-validation and shrinkage formulas is further discussed later in this chapter.

Assume that following derivation and empirical or estimated cross-validation, the predictors are found to adequately predict the criterion. Data on the predictors could then be obtained in applied settings from new subjects (e.g., clients, patients, students, or employment applicants) and entered into the regression equation to predict these subjects' unknown scores on the criterion. Of course, the determination of adequate prediction must be made by the users in the applied setting, on the basis of the potential consequences of their decisions.

These steps in the use of MRC for purposes of prediction are illustrated with the following example from the literature. First, the example is presented with a minimum of explanation. Then the example is used to explain the results and interpretations of MRC analyses for purposes of prediction.

Example 1: Estimating WAIS–R IQ

Willshire, Kinsella, and Prior (1991) reported the use of MRC to derive and cross-validate an equation to predict Wechsler Adult Intelligence Scale–Revised (WAIS–R) IQ scores—the criterion—from scores on the National Adult Reading Test (NART) and from demographic data (age, sex, occupation, and education level)—the predictors. The purpose of their investigation was to provide evidence for or against the use of these predictors for predicting premorbid intelligence in people suspected of having dementia and, if the evidence was supportive, to provide an optimal regression equation for making these predictions in clinical situations. Only select aspects of this investigation are presented here, to minimize complexity and emphasize the points relevant to our discussion of the use of MRC for purposes of prediction.

Willshire et al. (1991) collected criterion and predictor data from two samples of subjects. The first sample included 104 subjects, and the second, 49 subjects. Although the specific type of MRC analysis used was not indicated for each of the several MRC analyses performed, it seems likely that either *simultaneous* or *stepwise* MRC was used in all cases (these and other types of MRC analyses are described and discussed later in this chapter, under General Methodological Considerations and Assumptions).

First, Willshire et al. used the data from the first sample for their derivation study. Thus, they submitted these data to MRC analysis. The resulting regression equation is presented in Exhibit 1. This equation specifies that to predict a person's IQ score, one should proceed as follows: (a) multiply the person's NART score by 0.7, (b) subtract that product from 104.3, (c) multiply the person's education score (number of years

Exhibit 1

Example 1 Results: First Study

First Derivation Study

$N = 104$

Multiple regression equation

$$IQ = 104.3 - (0.7)\ (NART) + (4.6)\ (education)$$

Multiple correlation coefficient and multiple coefficient of determination

$R = .68$ $R^2 = .46$ $p < .05$

First Cross-Validation Study

$N = 49$

Cross-validation multiple correlation coefficient and multiple coefficient of determination

$R = .69$ $R^2 = .48$ $p < .05$

Note. Data are from Willshire et al. (1991). NART = National Adult Reading Test, education = education score.

of formal education) by 4.6, and (d) add that product to the result of Step b.

Also presented in Exhibit 1 are the multiple correlation coefficient (R) and multiple coefficient of determination (R^2) associated with this equation. These were statistically significant at well under the traditional alpha level of $p < .05$. As described in more detail after this example, R indicated that the degree of relationship between this linear combination of NART and education score, on the one hand, and IQ, on the other hand, was .68. Furthermore, R^2 indicated that 46% of the variance in IQ scores was predictable from this linear combination of NART and education score.

Note that the regression equation does not include all of the predictor variables; that is, age, sex, and occupation are missing. These predictors were dropped because they did not provide statistically significant contributions to the prediction of IQ above and beyond that provided by NART and the education score.

Also note that the weight for NART is negative (i.e., -0.7) whereas that for the education score is positive (i.e., 4.6). This is because NART scores indicate the number of errors made on the test rather than the number of correct responses. Thus, larger scores on NART suggest lower levels of reading. In contrast, the education score is simply the number of years of formal education. Therefore, larger scores for education mean more education. Thus, the signs of their weights reflect that the relationship between NART and IQ is negative whereas that between education and IQ is positive.

The data from the second data collection were then used for empirical cross-validation. The predicted IQ score for each of these 49 subjects were computed by inserting the subject's obtained scores for NART and education into the regression equation presented from the derivation study. These predicted scores were then correlated with the empirically obtained IQ scores for the 49 subjects in the second data collection to get an index of the relationship between predicted and actual IQ scores. This cross-validated R is an index of how good the prediction was when the regression equation from the derivation study was used to predict scores for the second sample of subjects. As indicated in Exhibit 1 under First Cross-Validation Study, the cross-validation R was .69, essentially identical to the R from the derivation study, and was statistically significant.

So far in this example we have seen a fairly classic derivation and cross-validation study. Willshire et al. (1991) went a step further by conducting a *double cross-validation study*. That is, in addition to what has already been described, they started over, this time treating the second data collection on 49 subjects as the derivation study and the first data collection on 104 subjects as the cross-validation study. Thus, they submitted the data on the criterion and on all of the predictors from the second sample to MRC analysis to obtain a new regression equation. This equation is presented in Exhibit 2 under Second Derivation Study. The R and R^2 for this equation were statistically significant and also are presented in this section of Exhibit 2.

Next, Willshire et al. (1991) cross-validated this new equation by using it to calculate predicted IQ scores for each of the 104 subjects in the first sample. These predicted IQ scores were then correlated with the actual obtained IQ scores for these 104 subjects to reveal a statistically significant cross-validation R (see Second Cross-Validation Study in Exhibit 2). Although statistical tests of the significance of the difference were not presented in the study, only a small degree of shrinkage is evident by the comparison of the second cross-validation study R (.68) to the R from the second derivation study (.75).

Exhibit 2

Example 1 Results: Second Study

Second Derivation Study

$N = 49$

Multiple regression equation

$$IQ = 123.7 - (0.8)\,(\text{NART}) + (3.8)\,(\text{education}) - (7.4)\,(\text{sex})$$

Multiple correlation coefficient and multiple coefficient of determination

$R = .75$ $R^2 = .56$ $p < .05$

Second Cross-Validation Study

$N = 104$

Cross-validation multiple correlation coefficient and multiple coefficient of determination

$R = .68$ $R^2 = .46$ $p < .05$

Note. Data are from Willshire et al. (1991). NART = National Adult Reading Test, education = education score.

Results and Interpretations for Purposes of Prediction

The Multiple Regression Equation, Partial Regression Coefficients, and Intercept

The regression equation is the most basic result of an MRC analysis. It indicates that to obtain a predicted score for the criterion, in this case IQ, the score on each predictor is multiplied by a number specific to that variable. These products are then summed (see the equations in Exhibits 1 and 2). The technical name for the number by which a predictor is multiplied is *partial regression coefficient* or *partial regression weight*. However, in many reports of MRC analyses, the term *partial* is omitted from this name.

There are two forms of the regression equation that are frequently reported in the literature. One form is the *raw score regression equation*, for which the scores on the criterion and on the predictors are in the original units of measurement. The equations in Exhibits 1 and 2 are of this form. The other form is the *standard score regression equation*, for which the scores for the criterion and predictors are presented in standard deviation units (i.e., z scores). When the standardized form of the equation is used to predict the criterion, the scores on the predictors must be transformed from their original raw score units to z scores. The resulting predicted scores will also be in z score form.

In addition to using either raw or standardized forms of the predictor and criterion variables, the raw and standardized forms of the regression equation differ in two other ways. As described below, the partial regression coefficients are different, both in their absolute values and in their potential interpretations. Also, the raw score form of the equation includes a numeric constant called the *intercept*. For example, the intercept in the equation from the second derivation study of Example 1 was 123.7 (see Exhibit 2). The intercept indicates the score on the criterion variable when all of the predictors are zero. When variables are transformed to standard scores, the intercept equals zero. Therefore, it is not included in the standard score form of the equation.

For prediction purposes, the raw score equation may be preferred because it does not require transforming variables. This is particularly the case when the raw score units are meaningful (e.g., years of education, number of occurrences, or dollars). However, the magnitudes of the raw score coefficients from a single equation cannot be directly compared with one another as an indication of the relative contributions of the corresponding predictors to the predicted value of the criterion. For example, the coefficients from the second derivation study regression equation in Example 1 (see Exhibit 2) do not indicate that sex makes the greatest contribution to the prediction of IQ, followed by education and then by NART. To some degree, comparisons of this sort are possible with the coefficients from the standardized equation. However, many of the complexities in interpreting the results of MRC occur when attempting to compare the contribution of various predictors. Because these types of comparisons are most common when explanation is the primary purpose of the study, further discussion of these issues is postponed until the use of MRC for purposes of explanation has been introduced.

To derive a regression equation, it is first necessary to obtain a sample of data for which values for each predictor and the criterion have been obtained for each subject. It is then the task of MRC to establish the values of the constants (partial regression coefficients and intercept) that, across subjects, produce predicted scores on the criterion that are as close as possible to the observed scores on the criterion (i.e., that have the smallest amount of error in prediction). That is, MRC selects the values of the constants that minimize the sum, across subjects, of the squared differences between their predicted and obtained scores on the criterion (i.e., that minimize the sum of the square errors). This criterion is known as the *least squares solution*.

The Multiple Correlation Coefficient (R) and Coefficient of Multiple Determination (R^2)

In addition to the regression equation, MRC provides an index of the degree of relationship between the criterion, on the one hand, and the weighted combination of predictors as specified by the regression equation, on the other hand—that is, R. Another way to think of R is as the bivariate correlation between the observed scores on the criterion and the scores predicted by the regression equation. R ranges from 0 to 1, with 0 indicating no relationship between predicted and actual criterion scores and 1 indicating a perfect relationship (i.e., perfect prediction).

The meaningfulness of multiple correlations might be more easily evaluated by examining R^2, which indicates the proportion of variance in the criterion that is shared by the weighted combination of predictors. In other words, it indicates the degree to which differences among individuals (i.e., variance) on the criterion are predictable from the set of predictors when those predictors are combined as specified in the multiple regression equation. It follows, then, that $1 - R^2$ is the proportion of variance that is *not* predictable. Thus, in the first derivation study of Example 1, the R of .68 and the R^2 of .46 indicate that (a) 46% of the variance in IQ scores was predictable from NART and education and (b) 54% (i.e., $1 - R^2$) of the variance in IQ scores was not predictable from these predictors.

To evaluate the potential for shrinkage of these results, a cross-validation study was conducted. The primary results of cross-validation studies are the *cross-validation* R and R^2. As described in Example 1, these are obtained by correlating the following: (a) the predicted criterion scores for a new sample of subjects that were predicted on the basis of the derivation study regression equation with (b) these subjects' actual observed scores on the criterion.

In Example 1, the values for R and R^2 from the first cross-validation analysis are virtually identical to the corresponding values from the derivation study (see Exhibit 1). Because these values did not change even when the multiple regression equation was used with a new sample of subjects (i.e., there was no shrinkage), one can have a good deal of confidence that this equation will also predict slightly less than 50% of the variance in IQ scores for other samples from the same population.

The investigators in Example 1 decided to obtain even more information on which to base their conclusions concerning the expected degree of prediction using these predictor variables. Therefore, they conducted a double cross-validation study. The R and R^2 from the new derivation study were slightly larger than those in the first derivation study (see Exhibits 1 and 2). Cross-validation in this second study did reveal a small degree of shrinkage, from a derivation R of .75 to a cross-validation R of .68. However, this shrinkage resulted in an R and R^2 that were virtually identical to those from the first cross-validation study. Given this consistency of findings, a high degree of confidence can be placed in the conclusion that somewhat less than 50% of the variance in IQ scores can be predicted from this set of predictor variables in this population.

Of course, drawing a conclusion about the utility of being able to predict slightly less than 50% of the variance in the criterion, and the corresponding inability to predict slightly more than 50% of this variance, is a judgment that must be made on the basis of the consequences of the various outcomes of the decisions in the applied setting. A discussion of procedures for examining the utilities of applied decisions is well beyond the scope of this chapter. However, the reader is referred to the excellent presentations of this topic in Cronbach and Gleser (1965) and Wiggins (1973).

Given that the multiple regression equations from both derivation studies indicated virtually identical cross-validated levels of prediction, the question remains as to which equation to use. According to the results presented in Example 1, it should make no difference which equation is used. Other considerations for deciding on which multiple regression equation to use in applied settings are discussed later in the chapter.

The Standard Error of Estimate (SE_{est})

Assume that a score is calculated for each subject by subtracting her or his predicted score from her or his actual observed score on the criterion. Each of these scores, commonly referred to either as *error scores* or *residual scores*, indicate the error in predicting or estimating a person's score on the criterion using the multiple regression equation. The SE_{est} is the standard deviation of the distribution of these error scores. Obviously, a multiple regression equation with a small SE_{est} is preferable because it indicates that on average, there is only a small amount of error in prediction. As with other standard deviations, SE_{est} can be used to create confidence intervals (in this case, frequently referred to as *prediction intervals*). When certain assumptions are met (described later under Error Score Assumptions), these intervals indicate the limits within which a criterion score is expected to fall (Wiggins, 1973). Given the absence of perfect prediction (i.e., $R < 1.00$), it is often more reasonable to determine the confidence that can be placed in a range or interval of predicted scores than to attempt to use a single specific predicted score (for more on the use of confidence intervals, see, e.g., Cohen & Cohen, 1983, or Wiggins, 1973). Despite its utility for prediction purposes, the SE_{est} often is not presented in published reports, as it was not in Example 1.

Actually, SE_{est} is not the most accurate statistic for computing confidence intervals around predicted scores. This is because SE_{est} is the average error of prediction. Thus, it is a constant for all values of the predictors; however, actual error in predicting is not constant. Rather, prediction error is greater, the further the values of the predictors are from their respective means. In other words, predictions for subjects with extreme scores on the predictors are less accurate than predictions for subjects who have scores close to the means of the predictors. More precise confidence limits are achieved by using the standard error of a predicted score, or $SE_{y'}$, which has a different value for each combination of values on the predictors.

It is unreasonable to expect published reports of prediction studies to include the many potential values of $SE_{y'}$. Rather, they should present the SE_{est} as an indication of the average error of prediction that potential users might expect in their applications. Users themselves will need to obtain the information to compute $SE_{y'}$ and confidence limits for their applications. (See, e.g., Pedhazur, 1982, for details of computing $SE_{y'}$ and confidence limits.)

Conclusions Concerning the Use of MRC for Prediction

MRC aids prediction by determining the degree to which a given set of predictors can empirically predict a criterion, as indexed by R, R^2, and SE_{est}. Furthermore, it provides a linear formula for combining the predictors to make the prediction. Given a certain degree of instability across samples, results either of cross-validation procedures or of shrinkage formulas should be inspected to evaluate the expected utility of these results for new samples.

Some have argued that the use of shrinkage formulas is preferred to empirical cross-validation because one's entire resources can be devoted to the derivation study (see Wiggins, 1973, for a summary of these arguments). The result would be a larger derivation sample and, consequently, more stable estimates of regression coefficients, multiple correlations, and standard errors. Therefore, less shrinkage would occur than if the sample was split to create separate derivation and cross-validation samples.

Theoretical considerations and empirical evidence appear to support this argument when there is a *fixed set of predictors*, that is, all predictors used in the derivation study are retained for future use (Wiggins, 1973). However, as further pointed out by Wiggins, empirical cross-validation procedures rather than shrinkage formulas should be used when the results of the derivation study are used not only to determine the regression coefficients, intercept, and other indexes but also to select a subset of the original predictors to be retained for future use. This is because shrinkage formulas *only* correct for overestimation of the degree of relationship and predictive accuracy that result from using the full regression formula from the derivation study. They do not account for the fact that the predictor variables that are most useful in the derivation sample may not be those most useful in subsequent samples. Therefore, when predictors are dropped from the equation on the basis of their empirical relationships with the criterion and with the other predictors, as in Example 1, empirical cross-validation with one or more truly independent samples is required.

The use of MRC for purposes of applied prediction is often considered a pure empirical process, because it doesn't really matter whether the prediction makes theoretical sense. Although the potential of MRC for empirically deriving useful prediction formulas is well established (e.g., Dawes, Faust, & Meehl, 1989; Wiggins, 1973), this potential is more likely to be realized when the results also make theoretical sense (Wiggins, 1973). When the specific predictors selected or the differential weighting of predictors fail to conform to any theoretical formulation of the nature of the phenomena, it is likely that they represent chance or idiosyncratic aspects of the samples on which they were derived. Such results should be viewed with skepticism, particularly in the absence of methodologically sound cross-validation.

Theoretical Explanation

On the basis of the axiom that one cannot draw causation from correlation, many people argue that regression and correlation are not useful for explaining the nature and causes of phenomena. However, the ability to make causative and explanatory interpretations is determined primarily by the design of the data collection and the logic of the reasoning rather than by the procedures for analyzing the data. A discussion of the philosophy and logic of causal and explanatory reasoning is beyond the scope of this chapter (see, e.g., Pedhazur, 1982). However, for the following discussion, two issues involved in this reasoning should be noted.

First, one can gain a better understanding of the nature of a phenomenon by identifying those factors with which it co-occurs. Although not conclusive, it is likely that co-occurring factors either are causally related to one another or have other causative factors in common. At least, information of co-occurrence helps to define the theoretical constructs involved in the phenomenon under study (see, e.g., Kerlinger, 1986). MRC addresses this issue by providing multiple indexes of co-occurrence.

Second, confidence in one's theory of causation can be improved by ruling out plausible alternative causal explanations. Preferably, these alternative causal explanations would be excluded by means of experimental controls in the design of the research. However, frequently, experimental control is not possible. In these cases, MRC can provide statistical control for specified alternative explanations of causation (i.e., third-variable explanations).

In this section of the chapter, I first present an example of the use of MRC for purposes of theoretical explanation. Immediately after the description of the study, I explain the results and use them to describe and clarify the following: (a) the MRC indexes of co-occurrence and (b) the nature of statistical control of potential third-variable explanations provided by MRC.

Example 2: Effects of Parenting Practices on Academic Performance

Steinberg, Elmen, and Mounts (1989) used MRC to test a series of hypotheses concerning the effects of parenting practices on academic performance.[2] One of these hypotheses was that each of three components of authoritative parenting practices (acceptance, support for psychological autonomy, and behavioral control), assessed in 1985, would make independent contributions to adolescents' academic success, as measured by grade point average (GPA) in 1986. The alternative explanation, that GPA in 1986 caused the authoritative parenting practices rather than vice

2. In their article, Steinberg et al. (1989) refer to these hypotheses as *path models*. This term comes from path analysis, which is a set of procedures for attempting to determine causative relationships among variables. In essence, a path model is a theory that specifies the nature of the relationships expected to occur among the predictors and between these and the criterion. Multiple regression and correlational analyses are commonly used to analyze the results in path analysis. A full discussion of path analysis is provided in chapter 3. Here, the discussion is restricted to the basic results of one of the multiple regression analyses conducted in this study rather than on all aspects of path analysis.

versa, was ruled out through experimental control; that is, GPA was measured 1 year after the measurement of parental practices.

MRC was used to provide statistical control for 6 third-variable alternative explanations of 1986 GPA: GPA in 1985, 1985 achievement test scores (CAT), age, socioeconomic status (SES), and family structure (FAM). This control was accomplished by including these six measures as additional predictors in the MRC analysis. For example, sex was included as a predictor to control for the possibility that any relationship found between the parental practices and academic performance might actually have been due to the students' sex rather than because the parental practices affected academic performance (e.g., girls might get better grades and also might elicit more authoritative parental practices than do boys, thereby producing a correlation between parental practices and academic performance even if these two variables are not causally related). As explained in more detail later (see *Independent Contributions and Statistical Control*), MRC controlled for sex by removing from the analysis any variability related to sex. What was left, then, were the relationships between the parental practices and 1986 GPA that were not attributable to sex.

The three parental practices and the potential third variables were entered as predictors into a simultaneous multiple regression analysis with 1986 GPA as the criterion. Simultaneous multiple regression examines the contributions of all predictors at the same time rather than adding or subtracting predictors one at a time or in sets, as do stepwise, hierarchical, or all-possible-subsets regressions. As previously described for prediction, MRC uses the least squares criterion to determine the best coefficients to use in a linear equation for predicting the criterion from the set of predictors. Results from Example 2 are presented in Table 1.

Results and Interpretations for Purposes of Theoretical Explanation

The Multiple Correlation Coefficient (R) and Coefficient of Multiple Determination (R^2)

In Example 2 (see Table 1), R is not only statistically significant but also rather large in magnitude, thereby indicating a good deal of co-occurrence between the combination of all predictors and the criterion. In fact, the obtained R^2 indicated that almost 62% of the variance in students' 1986 GPAs was shared by this combination of variables. These results suggest that a good deal of the individual differences (i.e., variance) among students' GPAs in 1986 were caused either by the combination of predictors themselves or by other factors that were casually linked to these predictors

Table 1

Example 2 Results: Predicting 1986 GPA

Predictors	Partial regression weight	
	Raw	Standardized
Parental practices		
Acceptance	0.039*	.127*
Support for autonomy	0.048**	.148**
Behavioral control	0.047**	.142**
Potential third-variables		
1985 GPA	0.367****	.363****
1985 CAT	0.011***	.300***
Sex	−0.007	−.004
Age	−0.002	−.041
SES_1	0.072	.034
SES_2	0.221	.118
FAM_1	0.159	.071
FAM_2	−0.268	−.107
Intercept	1.494	
Summary statistics: R = .787,* R^2 = .619.		

Note. From "Authoritative Parenting, Psychosocial Maturity, and Academic Success Among Adolescents," by L. Steinberg, J. D. Elmen, and N. S. Mounts, 1989, *Child Development, 60,* p. 1429. Copyright 1989 by the Society for Research in Child Development. Adapted with permission. GPA = grade point average, CAT = California Achievement Test, SES = socioeconomic status, FAM = family structure. There are two variables each for SES and FAM because of the need to dummy, or effect, code these three category, nominally scaled variables.
*$p < .10$. **$p < .05$. ***$p < .01$. ****$p < .001$.

(i.e., further third variables not included in the study). In either case, important information has been obtained about the criterion.

Independent Contributions and Statistical Control

As previously described, R and R^2 provide information on the contribution of all the predictors taken as a group. However, a primary purpose of the study in Example 2 was to determine the independent contributions of each of the three parental practice variables. The researchers wanted to know, for example, the effects on academic performance of different amounts of behavioral control above and beyond the effects of the other parental practices and potential third variables. Obviously, this information can give even greater insight into the nature and causes of the criterion than do R and R^2.

Among the most powerful and useful aspects of MRC are the indexes that provide information relevant to this idea of independent contribution. However, this is also the area in which the greatest complexity and misunderstanding of MRC results can occur. As is demonstrated, none of these indexes unequivocally represent the unique or independent contribution of a single variable. The best that one can do is examine all of the information and attempt to understand it in the context of a well-formulated theory and sound logical reasoning.

One way to consider the contribution of a single predictor is by examining its bivariate correlation (r) with the criterion. While r is certainly useful for the overall understanding of the criterion, it indicates the degree of relationship when all other predictors are ignored. This differs from the independent contribution of a predictor, for which the effects of all other predictors need to be controlled or eliminated. With r, some or all of the shared variance between the two variables, indexed by r^2, might also be shared by other predictors.

Before discussing the indexes from MRC that more directly reflect independent contribution than do r or r^2, the concept of statistical control should be clarified. Statistical control is also referred to as *partialing*, *controlling for*, *residualizing*, *holding constant*, and *covarying*.

As explained in basic research methods texts (e.g., Kerlinger, 1986), experimental control over confounding, or third, variables can be attained in several ways. Random assignment of subjects to groups is used to experimentally control for subject variables by equally distributing these variables across groups. Another method is to hold a potentially confounding factor constant. For example, if sex had been held constant in Example 2 by including only female subjects, then differences in sex could not have been an explanation for variability in parental practices or grades because all subjects would have been female. A second study could have included only male subjects. The results of these two studies could be presented separately or averaged together to indicate the relationships between parental practices and grades with sex held constant and, therefore, not a viable alternative explanation of the relationships.

However it is accomplished, experimental control is desirable because it allows for relatively straightforward interpretations of results. Unfortunately, experimental control is not always possible or ethical. Random assignment is not possible when predictors cannot be manipulated (e.g., degree or type of psychopathology). Furthermore, it is not feasible to experimentally hold constant (as just described for sex) certain types of variables because the large numbers of values would require too many separate investigations (e.g., continuous variables such as age or achievement test scores). It also is not feasible to experimentally hold constant in this way more than one or two variables at a time because the combinations of these variables also produce a large number of values that would have to be individually evaluated (e.g., experimentally holding constant both sex and age produces twice as many values and the need for twice as many separate investigations, as does holding only age constant). However, MRC can provide statistical control for the effects of these types of potential third variables. Conceptually, statistical control does what was described in the previous paragraph for sex: It conducts separate analyses at every level of the control variables and averages the separate results (Pedhazur, 1982).

Another useful way to conceptualize statistical control is by reference to residuals. A *residual* is the difference between a predicted score and the actual observed value of that score. As such, it represents the part of the actual score that is independent of and cannot be predicted from the combination of predictors. This implies that these predictors (or causes of these predictors) have no effect on the residual scores and, therefore, have been eliminated from, or controlled for, in the residuals.

For example, *autonomy* from Example 2 could be used as the criterion in an MRC analysis with the remaining predictors from Example 2 serving as the predictors. Residuals for autonomy could be computed by subtracting each subject's predicted score on the basis of this MRC analysis from her or his actual observed score. These residuals would represent the concept of autonomy when controlling for, or eliminating, acceptance, behavioral control, age, sex, and all of the other predictors. If these residual scores for autonomy were then correlated with the criterion of interest, 1986 GPA, the result would be interpreted as reflecting the relationship between 1986 GPA and autonomy, after first controlling for the effects of all the other predictors.

A few cautionary notes are in order at this time. The relationship just described is not between 1986 GPA and the construct of autonomy, as initially conceived and measured by the autonomy scale used in Example 2. Rather, it is between 1986 GPA and the concept of autonomy as measured by the residuals. These residual scores are indicants of a new, altered concept of autonomy that is different from the original concept in that it is independent of the other variables (i.e., acceptance, behavioral control, sex, and age). That is, these other predictors, and factors that are causally related to them, have no effect on the residual scores, whereas they did have an effect on the original autonomy scores (as indicated by the correlations). Ascribing meaning to this new construct is one of the complexities in interpreting MRC and emphasizes the need for explanatory interpretations to be based in sound theoretical reasoning.

Another issue that must be considered when interpreting independent contributions is that, as implied above, the contribution is only independent of the other variables included in the study. Thus, the appropriateness of interpreting this as an independent or unique contribution in some larger sense rests on having included all plausible third variables in the analysis. The values of the MRC indexes of independent contribution, described below, can change dramatically when even one new variable is added to the study.

There are two types of indexes provided by MRC that are typically used to indicate independent contribution: *partial regression coefficients* and *partial coefficients of correlation.* The term *partial* means that the effects of other predictors have been statistically controlled for or partialed out, in the sense just described. Thus, when interpreted within the framework of sound theoretical reasoning, these coefficients can provide information on the independent contribution to the criterion of the corresponding predictor.

Partial Regression Coefficients

As previously described, these coefficients are the weights, or multipliers, provided by the multiple regression equations. Because there are both raw and standard score equations, there also are both raw and standard score partial regression coefficients. The coefficients from the Steinberg et al. (1989) study are presented in Table 1. A partial regression coefficient (raw or standard) can be interpreted as the amount of change that is expected to occur in the criterion per unit change in that predictor when statistical control has occurred for all other variables in the analysis. The sign of the coefficient indicates the direction of the change. Thus, the raw coefficient for autonomy in Example 2 (.048) indicates that in general, when all other variables are partialed out and thereby have no effect, 1986 GPA increases by .048 when the score on autonomy increases by 1.0.

The desired implication of this finding is that if one could experimentally hold constant or otherwise control for student's age, sex, family structure, socioeconomic status, prior academic achievement, and the other two parental practices while concurrently increasing parental support for autonomy by a factor of 10, then GPA would be expected to improve by almost one half of a grade point (i.e., $10 \times .048 = .48$ or, for example, from approximately 2.5 to 3.0). This result might suggest a substantial independent contribution of autonomy to GPA that could have implications for parenting practices and for understanding why some students get higher grades than other students. However, it is difficult to determine the practical and theoretical meaningfulness of this finding because although most people understand units of GPA, autonomy is measured on an arbitrary scale with no commonly used real-world referents. Thus, the meaning of a 10-point change in autonomy is unclear. In cases like this, the standardized coefficients can be helpful.

Standardized partial regression coefficients are interpreted similarly to raw coefficients, except in terms of *z* scores rather than raw scores. As indicated in Table 1, the standardized coefficient for autonomy is .148. Therefore, to achieve an increase of only 0.148 standard deviations for 1986 GPA (e.g., an increase from the 50th percentile to just under the 56th percentile in a normal distribution), autonomy must be increased by a full standard deviation (e.g., an increase from the 50th to the 84th percentile in a normal distribution). These standardized coefficients suggest that it would undoubtedly require a substantial effort to increase autonomy even one unit and that the expected payoff in GPA would be rather small.

In general, when the units of measurement have some real-world meaning (e.g., grade points, dollars, or years), interpretations of the raw coefficients are generally easier to understand than are those for standardized coefficients. The opposite tends to be true when the units of measurement are on an arbitrary and relatively unknown scale (e.g., autonomy or personality test scores). Of course, before interpreting any result as meaningful for people outside the study sample, the statistical significance or confidence limits must be considered. For these purposes, MRC provides significance tests and standard errors for the partial regression coefficients. Note that the raw and standardized coefficients need not be tested separately because when one is significant, so is the other.

The statistical significance of partial regression coefficients means that within the probability of error represented by the *p* value, the magnitudes of these coefficients differ from zero. Therefore, the corresponding predictors would be considered to make statistically significant independent contributions to the prediction and understanding of the criterion. Furthermore, some confidence is gained in the hypotheses that each of these predictors has some unique causal effect on the criterion because these coefficients are significant even after controlling for the potential alternative casual explanations represented by the other predictors.

For example (see Table 1), it is reasonable to conclude that the parental practices represented by autonomy and behavioral control (and perhaps acceptance) each make unique contributions to academic success as measured by 1986 GPA. Furthermore, confidence in these being causative contributions is increased because important third variables (some of which make statistically significant unique contributions themselves) have been eliminated as explanations. At this point, the investigator and reader must rely on logic and theory rather than statistical analyses to address several unanswered questions. For example, what is the meaning of each altered (e.g., residualized) parental practice variable? What is the theoretical and practical importance of the relationship between the residualized parental practice variables and the 1986 GPA? Finally, are there other third variables that had they been included in the study, would have influenced the results?

The partial regression coefficients for most of the potential third variables included in this study were not statistically significant (see Table 1). This means that with the exception of 1985 GPA and CAT, these variables did not make significant unique contributions to the criterion. However, this does not mean that they can just be dropped from the results as if they were never included. In fact, it is possible to have a significant correlation between the combination of predictors and the criterion (i.e., R) when none of the predictors make significant independent contributions. Whether its partial regression coefficient was statistically significant, dropping a variable from an MRC analysis is likely to affect the values of coefficients for all other variables as well as the value of R. Thus, similar to the decision to include certain variables in the first place, careful theoretical reasoning must be used when deciding to drop predictors. Furthermore, whenever predictors are dropped or added, the MRC must be recomputed to obtain new estimates of all indexes.

It is tempting to interpret the relative magnitudes of statistically significant partial regression coefficients as indicating the comparative contributions of the various predictors to the criterion. It is even possible to test for the statistical significance of differences between regression coefficients, although these tests are seldom reported in the literature. However, caution is required when interpreting the comparative sizes of regression coefficients.

The magnitudes of raw regression coefficients cannot be directly compared across variables with different units of measurement. In Example 2, CAT has a mean of 59.53 and a standard deviation of 24.81, whereas autonomy's mean is 7.94 and standard deviation is 2.81. Obviously, a change of one CAT unit (i.e., a change of less than 0.05 standard deviations) does not have the same meaning for academic achievement as does a change of one autonomy unit (i.e., a change of greater than 0.35 standard deviations) for parental support of autonomy. These differences in units of measurement affect the magnitudes of the raw partial coefficients, thereby rendering their comparison meaningless.

Transforming the raw scores to *z* scores, however, puts all variables in the same measurement units, with means of 0 and standard deviations of 1. Therefore, the standardized coefficients are directly comparable. Thus, at least within this study and within the limitations of interpreting statistical control, it is reasonable to say that both 1985 GPA and prior achievement (as measured by CAT) appear to make larger independent contributions to 1986 GPA than do any of the three parental practice variables.

Note, however, that even if they are significantly different from zero, the absolute and relative magnitudes of standardized coefficients vary from one sample to another. This is because they are affected by sample variances, which tend to fluctuate (in contrast, the sizes of raw regression coefficients tend to be more stable). Therefore, interpretations of the absolute and relative sizes of standardized coefficients have limited generalizability to new samples.

In conclusion, statistically significant partial regression coefficients can indicate, with a good deal of confidence, whether specific predictors make contributions to the criterion that are unrelated to the contributions made by the other variables in the analysis. Furthermore, interpreting the magnitudes of the coefficients can provide insight into the meaningfulness of those contributions. Although caution is required when interpreting the magnitudes of regression coefficients (in particular, see the Multicollinearity section, below), the following guidelines might be useful:

1. Raw regression coefficients are most useful when the measurement scales of the variables have obvious intuitive meaning, such as dollars, GPA, or heart rate.

2. Standardized regression coefficients are most useful when comparing the relative contributions of each predictor to the overall effect.

3. Either raw or standardized regression coefficients can be useful in the context of the sample on which they were derived; however, do not expect standardized regression coefficients to replicate in another sample to the same degree as will raw regression coefficients.

Given the imprecision of the point estimations of size for both standardized and raw regression coefficients, statistical confidence intervals can be very helpful when interpreting the absolute or relative sizes of regression coefficients. Unfortunately, confidence limits are seldom provided in published reports.

Partial Coefficients of Correlation

Whereas regression coefficients indicate amount of change, coefficients of correlation (or their squares) indicate the degree to which individual differences (i.e., variance) on one variable correspond to individual differences on another variable. This notion of *shared variance* has been used extensively in the psychological research literature. The traditional index of the concept of shared variance is the square of the bivariate correlation coefficient (r^2). As previously described, r^2 does not index independent contribution because it ignores, rather than eliminates, the effects of other variables. However, MRC provides two indexes of correlation that do indicate independent contributions: the partial and semipartial correlation coefficients. There are formulas for calculating these indexes (e.g., Cohen & Cohen, 1983; Pedhazur, 1982), but the following conceptual descriptions should provide the nonstatistician with greater insight into their meanings and appropriate interpretations.

Semipartial, or Part, Correlation Coefficient **(r_{sp}).** This is the correlation between a specific predictor (e.g., X_1) and the criterion (Y) when all other predictors in the study have been partialed out of X_1 but not out of Y. This is exactly what I described when I explained the concept of statistical control by residualizing. When r_{sp} is squared (r^2_{sp}), it indexes the proportion of variance that is shared by the residualized predictor and the original, unchanged, criterion. Thus, only the meaning of the predictor has been altered. Although Steinberg et al. (1989) did not directly present r_{sp} for the variables used in Example 2, another form of this statistic was presented for another variable.

Steinberg et al. (1989) replicated the analysis reported in Table 1 but with one additional predictor included: psychosocial maturity. The difference between the R^2 in this new analysis (R^2 = .653) and the R^2 in the original analysis (R^2 = .620) is known as the change in R^2 associated with the newly added variable (R^2 change = .033). Because the only difference between the first and second analyses was the inclusion of maturity, the change in R^2 must have been due to this, and only this, variable (i.e., it was the independent contribution of maturity to 1986 GPA).

The change in R^2 is equal to, and is one method of calculating, r^2_{sp} between the criterion and the newly added predictor variable, partialing out all other predictors. When calculating r_{sp} by computing the square root of R^2_{change}, the sign of r_{sp} must be obtained by taking the sign of the partial regression coefficient for the newly added predictor.

***Partial Correlation Coefficient* (r_p).** This is the correlation between a specific predictor (e.g., X_1) and the criterion (Y) when all other predictors in the study have been partialed out of both X_1 and Y. Calculating r_p between 1986 GPA and behavioral control in Example 2 would be equivalent to (a) predicting behavioral control from all of the other predictors and computing the residuals for behavioral control, (b) predicting 1986 GPA from all predictors except behavioral control and computing residuals for 1986 GPA, and (c) correlating the residuals for behavioral control with the residuals for 1986 GPA. The square of this index, r^2_p, indicates the proportion of shared variance between behavioral control and 1986 GPA when the effects of all other variables have been statistically eliminated from both. Note that unlike r_{sp}, which would only change the meaning of behavioral control (i.e., the predictor variable), r_p changes the meaning of both behavioral control and 1986 GPA by removing everything that has to do with the other predictors from both the predictor and criterion variables involved in the partial correlation. Although both r_{sp} and r_p can be useful, the former is more frequently encountered in the literature. One probable reason is that typically, the purpose of explanatory studies is to understand a specific phenomenon as measured by the criterion. Therefore, researchers might avoid statistics that alter this variable, as does r_p.

Because the partial correlation, semipartial correlation, and partial regression weights are really only different indexes of the same basic concept (i.e., independent contribution), when one is statistically significant, so are the others. The significance test for the partial regression weight associated with a predictor typically serves as the test for all of these statistics.

Note that many of the caveats mentioned for partial regression coefficients also hold for the partial coefficients of correlation (i.e., the possibility of reverse causation and of additional third variables not included in the study and difficulty interpreting the altered construct). Furthermore, like standardized regression coefficients, magnitudes of r_{sp} and r_p for different predictors from a simultaneous MRC are comparable to one another because they use the same measurement scale. However, also like the standardized coefficients, the magnitudes of r_{sp} and r_p are affected by the sample variances and, therefore, can fluctuate across samples. Therefore, care is required when attempting to generalize interpretations of the magnitudes of r_{sp} and r_p or when attempting to compare their magnitudes across different studies.

Conclusions Concerning Theoretical Explanation

MRC aids our understanding of complex phenomena by providing multiple indexes of the degree of relationship between predictors and criteria while statistically controlling for certain alternative explanations of those relationships. Thus, MRC addresses both points previously described as important for explanatory and causal reasoning: identifying factors with which the phenomenon co-occurs and ruling out plausible alternative causal explanations. Although statistical control cannot be equated with experimental control, when interpreted with proper regard to its limitations, MRC can be a powerful scientific tool for the many important investigations in which experimental control is not possible.

Different information and a different conceptualization about the nature of the relationships between predictors and criteria are provided by the various MRC indexes. Although arguments have been made for the use of certain of these indexes over others (see, e.g., Achen, 1982; Cohen & Cohen, 1983; Pedhazur, 1982), they all have benefits and limitations, as described above. The use of specific indexes should be dictated by the nature of the variables and the purpose of the investigation. Frequently, comparisons of several indexes can be useful. In any case, consumers of MRC research need to be able to intelligently interpret whichever indexes are provided in the published reports. Therefore, consumers need to understand all of the indexes and their limitations.

General Methodological Considerations and Assumptions

The discussion so far has avoided some of the more complex methodological considerations and the statistical assumptions associated with MRC. However, appropriate use of MRC requires some understanding of these issues. This section of the chapter provides a general description of the most important of these issues, including the following: multicollinearity; assumptions involving residual scores, specification errors, and measurement errors; how to handle categorical variables; and the differences among the most commonly encountered variations of MRC.

Multicollinearity

The term *multicollinearity* is used when discussing the intercorrelations among the predictors in an MRC analysis. However, readers can become confused when encountering this term because, as described by Pedhazur (1982), the term has no consensually agreed on definition. Some authors use the term in a descriptive sense to indicate the degree to which the predictors are intercorrelated. Other authors use the term to indicate that some critical level has been surpassed—that is, that these intercorrelations are *too* high (e.g., "MRC is not applicable due to multicollinearity"). Because the results and interpretations of MRC analyses are affected by any level of intercorrelation among the predictors, not just when these relationships are extremely high, I prefer the descriptive use of *multicollinearity*. In this section, I attempt to provide a general understanding of the effects of multicollinearity to aid the reader in determining whether it might be a problem for any given MRC study.

In general, the greater the multicollinearity, the more problems exist in terms of technical aspects of MRC (e.g., mathematical solutions and statistical inference), as well as for practical prediction and theoretical interpretations. With regard to the technical aspects, when one or more predictors are perfectly correlated with one or more other predictors (i.e., a predictor can be totally accounted for or perfectly predicted by one or more of the other predictors), there is no possible mathematical solution in MRC. Furthermore, the greater the multicollinearity, the more unstable the partial regression coefficients. Therefore, the standard errors and confidence intervals of these coefficients are larger, and the likelihood that they are statistically significant is lower.

Given these technical problems, it is clear that any two predictors should not be perfectly intercorrelated—a condition that is easily detected by examining the bivariate correlations between all pairs of predictors. But it also follows that no predictor should be totally accounted for, or predicted by, any combination of other predictors. This latter requirement is not easily detected, particularly when there are many predictors. Although extremely high, yet imperfect, relationships also can cause these technical problems, there is no universally accepted rule of thumb concerning how high is too high. Still, most investigators would probably agree that correlations of $r > .80$ between predictors should be considered very problematic. Correlations of this magnitude might suggest that the two variables largely measure the same construct and that only one, or a combination of the two, be used: a solution that is discussed later in the chapter (see *Measurement Errors*).

For these technical reasons alone, smaller rather than larger correlations among predictors are preferred. However, there are also practical and interpretive reasons for preferring small correlations between the predictors. A practical advantage of low multicollinearity when MRC is used for purposes of applied prediction is that the most efficient prediction will occur when each predictor is highly correlated with the criterion but uncorrelated with other predictors. In this case, each predictor is uniquely important, there is no redundancy or duplication of effort, and prediction can thus be less costly.

An exception to the preference for low multicollinearity when MRC is used for applied prediction is the use of *suppressor variables*. These are predictors that are highly correlated with one or more of the other predictors but have no or low correlations with the criterion. Given the partialing procedures that occur in MRC, suppressors cause the elimination of variance in the other predictors that is irrelevant to, or not shared with, the criterion. This results in larger partial coefficients for these other predictors because irrelevant variance has been eliminated, that is, suppressed. The overall outcome is better prediction of the criterion. Unfortunately, the search for variables that have a primarily suppressor effect, or at least a large enough suppressor effect to be worth the cost of their measurement, has been disappointing (Wiggins, 1973).

Multicollinearity also poses problems for theoretical interpretations of MRC results. The larger the correlations between predictors, the more likely it is that they will share the same variance in the criterion (Y). The problem is deciding which predictor should be credited with contributing this shared, or *redundant*, variance in Y. MRC, or any other statistical procedure, cannot make this decision. In MRC, this redundant variance does not appear as the independent contribution of any of the predictors (as indexed by semipartial correlations) even though, in reality, this shared variance may be caused solely by one of the predictors and is merely correlated with the others. Rather than blind reliance on statistics, the decision of which, if any, of the predictors is ultimately responsible for redundant variance in Y must be based on careful theoretical reasoning and, if possible, experimental investigations.

The degree of multicollinearity also affects the absolute and comparative magnitudes of partial regression weights and, thus, their interpretations. Conceptually, it might be expected that the relative magnitudes of standardized regression coefficients would approximately parallel the relative sizes of their bivariate correlations with the criterion. This is the case when multicollinearity is small. However, the relative sizes of these regression coefficients can differ greatly from those of the bivariate correlations as the relationships among predictors increase.

An example presented by Pedhazur (1982, pp. 244–245) illustrates this point. That example included a total of three predictors (X_1, X_2, and X_3), each with virtually identical bivariate correlations with the criterion, Y (rs = .50, .50, and .52, respectively). Thus, one might expect that these three variables are of about equal importance for understanding Y. However, if (a) the correlation between X_1 and X_2, and between X_1 and X_3, is relatively small (rs = .20) and (b) the correlation between X_2 and X_3 is relatively large (r = .85), then the sizes of the standardized regression coefficients for the three variables are quite different from one another. In fact, the standardized regression coefficient for X_1 is nearly 1.5 times larger than that for X_3 and nearly 2.5 times larger than that for X_2.

If one were to simply interpret the relative sizes of the standardized regression coefficients in this example, then one might conclude that the contribution to the criterion of X_1 was 2.5 times that of X_2 and 1.5 times that of X_3. However, the differences in the sizes of the standardized regression coefficients are primarily due to the correlations among the predictors and not due to their relative relationships with the variable that is being predicted or explained, that is, the criterion, Y. X_1 has the largest standardized regression coefficient because it has the smallest correlation with the other predictors. The standardized regression coefficients for X_2 and X_3 are smaller because their relationships with Y overlap. However, there is no guarantee that these differences in standardized regression coefficients will make theoretical sense or reflect the true causal contributions of the three predictors.

To illustrate this point, assume X_3 has a strong causative effect on both Y and X_2, thus accounting for its correlations with these two variables (r_{Y,X_3} = .52 and r_{X_2,X_3} = .85). Also, assume that X_2 has little, if any, causative effect on Y and that its correlation with Y (r_{Y,X_2} = .50) is entirely the result of its relationship with X_3. Further assume that X_1 also has a strong causative effect on Y that accounts for their intercorrelation (r_{Y,X_1} = .50) but that this effect is different from (i.e., independent of) the causative effect of X_3. In this scenario, X_3 would make at least an equally important contribution to Y as would X_1. However, because MRC has no way of determining the true underlying causality, the standardized regression coefficients will be as described above—suggesting a greater contribution of X_1. Thus, the complexity of interpreting MRC results for purposes of theoretical explanation increases with increasing multicollinearity because the relationships among predictors must be considered.

The problem of multicollinearity is the catch-22 of MRC. A major advantage of MRC is its ability to statistically control for the effects of potential third variables. The most plausible third variables are those that are highly correlated with both the criterion and the predictor in question. If one attempts to reduce the complexity of interpretation by using only uncorrelated predictors, one of the most powerful aspects of MRC, controlling for potential third-variable explanations, is negated. However, highly correlated predictors can confuse interpretations of partial regression coefficients and leave large amounts of variance in Y unexplained by partial coefficients of correlation.

Multicollinearity can be a frustrating problem, and for technical reasons, predictors cannot be *too* highly correlated (e.g., $rs > .80$). However, selecting predictors only because they have relatively small intercorrelations is not the solution. As discussed later, under *Specification Errors*, serious problems can occur when important variables are not included in MRC analyses. The decision with respect to which predictors to include in the study must be based on theoretical considerations involving the hypotheses being tested. These theoretical considerations about the direction of causality among predictors can direct the use of MRC analyses in such a way as to help minimize the interpretive problems resulting from multicollinearity (e.g., see the discussion of path analysis in chapter 3). Thus, in the absence of the ability to exert experimental control, MRC's statistical control can still provide important information for understanding the nature and causes of complex phenomena as long as it is designed and interpreted on the basis of sound theoretical considerations and with knowledge of its limitations.

Assumptions

The primary assumptions for the use of MRC can be classified into three categories: (a) those involving the error, or residual, scores; (b) those involving *specification errors*; and (c) those involving *measurement errors*.

Error (Residual) Score Assumptions

As previously described in the *Standard Error of Estimate* section, an error, or residual, score is the difference between a subject's actual observed score on the criterion and the score predicted for that subject using the regression equation. For a variety of statistical and interpretive reasons, it is best if these error scores (a) have a mean of zero; (b) are homoscedastic (i.e., have equal variances at all values of the predictors); (c) are uncorrelated with each other and with the predictors; and (d) are normally distributed. Furthermore, the existence of outliers, or extreme residual scores, can have a number of undesirable effects (see, e.g., Cohen & Cohen, 1983). Although these characteristics of error scores should be considered when evaluating MRC studies, moderate violations of these assumptions tend not to be very problematic (see Lewis-Beck, 1980; Pedhazur, 1982). Furthermore, published reports seldom provide the necessary information to allow one to evaluate these issues. Therefore, they are not discussed at length here (readers are referred to the suggested readings presented at the end of the chapter for further details on these assumptions). However, the other two categories of assumptions, specification and measurement errors, must be very carefully considered because they are involved in most of the important caveats mentioned in this chapter. The rest of this section is devoted to these two categories of assumptions.

Specification Errors

Violations of any of the following requirements constitute a specification error: (a) The relationships among variables must be linear, (b) all relevant predictors must be included, and (c) no irrelevant predictors can be included. With regard to the first requirement, it was previously suggested that MRC could handle linear or curvilinear relationships between predictors and criteria. Although only linear relationships will be detected by MRC, nonlinear relationships and interactions among predictors can be accommodated by certain transformation strategies (see, e.g., Cohen & Cohen, 1983; Pedhazur, 1982).

The primary reason for the last two specification requirements is that the values of all indexes from MRC can change dramatically when even one important predictor is added to the analysis. Therefore, if an important variable is excluded, the obtained MRC indexes can be misleading. Although the inclusion of irrelevant variables also can affect the values of the indexes, the effect will be small if the variables are truly irrelevant. However, if too many irrelevant variables are included, it will be more difficult to achieve statistical significance or acceptable cross-validation.

Empirical and theoretical considerations can be used to select predictors for use in an MRC study. Some objections to pure empirical procedures, that is, procedures based solely on empirical relationships without regard to theoretical considerations, have already been discussed. These objections are even more critical when MRC is used for purposes of explanation. Pure empirical selection of predictors is not likely to include all theoretically relevant, nor exclude all irrelevant, predictors and, thus, is likely to produce misleading and nonreproducible results. Pure empirical selection should probably only be used for generating hypotheses when little is known about the subject matter, and the results should be interpreted with extreme caution.

The most effective means of avoiding specification errors in MRC is to use formal theories about the phenomena of interest to direct both the hypotheses to be tested and the variables to be included in the study. It is expected that specification errors will be most problematic when the theories about the phenomena are weak. In these cases, MRC results should be considered more suggestive than definitive. However, as the theories and hypotheses become more sophisticated and empirically substantiated, improvement should occur in the specification of variables to be included in further studies and confidence in interpretations of results should increase accordingly.

Measurement Errors

Unreliability and invalidity of measures are problems that plague all types of research, not only those that use MRC. Statistical corrections (e.g., corrections for attenuation) have been suggested for the effects of some measurement errors on certain MRC indexes. However, these have limited utility (e.g., only when errors of measurement are truly random) and, even when appropriate, pose additional problems for MRC analyses (Pedhazur, 1982). Rather than attempt to statistically overcome poor measurements, researchers need to extend more effort in the selection and development of measurement procedures. Furthermore, they must present evidence in support of the measures of the constructs of interest. Consumers of this research literature must then carefully consider this evidence. Note that issues concerning the reliability and validity of measurements involve theoretical considerations and, therefore, further emphasize the role of theory in the interpretation of MRC results.

Consideration of these measurement issues can also help minimize problems of multicollinearity. If two or more predictors intended to measure separate constructs have high empirical intercorrelations, than perhaps some are not measuring the intended constructs and need to be replaced. If, on the other hand, they are intended as measures of the same construct, then they might be combined into a single score, or some might be eliminated. Careful theoretical reasoning and empirical tests of construct validity can help to resolve these issues (see, e.g., Linn, 1989).

Categorical Variables

A common misconception about MRC is that it can only be used with quantitative variables measured on continuous interval or ratio scales. However, as previously stated, MRC can accommodate many types of categorical, or nominally scaled, variables as well. To be used in MRC, categorical variables need to be transformed (referred to as *dummy coding* or *effect coding*).

For instance, in Example 1, sex was a dichotomous categorical predictor variable with two categories: male and female. Willshire et al. (1991) reported representing (or dummy coding) men with the value of 1 and women with the value of 2. By giving women a larger number, they did not mean to imply any quantitative interpretation (e.g., that women are greater than men). Rather, they simply were using the numbers to indicate which subjects were men and which were women. When entered into the MRC analysis for the second study, the partial raw regression coefficient for sex turned out to be −7.4 (see Exhibit 2). This meant that there was a negative correlation between sex and IQ, indicating that on the whole, men (represented by the number 1) in the sample had higher IQs than did the women (represented by the number 2). When the predicted IQ scores were computed by means of the multiple regression equation in Exhibit 2, for each woman, 14.8 (the partial regression coefficient of −7.4 multiplied by the value on the sex predictor variable, which for women was 2) was subtracted from the total score. For each man, on the other hand, only 7.4 was subtracted from the total score. Thus, all other things being equal (i.e., equal NART and education scores), men would have higher predicted IQ scores than would women.

Examples of categorical predictors with more than two categories were provided by Example 2, in which SES and family structure (FAM) were both categorical with three levels, or categories, each (e.g., for FAM, the three levels were biologically intact family, single-parent family, or stepfamily). The three categories were not represented as a single predictor variable with the values 0, 1, and 2 because this would imply a specific quantitative order (e.g., stepfamily representing more of something than did the other two). Rather, FAM was transformed into two dichotomous dummy predictor variables coded 0 for the absence and 1 for the presence of a given category. For example, FAM_1 might be 1 if the student's family was biologically intact and 0 if it was either single-parent or stepfamily. Similarly, the value of FAM_2 might be 1 if the student had a single parent and 0 if it was either of the other two categories. Because there were only three categories of family structure in this study, there could not have been a FAM_3 because, taken in combination, scores on FAM_1 and FAM_2 already indicate the third category. That is, a 0 on FAM_1 combined with a 0 on FAM_2 must mean a stepfamily, whereas a 1 on either FAM_1 or FAM_2 must mean that it is not a stepfamily. Including FAM_3 would lead to perfect multicollinearity. Therefore, only the two dummy variables, FAM_1 and FAM_2, are entered into the MRC analysis as separate predictors.

Types of MRC Analyses

The most basic form of MRC is known as *simultaneous regression* because the regression equation and multiple correlation are determined by analyzing all predictors at the same time. This was the procedure used in the primary analysis described for Example 2.

Another form of MRC commonly reported in the literature is *hierarchical regression*, which actually is a series of simultaneous analyses, all of which use the same criterion. The first analysis in the series contains one or more predictors. The next analysis adds one or more new predictors to those used in the first analysis. The next analysis adds new predictors to those used in the second analysis and so on. The change in R^2 between consecutive analyses in this series represents the proportion of variance in the criterion that is shared exclusively with the newly added variables. Thus, hierarchical analysis is one way of calculating semipartial correlations, as was previously described for Example 2 wherein a second analysis was used to add a new predictor (maturity) to those used in the first analysis.

A very important consideration in the use of hierarchical regression is the order in which variables are entered into the series of analyses because it is this order that determines the variables that are being partialed, or controlled. As described above, the effects of the variables entered in earlier steps are partialed from relationships involving variables entered in later steps. Consequently, the partial indexes from different steps in the hierarchical regression do not involve the exact same sets of variables and are not directly comparable to one another.

Another type of MRC that is frequently used is stepwise regression. One variant of stepwise regression, *forward inclusion*, follows the same basic process as described for hierarchical regression. However, in stepwise regression, the order in which predictors are included is determined solely by their empirical relationships with the dependent variable and other predictors (e.g., the predictor that will produce the greatest increase in the R^2 at that step is included next). In hierarchical regression, the order of entry is determined by the researcher, presumably on the basis of theory.

Another variant of stepwise regression is *backward elimination*, in which the first analysis of the series includes all of the predictors. Each successive analysis involves fewer, rather than additional, predictors. Again, the order of elimination is based solely on empirical relationships among the variables (e.g., the variable that produces the smallest decrement in the R^2 at that step is removed).

In hierarchical regression, the researcher determines the number of variables to be added and, thus, the number of analyses in the series. In contrast, stepwise regression terminates when no additional variable significantly increases R^2 in forward inclusion or when the elimination of any additional variable significantly reduces R^2 in backward elimination. Thus, stepwise regression is one approach to empirically selecting a set of predictors from a larger pool of potential variables.

Another type of MRC that is used for empirical selection of variables from a larger pool is *all possible subsets*. Although it is less frequently used than the other types of MRC discussed here, one is likely to find it in the literature. Actually, there are several variations of this type of MRC, and computer programs usually use various algorithms to produce the results. However, conceptually, it is as if separate simultaneous MRC analyses are conducted for all possible combinations and permutations of the potential predictors in the larger pool. Then, the one solution is selected that produces the largest R^2 with the criterion.

Extreme care must be used when using either stepwise or all-possible-subsets regression because, as previously discussed, pure empirical selection of predictors is likely to be highly sample specific and is not likely to include all theoretically relevant, or to exclude all irrelevant, predictors. Thus, these procedures are likely to produce misleading and nonreproducible results.

Controlling Type I and Type II Errors and Preventing Alpha Inflation in MRC

Whenever MRC is used for statistical inference, it is important to control and balance the probabilities of *Type I* and *Type II* errors. As defined in any basic statistics text, a Type I error is the probability of rejecting the null hypothesis when it is, in fact, true, whereas a Type II error is the probability of failing to reject the null hypothesis when, in fact, it is false; or as described by Cohen and Cohen (1983), a Type I error is "finding things that are not there" and a Type II error is "failing to find things that *are* there" (p. 166).

The statistical probability of a Type I error when testing a single hypothesis is determined by the investigator when she or he sets the alpha level for the analysis, which is typically $p < .05$. However, when multiple hypotheses are being tested, the question arises as to whether alpha should be set at .05 for each hypothesis being tested or for some larger group of hypotheses (often referred to as *experimentwise* or *investigationwise*). This is an important consideration with regard to MRC because each MRC analysis involves the testing of multiple hypotheses.

For example, included in a simultaneous multiple regression are tests of the statistical significance of R and of the partial coefficients associated with each of the predictors. Thus, with five predictors there are at least six hypotheses being tested (one need not be concerned with differentiating tests of the statistical significance of semipartial correlations, partial correlations, raw partial regression coefficients, and standardized partial regression coefficients because the tests of these various types of partial coefficients are identical and, therefore, do not constitute separate hypotheses). Even more tests are associated with hierarchical, stepwise, and all-possible-subsets regressions because they involve multiple simultaneous regressions with separate tests occurring at each step.

It is difficult to determine the exact degree to which experimentwise alpha might be inflated over .05 when $p < .05$ is used for each individual test because this is, in part, determined by the degree to which the hypotheses are mutually dependent. However, the increased probability of committing Type I errors is very likely to be large enough to be of major concern, particularly when there are multiple steps in the analysis and there is minimal theoretical guidance used, as in stepwise and all-possible-subsets regressions. A variety of procedures for controlling experimentwise Type I error rates have been described with regard to ANOVA types of statistical procedures (e.g., Bonferroni, Dunn's, Newman-Keuls, Duncan, and Scheffé tests—see, e.g., Cohen & Cohen, 1983; Hays, 1988). For example, the Bonferroni procedure usually involves dividing the desired experimentwise alpha by the number of individual tests to determine the alpha to be used for each hypothesis. However, it is unusual to see a detailed discussion of this issue with regard to MRC. A notable exception is the excellent presentation by Cohen and Cohen (1983, pp. 166–176), which I attempt to briefly summarize in the remainder of this section.

First, Cohen and Cohen (1983) argue the general principle that "less is more" (p. 169; also see Cohen, 1990). That is, for a number of reasons (including minimizing multicollinearity, as previously discussed), reducing the complexity of the investigation by minimizing the number of predictors is likely to result in more meaningful and comprehensible results. With specific regard to the current discussion, the use of fewer predictors reduces the number of hypotheses being tested and thus results in lower experimentwise error rates. Also, power is increased, and thus the probability of Type II errors is decreased, with fewer predictors (the reasons for this are straightforward but their explanations go beyond the scope of this chapter; see, e.g., Cohen & Cohen, 1983; Pedhazur, 1982).

Thus, investigators are encouraged to use careful theoretical reasoning in selecting only the important and necessary variables to include in MRC studies. Furthermore, they should minimize the redundancy in the predictors used in MRC analyses by choosing as few alternate methods of measuring a specific concept as possible—preferably choosing or developing the method of assessment that has the greatest reliability and validity for the purposes of the investigation. (The suggestion that multiple measures of each construct not be used is made for MRC analyses and not for techniques that include specific tests of measurement models, such as structural equation modeling.)

In general, "fishing expeditions," in which variables are included because they might be useful, are discouraged because they are likely to result in highly inflated Type I error rates; although, in preliminary stages of the study of a phenomenon, these types of exploratory investigations can prove useful. When they are used, however, they should be clearly labeled as exploratory, statistical significance should be interpreted with extreme caution, and results should be replicated in more carefully designed confirmatory studies.

Cohen and Cohen (1983) also suggested that MRC analyses use an extension of Fisher's protected *t* test, designed for ANOVA procedures, on the grounds that it is simple, practical, and enjoys strong empirical support. In ANOVA, this procedure involves first inspecting the overall F for statistical significance. Then, only if this overall F is significant at the desired level of alpha (e.g., $p < .05$) are the means of individual groups statistically compared. These individual tests are "protected from the mounting up of small per-comparison [alpha] to large experimentwise error rates [because we are prevented] from comparing the sample means 95% of the time when the overall null hypothesis is true" (Cohen & Cohen, 1983, p. 172). Cohen and Cohen (1983) suggested that this strategy be generalized to MRC by allowing the examination of the statistical significance of partial coefficients, and thus of the contributions of specific predictors, only when the overall R is statistically significant. Further elaboration of this strategy is provided by Cohen and Cohen (1983), and readers are encouraged to consult this source.

General Conclusions

The complexity of the subject matter in the social and behavioral sciences requires that to adequately predict or explain phenomena of interest, multiple determinants must be used. MRC provides powerful tools for analyzing the combined and independent contributions of multiple potential determinants, particularly when experimental control is not possible. However, MRC is not without its limitations and complexities, although most of these have greater relevance for explanation than for prediction.

Statistical control should never be equated with experimental control. MRC's partialing procedures cannot rule out the possibility of reverse causation and may result in altered variables that have no reasonable theoretical meaning or representation in practical reality. Also, third variables not included in the analysis, or controlled for experimentally, remain as alternative explanations of results. Furthermore, the values of MRC indexes are specific to the variables, and in some cases the subjects, included in the analysis. Therefore, it is important to evaluate the degree to which all relevant, and no irrelevant, variables have been included. As in all research, the reliability and validity of the measures for the purposes of the investigation, as well as the representativeness of the samples, should be evaluated with great care.

These caveats and complexities might appear formidable and bring into question the general utility of MRC. However, the subject matter in the social and behavioral sciences is also highly complex and our ability to exert experimental control is limited. When formal theory and knowledge of MRC's limitations are used to guide the interpretation of results, MRC is a powerful scientific tool for use with experimental, quasi-experimental, and correlational research designs.

Suggestions for Further Reading

Several useful texts and articles provide further explanation and greater detail about MRC analyses and interpretations than was possible in this chapter. Foremost among these are the thorough and very readable texts by Cohen and Cohen (1983) and Pedhazur (1982). Although some mathematical derivations and explanations are provided in these texts, their emphases are on conceptual understanding of the procedures and interpretations of MRC. They are highly recommended.

Cohen and Cohen (1983) and Pedhazur (1982) focus primarily on the use of MRC for purposes of theoretical explanation. An excellent conceptual presentation of the use of MRC in personality, clinical, and industrial psychology for applied prediction, as well as for theoretical explanation, is provided by Wiggins (1973). Also, for more comprehensive discussions of the use of empirical cross-validation procedures versus shrinkage formulas, see Fowler (1986), Mitchell and Klimoski (1986), and Wiggins (1973, pp. 46–49).

Glossary

ALL-POSSIBLE-SUBSETS MULTIPLE REGRESSION/CORRELATION A form of MRC that, by means of one of several algorithms that have been derived for these purposes, attempts to select from a larger set of predictors those that produce the largest R with, and best prediction of, the criterion. The selection of variables is based solely on the empirical relationships among the variables in the analysis.

BIVARIATE OR ZERO-ORDER CORRELATION COEFFICIENT A coefficient (usually denoted symbolically as r) that indicates, on a scale from -1 to 1, the degree and direction of linear relationship between two variables. When squared (r^2), it is called the *coefficient of determination*; it indicates the proportion of variance that is shared by the two variables.

BIVARIATE REGRESSION/CORRELATION (BRC) Data-analytic procedures that examine the relationship between two variables and the prediction of each of these variables from the other.

CHANGE IN R^2 The difference between the R^2 from two consecutive steps in a hierarchical or a stepwise MRC analysis. It equals the squared semipartial correlation coefficient between the criterion and the additional predictors (or set of variables) that were included in one, but not the other, step in the analysis.

COEFFICIENT OF MULTIPLE DETERMINATION The square of the multiple correlation coefficient (usually denoted symbolically as R^2) indicating the proportion of variance in the criterion that is shared by the combination of predictors in an MRC analysis.

CRITERION OR DEPENDENT VARIABLE The variable to be explained or predicted in a regression/correlation analysis. The *observed criterion* or *criterion score* is the actual empirical score obtained for a subject (usually denoted symbolically as Y for raw scores and as z_Y for standard scores). The *predicted criterion* or *criterion score* is the score predicted for a subject on the basis of the regression equation (usually denoted symbolically as Y' or $\hat{Y}$ for raw scores and as $z_{Y'}$ or $z_{\hat{Y}}$ for standard scores).

CROSS-VALIDATION STUDY A study in which (a) empirical scores on the criterion and the predictors are obtained from a different sample of subjects than was used in the derivation study—this new sample is known as the *cross-validation*, *calibration*, or *hold-out sample*; and (b) these data are used to evaluate the results of the derivation study. This evaluation is accomplished by actually using the multiple regression equation from the derivation study with the cross-validation sample to predict their scores on the criterion from their empirically obtained scores on the predictors. These predicted scores for the cross-validation samples are then correlated with their empirically obtained criterion scores to obtain the *cross-validated multiple correlation* (usually denoted symbolically as R^2_{cv}); an empirical form of a *shrunken R*.

DERIVATION STUDY A study in which (a) empirical scores are obtained on the criterion and the predictors—the sample of subjects used in this study is known as the *derivation*, *training*, or *screening sample*; and (b) these empirical scores are used to derive the regression equation and other indexes of association between predictors and the criterion.

DIFFERENTIAL WEIGHTING Obtaining a predicted score for the criterion by multiplying the scores for each predictor by that variable's regression coefficient and then summing these products. Because each predictor has a unique regression coefficient, there is differential weighting of predictors. *Also see* UNIT WEIGHTING

DUMMY OR EFFECT CODING A method of including categorical variables in an MRC analysis. Instead of using the original scale, these variables are transformed into dichotomous variables (there will be one less variable than number of categories in the original scale), indicating the presence or absence of a specific category.

EXPERIMENTAL CONTROL Controlling for potential third-variable explanations by actually, as opposed to statistically, holding a variable constant or equating on that variable. For example, random assignment of subjects to groups attempts to equate, or equally distribute, subject characteristics across groups. Also, using only females in that study holds sex constant.

HIERARCHICAL MULTIPLE REGRESSION/CORRELATION A form of MRC that consists of a series of simultaneous MRC analyses in which one or more new predictors are added to those used in the previous analysis. The decision concerning which variables to add at each point in the series is made by the investigator.

HOMOSCEDASTIC/HETEROSCEDASTIC The degree to which the variance of criterion scores is the same across the various values of the predictors. For example, homoscedasticity would exist if the variance of happiness scores (i.e., the criterion) was essentially the same for 50-, 60-, and 70-year olds (i.e., the predictor of age). If the variance of happiness scores differed for people of different ages, then the data would be heteroscedastic. It is important because summary statistics from regression/correlation analyses, such as correlations and standard errors, average across levels of the predictors. Thus, they only give reasonable estimates for all levels when the data are relatively homoscedastic.

INDEPENDENT CONTRIBUTION The relationship of a predictor to the criterion after the relationships of all other predictors in the study have been statistically controlled for.

INTERCEPT An element of the raw score regression equation (usually denoted symbolically as a) that indicates the criterion score when all of the predictors equal zero. It is not an element of the standard score regression equation because it always equals zero in standard form.

LEAST SQUARES SOLUTION OR CRITERION The criterion used to determine the values of the regression coefficients and intercept in bivariate and multiple regression equations. The values are chosen that minimize the sum, across subjects, of the squared differences between the predicted and observed scores on the criterion, that is, the values that minimize the sum of the squared errors in prediction.

MEASUREMENT ERRORS The existence of problems with the reliability or validity of the procedures used to measure the variables in an MRC study.

MULTICOLLINEARITY This term refers to the intercorrelations among the predictors in an MRC analysis. Either it indicates that these correlations have surpassed some arbitrary cutoff level, and thus are too high for MRC to be applicable, or it descriptively refers to the degree of intercorrelation among the predictors (e.g., "high, medium, or low multicollinearity exists in this study").

MULTIPLE CORRELATION COEFFICIENT A result of an MRC analysis (usually denoted symbolically as R) that, on a scale from 0 to 1, indicates the degree of linear relationship (a) between the criterion and the combination of predictors and (b) between the actual observed and predicted values of the criterion.

MULTIPLE REGRESSION/CORRELATION (MRC) Data-analytic procedure, based on the least squares criterion, that determines the linear relationships between a set of predictors and a single criterion and determines the best combination of the set of predictors for predicting the single criterion.

MULTIPLE REGRESSION EQUATION Result of an MRC analysis denoting the weighted linear combination of predictors that, based on the least squares criterion, provides the best prediction of the criterion. There are both *raw score* and *standard score* forms of these equations.
Raw Score Multiple Regression Equation:

$$Y' = a + B_1X_1 + B_2X_2 + \cdots + B_KX_K$$

Standard Score Multiple Regression Equation:

$$z_{Y'} = \beta_1z_1 + \beta_2z_2 + \cdots + \beta_Kz_K,$$

where Y' and $z_{Y'}$ denote the predicted criterion, X and z denote the predictors, B is the raw partial regression coefficient, β is the standardized partial regression coefficient, and a is the intercept. *Also see* CRITERION, PREDICTOR, PARTIAL REGRESSION COEFFICIENT, and INTERCEPT. (In *bivariate regression*, there is only one predictor and one weight.)

PARTIAL CORRELATION COEFFICIENT A coefficient (usually denoted symbolically as $r_{Y1 \cdot 2,3 \ldots k}$ or r_p) that indicates, on a scale from -1 to 1, the degree and direction of linear relationship between two variables (Y and X_1) when the effects of one or more other variables have been removed from both variables Y and X_1. When squared ($r^2_{Y1 \cdot 2,3\,..\,k}$ or r^2_p), it indicates the proportion of variance in Y shared by X_1 after variance shared with the other predictors has been removed from both Y and X_1.

PARTIAL REGRESSION COEFFICIENT OR WEIGHT The number specified in the regression equation by which the score on the predictor is multiplied when predicting scores on the criterion. Each predictor has its own, unique, coefficient. Although the *partial* is often dropped when referring to these coefficients, they are *partial* coefficients because they represent the effects of a predictor after partialing out the effects of all other predictors. (In bivariate regression, the regression coefficient is not a partial because with only one predictor, nothing is partialed out.) There are two forms of these coefficients: *raw score partial regression coefficients* (usually denoted symbolically as B) and standard score partial regression coefficient (usually denoted symbolically as β).

PREDICTORS OR INDEPENDENT VARIABLES The variables that are used to explain or predict the criterion in an MRC analysis (usually denoted symbolically as X_K for raw scores and z_K for standard scores, where K is a number label indicating a specific predictor). Scores for these variables are always obtained empirically.

REDUNDANT VARIANCE Variance in the criterion that is shared by two or more predictors. Thus, this variance can be predicted or accounted for by any of these predictors, that is, they are redundant for the purpose of predicting this overlapping variance.

RESIDUAL OR ERROR SCORES Scores computed by subtracting subjects' criterion scores predicted on the basis of the regression equations from their actual observed criterion scores.

SEMIPARTIAL OR PART CORRELATION COEFFICIENT A coefficient (usually denoted symbolically as $r_{Y(1 \cdot 2,3\,..\,k)}$ or r_{sp}) that indicates, on a scale from -1 to 1, the degree and direction of linear relationship between two variables (Y and X_1) when the effects of one or more other variables ($X_2, X_3, .. X_k$) have been removed from variable X_1 only. When squared ($r^2_{Y(1 \cdot 2,3\,..\,k)}$ or r^2_{sp}), it indicates the proportion of variance in Y shared by X_1 after variance shared with the other predictors has been removed from X_1 only.

SHRINKAGE, SHRINKAGE FORMULA, SHRUNKEN OR ADJUSTED R The results of an MRC analysis reflect relationships among variables that are, in part, specific to the sample of subjects used in the derivation study. Therefore, the measures of association and the accuracy of prediction are expected to be lower (i.e., to shrink) when the regression equation is used with another sample of subjects. This is known as *shrinkage.* Formulas intended to estimate expected shrinkage are known as *shrinkage formulas*, and an estimated multiple correlation coefficient from one of these formulas is known as a *shrunken* or *adjusted R* (usually denoted symbolically as R_a) and usually reported in the form of a *shrunken* or *adjusted* R^2_a.

SIMULTANEOUS MULTIPLE REGRESSION/CORRELATION A form of MRC that examines the contributions of all predictors at the same time rather than by adding or subtracting variables one at a time.

SPECIFICATION ERRORS Violations of any of the following three assumptions of MRC: (a) The relationships among variables are linear, (b) all relevant predictors have been included in the analysis, and (c) no irrelevant predictors have been included in the analysis.

STANDARD ERROR OF A GIVEN PREDICTED SCORE This standard error (usually denoted symbolically as $s_{y'}$) indicates the error when predicting from a given set of values for the predictors. It is more precise than the standard error of estimate (which is an average error) for building confidence intervals around a specific predicted score.

STANDARD ERROR OF ESTIMATE OR PREDICTION A result of an MRC analysis (usually denoted symbolically as SE_{est} or $SE_{Y\text{-}Y'}$) that is the standard deviation of the distribution of error, or residual, scores. It indicates the average magnitude of error in predicting values of the criterion.

STATISTICAL CONTROL Controlling for potential third-variable explanations by using statistical, rather than experimental, methods. That is, in the empirical data, the third variable does pose a potential confound, but statistical procedures (such as MRC) are used to attempt to statistically eliminate this confound. Also known as *partialing*, *controlling for*, *residualizing*, *holding constant*, and *covarying*.

STEPWISE MULTIPLE REGRESSION/CORRELATION A form of MRC that consists of a series of simultaneous MRC analyses, each of which constitutes a step. At each step, one or more new predictors are either added to (in *forward inclusion*) or subtracted from (in *backward elimina-*

tion) those used in the previous step. The decision concerning which variables to add or subtract at each step is determined entirely on the basis of the empirical relationships among the variables in the analysis.

SUPPRESSOR VARIABLE A predictor that is highly correlated with other predictors but that has a small correlation with the criterion. The effect of a suppressor variable is to partial out of the other predictors variance that is irrelevant to the criterion, resulting in larger relationships with, and prediction of, the criterion.

THIRD-VARIABLE EXPLANATIONS When the causal explanation of the relationship between two variables is attributed to another (third) variable, to which each of the two variables is also related. For example, when a relationship between age and happiness in adults is caused by differential physical health.

UNIT WEIGHTING Obtaining a predicted score for the criterion by simply summing the scores on the predictors without first multiplying each predictor score by that variable's partial regression coefficient. This is the equivalent of setting all regression coefficients or weights to 1 (i.e., unity). *Also see* DIFFERENTIAL WEIGHTING

References

Achen, C. H. (1982). *Interpreting and using regression*. Beverly Hills, CA: Sage.

Cohen, J. (1990). Things I have learned (so far). *American Psychologist, 45*, 1304–1312.

Cohen, J., & Cohen, P. (1983). *Applied MRC analysis for the behavioral sciences* (2nd ed.). Hillsdale, NJ: Erlbaum.

Cronbach, L. J., & Gleser, G. C. (1965). *Psychological tests and personnel decisions* (2nd ed.). Urbana: University of Illinois Press.

Dawes, R. M., Faust, D., & Meehl, P. E. (1989). Clinical versus actuarial judgment. *Science, 243*, 1668–1674.

Fowler, R. L. (1986). Confidence intervals for the cross-validated multiple correlaton in predictive regression models. *Journal of Applied Psychology, 71*, 318–322.

Hays, W. L. (1988). *Statistics* (4th ed.). Chicago: Holt, Rinehart & Winston.

Kerlinger, F. N. (1986). *Foundation of behavioral research* (3rd ed.). New York: Holt, Rinehart & Winston.

Lewis-Beck, M. S. (1980). *Applied regression: An introduction*, Beverly Hills, CA: Sage.

Linn, R. L. (Ed.). (1989). *Educational measurement* (3rd ed.). London: Cassel & Collier Macmillian.

Mitchell, T. W., & Klimoski, R. J. (1986). Estimating the validity of cross-validity estimation. *Journal of Applied Psychology, 71*, 311–317.

Pedhazur, E. J. (1982). *Multiple regression in behavioral research: Explanation and prediction* (2nd ed.). New York: Holt, Rinehart & Winston.

Steinberg, L., Elmen, J. D., & Mounts, N. S. (1989). Authoritative parenting, psychosocial maturity, and academic success among adolescents. *Child Development, 60*, 1424–1436.

Tatsuoka, M. M. (1975). *The general linear model: A new trend in analysis of variance*. Champaign, IL: Institute for Personality and Ability Testing.

Wiggins, J. S. (1973). *Personality and prediction: Principles of personality assessment*. Reading, MA: Addison-Wesley.

Willshire, D., Kinsella, G., & Prior, M. (1991). Estimating WAIS–R IQ from the National Adult Reading Test: A cross-validation. *Journal of Clinical and Experimental Neuropsychology, 13*, 204–216.

3 Path Analysis

Laura Klem

Path analysis can be viewed as an extension of *multiple regression*. In multiple regression, the researcher is interested in predicting to a single dependent variable. In a path analysis, there is more than one dependent variable. The concern in path analysis is with the predictive ordering of variables. The model "*X* causes *Y*" is a regression model, whereas "*X* causes *Y* and *Y* causes *Z*" is a path analysis model. Path analysis allows a researcher to test a theory of causal order among a set of variables.

I aimed in this chapter to provide enough information so that the reader could follow (and perhaps even perform) a sensible path analysis. In addition, I tried to alert the reader to the limitations and assumptions of the technique. Those were my two main goals. I also had in mind a possible fringe benefit to the reader. Path analysis is a special case of *covariance structure analysis*: If the chapter encourages the reader to investigate covariance structure analysis further, that is yet another gain.

The chapter is divided into three main sections. The first section, the Overview, contains the basic elements of path analysis models and diagrams and explains how to obtain path coefficients. I begin the Overview with some background, and I then use two examples to illustrate these ideas. The second section, Statistical Concepts, is a bit more technical. It explains how to manipulate path coefficients (by multiplying and adding them according to special rules) to assess indirect effects of variables and test models. The two examples previously introduced, plus a third example, are analyzed using the new tools. The final section, Assumptions and Issues, deals with the assumptions of path analysis and offers some ways of thinking about them. Also, a short discussion of the standardization issue is included in this section. Brief Concluding Comments consist of a recommendation concerning actually doing path analyses.

Overview

History and Applications

Path analysis was first used in 1918 by Sewall Wright and described by him in papers in the 1920s and 1930s. Wright, a geneticist, used path analysis primarily to study population genetics. Many of his papers have intriguing titles, for example, there was one involving the piebald pattern of guinea pigs (1920) and one about corn and hog correlations (1925). Wright himself (1960) singled out his 1921 article published in the *Journal of Agricultural Research* as the first general account of path analysis. However, because his work was published in journals seldom read by social scientists, it went virtually unnoticed by them for 40 years.

Then, in the 1960s, sociologists, psychologists, economists, and political scientists began writing about the technique, bringing it to the attention of social scientists. H. M. Blalock, Jr., and O. D. Duncan, both sociologists, were particularly influential in this respect. *Causal Models in the Social Sciences* (1971), a volume edited by Blalock, is a general overview of causal modeling, with an interdisciplinary emphasis; it captures the enthusiasm of social scientists as path analysis became widely used just over 20 years ago. Since then path analysis has been applied to a wide variety of problems. A sense of the diversity can be gleaned from a sample of topics: self-esteem, health, teacher expectations, moral judgment development, organizational commitment, transit system maintenance performance, medical residency, attitudes toward microcomputers, and ethical-decision-making behavior (Bachman & O'Malley, 1977; Boldizar, Wilson, & Deemer, 1989; Curry, Yarnold, Bryant, Martin, & Hughes, 1988; DeCotiis & Summers, 1987; Igbaria & Parasuraman, 1989; Jussim, 1989; Obeng, 1989; Roth, Wiebe, Fillingim, & Shay, 1989; Trevino & Youngblood, 1990).

The most recent, quite exciting, developments in path analysis began in the mid-1970s, when computer programs for general covariance analysis became available. Covariance analysis programs can be used to perform elegant path analyses among other things. (The advantages of using such software are discussed later in this chapter.) The new software has brought with it a resurgence of interest in the technique: Path analyses are being performed and published, and they are more thorough and sophisticated than ever before.

The Model

The starting point for a path analysis is a researcher's theory about the causal relationships among a set of variables. The theory is then expressed formally and explicitly by a model: The model is usually presented both in words and by a *path diagram* (and sometimes by a set of mathematical equations). A path diagram can be viewed as a compact statement of a set of hypotheses. Explicitness (required to lay out a model) and compactness (achieved by the conventions of a path diagram) are two of the characteristics that make path analysis so attractive.

Data for a Path Analysis

To perform a path analysis, one needs data for each of the variables in the model. Each variable should be measured on an interval scale, or possibly an ordinal scale that one believes can be treated as interval. The number of cases needed depends on the complexity of the model; most models require at least 200 or 300 cases.

Results of a Path Analysis

There are two major kinds of results. First, a path analysis provides estimates of the magnitude of the hypothesized effects. It is important to appreciate that the estimates obtained are conditional on the model being correct: In other words, they are estimates of the size of the effects, under the assumption that the model is correct. Second, a path analysis (usually) allows one to test that the model is consistent with the observed data. If the model is not consistent with the observed data, one can reject the model as being very unlikely. On the other hand, if the model and data are consistent, one can say that the model is plausible. One cannot prove that a path analysis model is correct because different models can be consistent with the same observed data.

Example One: Recalled Symptoms

Many researchers have found that neurotic individuals tend to overreport physical symptoms. That is, there is a positive correlation between neu-

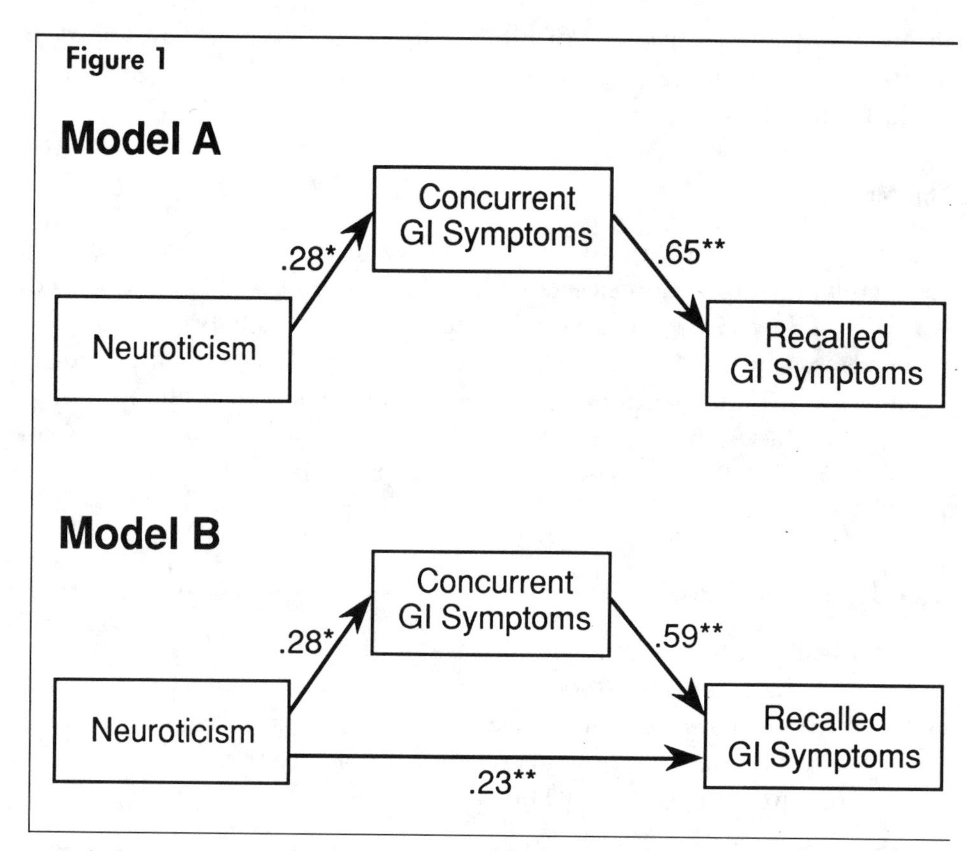

Path diagrams depicting the relation between neuroticism and recall with standardized coefficients. GI = gastrointestinal. The coefficients shown were calculated using the correlations in Table 1. Due to rounding, one coefficient differs from Larsen's (1992) by .01. *$p < .05$, one-tailed. **$p < .01$, one-tailed. From "Neuroticism and Selective Encoding and Recall of Symptoms: Evidence from a Combined Concurrent–Retrospective Study" by R. J. Larsen, 1992, *Journal of Personality and Social Psychology, 62,* Figure 1, page 481, and Figure 2, page 484. Copyright 1992 by The American Psychological Association. Adapted by permission of the author.

roticism and recall of symptoms. Larsen (1992) conducted a study to compare two models to account for the correlation. His two models are shown in Figure 1; the arrows represent hypothesized effects.[1] There are three variables in each of these models: (a) neuroticism, a score from a scale consisting of 24 items; (b) concurrent symptoms, an average based on reports of gastrointestinal symptoms three times a day for 2 months; (c) recalled symptoms, a score based on questions at the end of the study

1. Larsen (1992) studied four kinds of symptoms, one psychological and three physical. Only the gastrointestinal data are used for this example.

Table 1

Bivariate Pearson Product–Moment Correlations for Recalled-Symptoms Data

Variable	1	2	3
1. Neuroticism	–		
2. Concurrent GI symptoms	.28	—	
3. Recalled symptoms	.39	.65	—

Note. GI = gastrointestinal. From "Neuroticism and Selective Encoding and Recall of Symptoms: Evidence From a Combined Concurrent–Retrospective Study" by R. J. Larsen, 1992, *Journal of Personality and Social Psychology, 62,* p. 483. Copyright 1992 by the American Psychological Association. Adapted by permission of the author.

about gastrointestinal symptoms during the last 2 months. The bivariate Pearson product–moment correlations between these variables are given in Table 1. The subjects were 43 undergraduate students. (Although 43 cases would be too few for most path analyses, they were sufficient for this simple model.)

In Model A, the effect of neuroticism on recall is all *indirect*: Neuroticism affects how symptoms are perceived at the time they occur, and that perception affects the frequency later recalled. In Model B, the effect of neuroticism on recall is both indirect, through perception as symptoms occur, and also *direct*: The direct effect expresses an additional hypothesis that neurotic individuals are more likely to recall symptoms as being worse than they were perceived at the time. The numbers above the arrows are standardized path coefficients. They are estimates of the size of the effects, if the model is the correct model. (Note that the magnitude of the effect of concurrent symptoms on recall is different in the two models: Although the difference is not pronounced in this example, it is a reminder that estimates are conditional on the model.) Standardized path coefficients are on the same scale as Pearson product–moment correlations; however, they differ from Pearson product–moment correlations in that they occasionally fall outside the range of -1.0 to 1.0.

In both of these models, all the effects are significant. Given the research question, it is of particular interest that the direct effect of neuroticism on recall in Model B is significant. Regarding the plausibility of the models, when I later discuss *implied correlations*, I show that Model A does not fit the data very well and is almost certainly incorrect. When I discuss *fully recursive models* (models that include all of the direct links allowed by the causal ordering), I show that although Model B is consistent with the data, and may well be the correct model, it cannot be tested.

Endogenous and Exogenous Variables

There are two kinds of variables in a path model, *endogenous* and *exogenous*. The values of endogenous variables are explained by one or more of the other variables in the model. The values of exogenous variables are taken as given; the model does not attempt to explain them. The distinction is similar to that between dependent variables (endogenous) and independent variables (exogenous). However, in a path model, a variable can be both independent and dependent. For example, in the recalled-symptoms models, concurrent symptoms are caused by neuroticism and in turn cause recalled symptoms. Concurrent symptoms are endogenous; the rule is that if a variable is dependent in any part of the model, the variable is endogenous. Recalled symptoms are also endogenous. On the other hand, neuroticism is an exogenous variable in this model because the model does not undertake to explain it. The variables that explain neuroticism are external to the model.

Example Two: Remarriage and Well-Being

Greene (1990) used path analysis to study the positive relationship between remarriage of older widowers and their feelings of well-being. A sample of 335 men who became widowed was drawn from a larger data set. The study was longitudinal: Men were interviewed once before they became widowed and twice afterwards. One of the models Greene examined she called the *expanded selectivity model*. The theory behind this particular model, presented in Figure 2, is that differences in well-being between remarried and not remarried widowers can be accounted for by prior well-being: widowers who remarry are men with higher well-being in the first place.[2] This path model has three exogenous variables: prior health, prior wealth, and education. It has five endogenous variables: prior well-being, health, wealth, remarriage, and well-being. Note that there are no direct or indirect effects of remarriage on well-being. The hypothesis being tested by the model is that the relationship between

2. Time 2 and Time 3 data were used for this model. I have used the word *prior* to refer to Time 2 data.

Figure 2

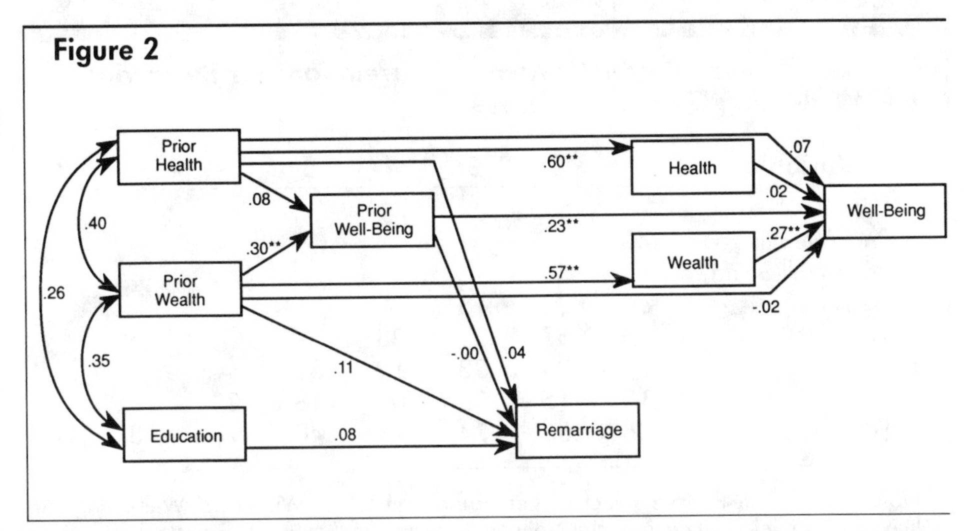

Path diagram depicting the expanded selectivity model of the relation between remarriage and well-being with standardized coefficients. The coefficients were calculated using the correlations in Table 2. Due to rounding, four coefficients differ from Greene's (1990) by .01. $^*p < .05$, one-tailed. $^{**}p < .01$, one-tailed. From "The Positive Effect of Remarriage on Older Widowers' Well-Being: An Integration of Selectivity and Social Network Explanations" by R. W. Greene, 1990 (Doctoral dissertation, University of Michigan, 1990), *Dissertation Abstrates International, 52,* 297A, p. 71. Copyright 1990 by R. W. Greene. Adapted by permission.

remarriage and well-being can be accounted for by common antecedents. The bivariate Pearson product–moment correlations for the variables are given in Table 2.

Calculation of Path Coefficients

The traditional and simplest way to estimate path coefficients is to use multiple regression (see chapter 2). One regression analysis is needed for each endogenous variable in the path model. (An easy way to spot endogenous variables is to look in the diagram for variables that have one or more straight arrows pointing at them. The model is intended to explain variance in those variables.) For each regression, the dependent variable is the endogenous variable, and the predictors are all of the variables that have arrows pointing directly at them. Two regression analyses were required to obtain the path coefficients for the recalled symptoms model in Figure 1. The variables for each of these regressions are listed in the top portion of Table 3. Five regression analyses were required to obtain the path coefficients for the remarriage and well-being model

Table 2

Bivariate Pearson Product–Moment Correlations for Remarriage and Well-Being Data

Variable	1	2	3	4	5	6	7	8
1. Prior Health	–							
2. Prior Wealth	.40	–						
3. Prior Well-Being	.20	.33	—					
4. Health	.60	.32	.18	—				
5. Wealth	.36	.57	.33	.33	—			
6. Well-Being	.21	.24	.33	.18	.36	—		
7. Remarriage	.10	.15	.06	.07	.18	.33	—	
8. Education	.26	.35	.24	.18	.40	.18	.13	—

Note. From "The Positive Effect of Remarriage on Older Widowers' Well-Being: An Integration of Selectivity and Social Network Explanations" by R. W. Greene, 1990 (Doctoral dissertation, University of Michigan, 1990), *Dissertation Abstracts International, 52,* 297A, p. 60. Copyright 1990 by R. W. Greene. Adapted by permission.

Table 3

Variables Used in Regression Analyses to Obtain Path Coefficients

Model	Dependent variable	Independent variables
Recalled symptoms	Concurrent symptoms	Neuroticism
	Recalled symptoms	Neuroticism Concurrent symptoms
Remarriage and well-being	Prior well-being	Prior health Prior wealth
	Remarriage	Prior health Prior wealth Prior well-being Education
	Health	Prior health
	Wealth	Prior wealth
	Well-being	Prior health Prior wealth Prior well-being Health Wealth

in Figure 2.[3] The variables for these regressions are listed in the bottom portion of Table 3.

Path coefficients are usually, as in these examples, displayed on a path diagram. Each path coefficient is the regression coefficient from the appropriate regression analysis. The other regression results, most notably the standard errors and significance of each coefficient and the amount of variance explained by the predictors jointly (the R^2), are also relevant to the path analysis.

Usually multiple regression is performed by means of a computer program. Most multiple regression computer programs can start either from raw data (scores for each case on each variable) or from a correlation matrix (a matrix that contains the correlation between each pair of variables). If raw data is the starting point, the program calculates a correlation matrix as the first step in the regression analysis. Path coefficients for this chapter were calculated by means of multiple regression that started from correlation matrices.

Conventions for Drawing Path Diagrams

The following three practices are customary when drawing path diagrams: (a) Diagrams are drawn so that the causal flow is from left to right. (b) Causal relationships among variables in the model are shown by one-headed arrows: The arrows indicate the direction of the hypothesized causal influence. (c) Relationships among exogenous variables (at the left of the diagram) are shown by two-headed arrows: These relationships are hypothesized to exist, but their causal structure is not explicated by the model.

Sometimes, because of printing constraints or to keep a diagram from appearing cluttered, the conventions are modified. For example, curved two-headed arrows, which represent relationships between exogenous variables, may be shown as straight two-headed arrows or may be omitted from the diagram altogether. Sometimes, as in Figures 1 and 2, the arrows and coefficients for the residual paths are omitted. (Residual paths are paths for causes that are omitted from the model. For the path diagrams in Figures 1 and 2 to be complete, each endogenous variable

---

3. Because remarriage was a dichotomous variable, Greene (1990) supplemented her regression analysis with a logistic regression analysis (see chapter 7). The results of the two analyses were in general agreement.

needs an additional arrow pointing at it that is the path for a residual variable that represents unknown causes that are omitted from the model.)

Each path coefficient is represented with p. In a report, letters or numbers are used for variables. By convention, particular path coefficients are identified by subscripts, with the subscript for the dependent variable coming first. If, for example, remarriage was identified by r and prior wealth was identified by w, then p_{rw} would refer to the arrow in Figure 2 that goes from prior wealth to remarriage.

In summary, this first section of the chapter has presented the basic ideas of path analysis. The crucial element, that a path analysis begins with a theoretical model, was discussed. Path diagrams and some specialized vocabulary were introduced. Also discussed was the use of multiple regression to estimate the parameters of the model. The next section examines ways in which those parameter estimates, that is, the path coefficients, can be used to understand and test the model.

Statistical Concepts

Estimating Direct and Indirect Effects

Direct effects are shown in path diagrams by straight arrows from one variable to another. For example, the model portrayed in Figure 2 posits a direct effect of prior wealth on remarriage. As previously indicated, direct effects are estimated by regression coefficients, typically read directly from a computer printout; for example, the direct effect of prior wealth on remarriage was estimated at .11. The model in Figure 2 also posited an indirect effect of prior wealth on remarriage: Prior wealth affects prior well-being, and prior well-being in turn affects remarriage. Indirect effects involve chains of straight arrows, where the path along the arrows is always forward (in the direction of the arrow). These are called *compound paths*. To estimate the magnitude of an indirect effect of one variable on another, locate all of the indirect routes, that is, all of the compound paths, by which influence can flow from one variable to the other. Then, for each route, multiply the path coefficients to obtain their product. Finally, add the products to get the indirect effect. The indirect effect of prior wealth on remarriage is $-.00$ ($.30 \times -.00$). (The path coefficient from prior well-being to remarriage, $-.00$, may be confusing. The coefficient was a small, negative number that rounded to $-.00$.) The effect is through prior well-being. The indirect effect of prior wealth on well-being, .22, involves two routes (.30 × .23) + (.57 × .27). One route is through prior well-being, and one route is through wealth. A variable that intervenes between two variables in a path model is hypothesized to function as a *mediator* between the two variables. The variable does function as a mediator if the path coefficients are sizable enough to establish that some of the causal influence is indeed traveling on the indirect route. Both prior well-being and wealth function as mediators between prior wealth and well-being. That is, prior wealth affects prior well-being and wealth, and these variables affect well-being.

A variable in a path model may have only a direct effect on another variable, only an indirect effect, both, or neither. The sum of the direct and indirect effects is sometimes called the *total effect*; some authors term it the *effect coefficient*.

Calculating and Using Implied Correlations

For each pair of variables in a model, there is a correlation that is implied by the model. The implied correlation is a sum of four components: a direct effect (if any), the sum of indirect effects (if any), and two possible further components, the sum of *spurious effects* and the sum of *unanalyzed effects*. An implied correlation need not have all of the components.[4]

A spurious effect between two variables arises because they have a common cause: In a path diagram, a spurious effect is characterized by a path that goes against the direction of the arrows. In the remarriage and well-being model depicted in Figure 2, all of the relationship between remarriage and well-being is hypothesized to be spurious. That's really the whole notion behind the model: Those two variables are correlated because they have common causes (prior health, prior wealth, and prior well-being). Note that in Figure 2 all of the paths from remarriage to well-being (or vice versa) start by going backward along an arrow (against the direction of an arrow). For example, one path from remarriage to well-being is backward to prior wealth and then forward (in the direction of the arrow) to well-being. The magnitude of a spurious effect is the product of the coefficients for the individual links.

An unanalyzed effect is one that involves a two-headed curved arrow, that is, a correlation between exogenous variables. For example, there is an effect of prior wealth on remarriage that flows through education.

4. There is minor variation among authors in nomenclature for effects. The names used here follow Pedhazur (1982).

This effect, the product of two coefficients, is unanalyzed from a causal perspective because it involves a correlation (between prior wealth and education) for which a causal order is not specified. If a path from one variable to another contains any common cause, the effect is said to be *spurious*. Thus, for example, the path from wealth to prior wealth to education to remarriage is termed *spurious* rather than *unanalyzed*.

The implied correlation between two variables in a model can be analyzed by computing all of the effects and summing them.[5] Often it is interesting to examine the contribution of each of the four types of effect. Always it is interesting to compare the sum, the implied correlation, with the observed correlation: If the implied correlation is very different from the corresponding observed correlation, the model isn't plausible. For example, the observed correlation between remarriage and well-being is .33 (from Table 2). The correlation implied by the model is .04. (There are 27 pathways between the variables. The calculations are given in the Appendix.) The inescapable conclusion is that the expanded selectivity model does not account for the observed relationship between remarriage and well-being. It is not a plausible model.

Similar logic leads to the rejection of Model A for the recalled-symptoms data. The implied correlation between neuroticism and recalled symptoms is .18 (.28 × .65), whereas the observed correlation .39 (from Table 1) is more than twice as big.

Fit of a Model

In the preceding section, I rejected the expanded selectivity model because the single correlation between remarriage and well-being that was implied by the model was not at all close to the actual correlation. For that model, that particular correlation was crucial for theoretical reasons. However, more generally, a model implies a correlation between every pair of variables in the model. Sometimes the observed and implied correlations are displayed in a square matrix, with all of the actual correlations in the triangle below the diagonal and all of the implied correlations in

5. There are rules for determining the legitimate paths, or tracings. The rules can be expressed in various ways. Following Kenny (1979), for the types of models discussed in this chapter, the set of tracings between two variables includes all the possible routes between the variables, given that the following two rules are followed. First, a path may not pass through the same variable twice. Second, a path may not enter a variable on an arrowhead and leave the variable on an arrowhead. The second rule, among other things, means that on any given tracing, there cannot be more than one unanalyzed effect.

Table 4

Hypothetical Data

Variable	A	B	C
A	—		
B	.50	—	
C	.10	.20	—

the triangle above the diagonal. Sometimes the difference between observed and implied correlation for each pair of variables is displayed in a triangular matrix.

A comprehensive measure of fit involves comparing all of the implied correlations to all of the actual correlations, rather than checking just one or two. The average of the absolute values of all of the differences is one measure of the fit of the model.[6] In practice, when multiple regression is used to estimate the parameters, researchers usually calculate the implied correlations only for relationships of particular theoretical interest. Another possibility, when multiple regression has been used, is to hand compute *Q*, a measure of fit that answers the question of how well a model reproduces the complete matrix.[7] *Q* can vary from zero to one; values near one signify a good fit of the model to the data. However, the use of *Q* is somewhat old-fashioned. As discussed later in the chapter, one benefit of using software for covariance structure models to estimate path coefficients is the computer printout, which provides overall measures of fit. If overall fit of the model is of interest—and, unless a model is fully recursive, it usually is—the use of such software is distinctly advantageous.

Finally, note two things that have nothing to do with fit. First, fit has nothing to do with the magnitude of coefficients or the amount of variance explained in the endogenous variables. To illustrate this, consider the hypothetical data in Table 4 and the model at the top of Figure 3. The correlation between *A* and *C* that is implied by the model is small, $(.5 \times .2) = .10$. But it is exactly equal to the observed correlation. Thus,

6. I know of no guideline for an acceptable level for the average of the absolute differences. However, such averages can be used to compare the fits of different models to the same data, with the smaller number indicating the better fit. To get a sense of fit for a single model, it is helpful to compare the magnitude of the average absolute difference with the range of the original correlations.
7. The formula for *Q* can be found in Pedhazur (1982, p. 619).

Figure 3

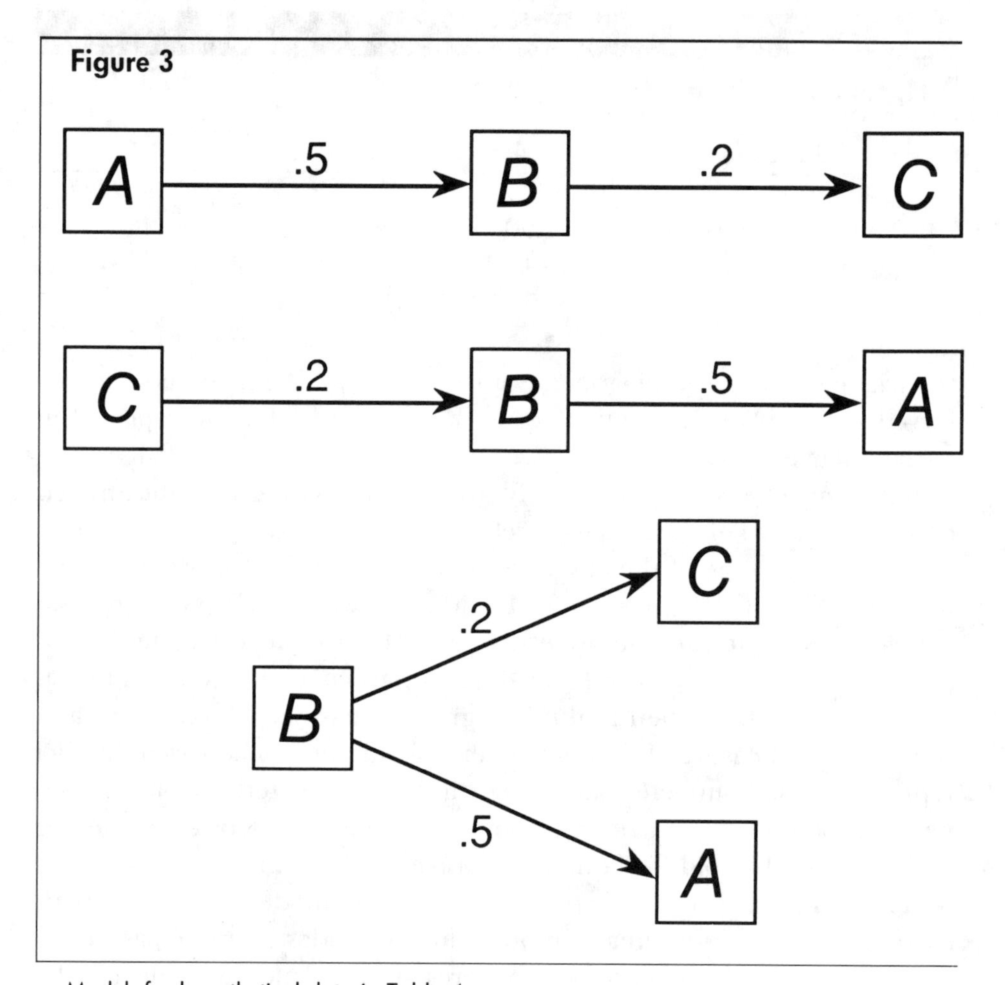

Models for hypothetical data in Table 4.

the model perfectly fits the data, although the percentage of variance in *C* that the model explains is slight (4%).

A second thing to note about fit is that it does not confirm the correctness of the model. The two lower models in Figure 3 also explain the observed data perfectly. It is not possible that all three models are correct; in fact, none of them may be correct.

Residual Path Coefficients and Error Variances

As noted above, in a path analysis, a multiple regression is performed for each endogenous variable. The R^2 on a multiple regression computer printout is the amount of variance explained in the endogenous variable

Figure 4

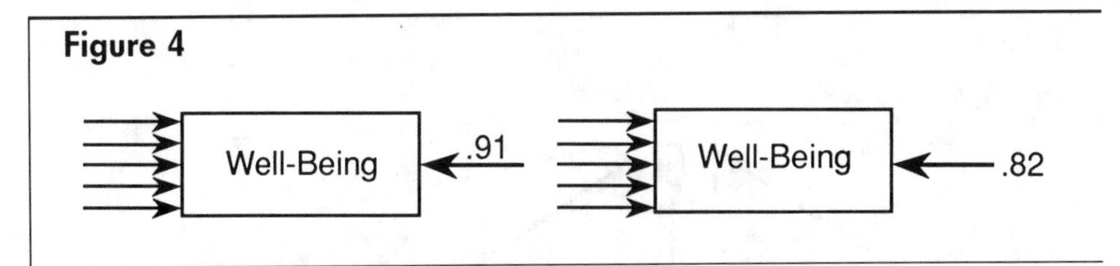

Alternative approaches for diagramming unexplained variance. The endogenous variable is well-being in the expanded selectivity model shown in Figure 2. On the left it is shown with its residual path coefficient; on the right, it is shown with its error variance.

by the variables that directly affect it. In a report of a path analysis, the R^2s are often reported in the text or in a table. In a path diagram (i.e., in the actual figure) the opposite perspective is taken, namely, the amount of variance left unexplained is indicated.

There are two, alternative, traditions for indicating unexplained variance in a path diagram. Recall that in a multiple regression, each case has a residual that is the difference between the observed value for the dependent variable and the predicted value; the residual is the error in the prediction. The older tradition is to standardize the residuals so that they have a variance of one (just as the other variables in the analysis) and report the effect of the residual variable on the endogenous variable. The residual path coefficient is printed above the arrow in the same manner as other coefficients. (In practice, it is not necessary to standardize the residuals: The value of a residual path coefficient is simply $\sqrt{1 - R^2}$ where R^2 is the R^2 from the relevant multiple regression.) Asher (1983) and Pedhazur (1982) gave examples of residual path coefficients. The second approach, which is being used more and more often, is to include the variance of the residuals in the diagram, that is, the variance of the errors, $1 - R^2$. If this is done, the variance is written at the end of the arrow (i.e., the end opposite the arrowhead). This approach is consistent with the way results from covariance analysis are displayed. The two alternative approaches to indicating unexplained variance are shown in Figure 4. Sometimes, to keep a diagram from being too complicated, the residual path coefficients or error variances are not shown in the diagram.

Types of Models

A useful distinction can be made between a fully recursive model and recursive models generally. In a fully recursive model, each variable has

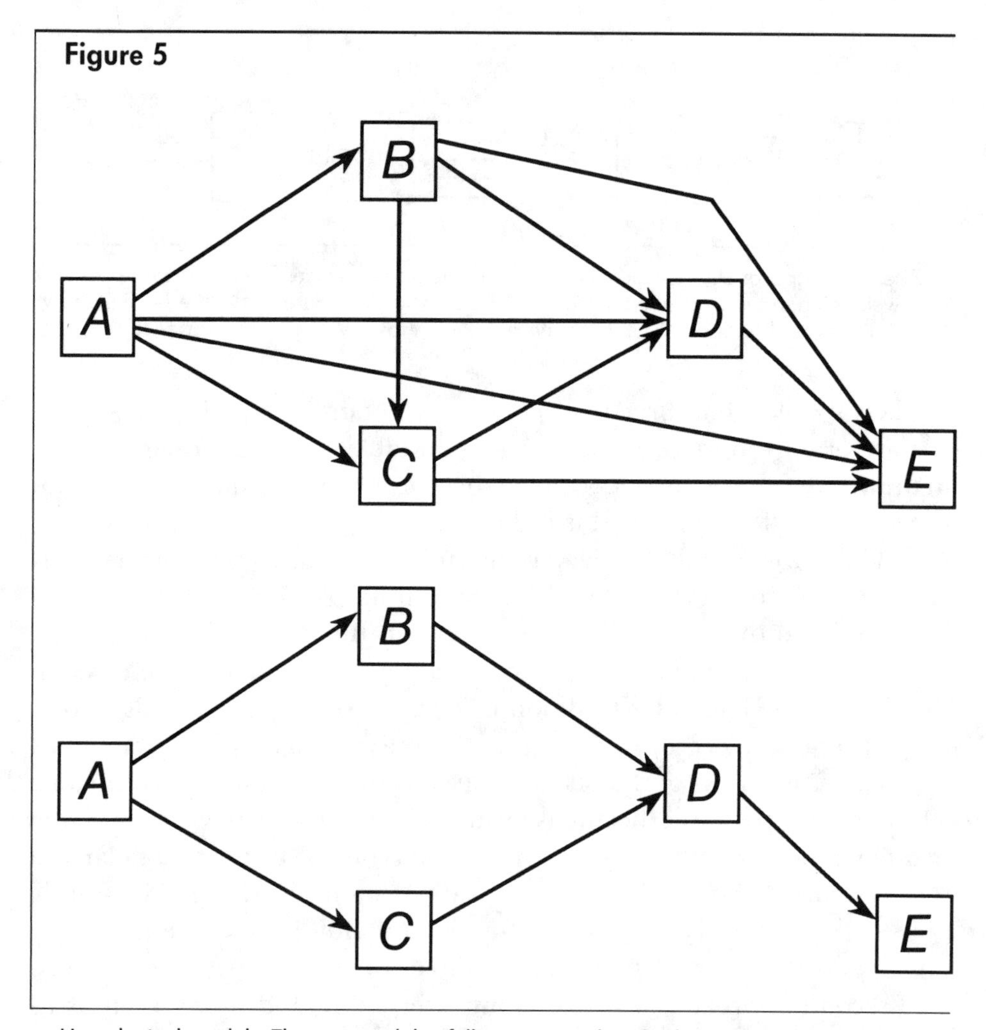

Hypothetical models. The top model is fully recursive, but the bottom model is not.

a direct effect on all of the variables further down the causal chain. The model at the top of Figure 5 is fully recursive: Variable *A* causes variable *B*, variable *A* and variable *B* cause variable *C*, variable *A*, variable *B*, and variable *C* cause variable *D*, variable *A*, variable *B*, variable *C*, and variable *D* cause variable *E*. In a model that is not fully recursive, one or more of the direct links allowed by the causal ordering are missing. The model at the top of Figure 5 would not be fully recursive if one or more of the links were omitted, such as in the model at the bottom of Figure 5. The recalled symptoms Model A is recursive, whereas Model B is fully recursive (Figure 1). The remarriage and well-being model is recursive (Figure 2).

One reason to note whether a model is fully recursive is that fully recursive models always fit the observed data perfectly. Thus, one should not be impressed that the model at the top of Figure 5 fit some set of data: That model would fit any data for five variables. If the model was the one with missing links shown at the bottom of Figure 5 and if the data fit the model, it might be impressive: The model assumes that the effects of variables *A*, *B*, and *C* on *E* are all indirect.

To repeat, if a model is fully recursive, it will always fit any data, and only the magnitude of the parameter estimates is of interest. However, if a model is not fully recursive, then the question of fit is also germane. Thus, for the recalled-symptoms data, the results of interest were the magnitude of the parameter estimates for both models and the rejection of Model A because it did not fit the data. The fit of Model B could not be tested because it was a fully recursive model.

A second useful distinction is between recursive and nonrecursive models. In a recursive model, all of the effects are unidirectional. That is, there is no reciprocal causation among variables (such as variable *A* causes variable *B* and variable *B* causes variable *A*) or loops (such as variable *A* causes variable *B* and variable *B* causes variable *C* and variable *C* causes variable *A*). Furthermore, in a recursive model, the error terms (residuals) are assumed to be uncorrelated with one another. All of the models discussed in this chapter are recursive. Nonrecursive models are often attractive (it is easy to believe that in the real world many variables have reciprocal effects), but it is difficult to estimate their parameters. From a purely mechanical standpoint, they should not be estimated by means of multiple regression. However, that is not much of a problem because there is a variety of appropriate techniques that are incorporated in computer software packages. A more troublesome problem with nonrecursive models is that they are often quite difficult to identify. That is, it is difficult to formulate a nonrecursive model so that there is a unique estimate for each parameter. There may be sets of estimates that fit the data equally well. For example, imagine a model with just two variables, *A* and *B*, that cause each other. That is, there is a path from *A* to *B* and a path from *B* to *A*. The single observed correlation between *A* and *B* does not provide enough information to estimate two unique path coefficients. Two references for nonrecursive models are given in the Suggestions for Further Reading.

Trimming a Model

Often, after the parameters of a model have been estimated, the researcher decides to simplify the model by removing links that are small, insignificant, or both. This is called *trimming* a model. There are two things to remember about a model that has been trimmed.

First, the parameters of the new, trimmed model should be reestimated. If the effect or effects that were removed were small, the new estimates will not vary much from the old estimates, but still it is necessary to reestimate them in order that the parameter estimates reported be the parameter estimates for the model presented. Second, if a model has been trimmed, the results are to some extent dependent on the observed data. That is, the trimming may capitalize on chance relationships in the data. The statistical significance of the parameters in the new model may not be quite as great as the new computer printout reports.

A related procedure, probably a more important threat to the validity of the final results, is snooping around in data before laying out the model. Prior snooping, like trimming, tends to capitalize on chance, should affect one's confidence in the final model, and is contrary to the model-driven perspective of path analysis.

There is a classic way to protect a study from the pitfall of capitalization on chance; this is to use a *hold-out* sample. If possible, that is, if there are enough observations, it is excellent practice to set aside a randomly selected part of the sample (say, one third) during the model development. Then the final model can be tested on the hold-out sample. This procedure is known as *cross-validation*.

Example Three: Two Quality-of-Life Models for Cardiac Patients

Romney, Jenkins, and Bynner (1992) used path analysis to study quality of life in cardiac patients. They reanalyzed quality-of-life data that were collected on patients 6 months after surgery for cardiovascular problems (Jenkins & Stanton, 1984). The data were collected by means of patient interviews, patient questionnaires, and from hospital charts; altogether there were 58 items and scales for each of 469 patients. The data were further analyzed by means of factor analysis by Jenkins, Jono, Stanton, and Stroup (1990), who were interested in discovering the number of factors that underlay the 58 measures and what the factors were. (See chapter 4 for a discussion of factor analysis.) The following five factors, subsequently used by Romney et al., emerged from their analysis: (a) low morale, (b) symptoms of illness, (c) neurological dysfunction, (d) poor interpersonal relationships, and (e) diminished socioeconomic status. The matrix of intercorrelations among the factors scores is given in Table 5.

Romney et al. (1992) hypothesized two competing causal models of

Table 5

Bivariate Pearson Product–Moment Correlations of Factor Scores for Quality of Life for Cardiac Patients Data

Factor	1	2	3	4	5
1. Low morale	—				
2. Symptoms of illness	.53	—			
3. Neurological dysfunction	.15	.18	—		
4. Poor interpersonal relationships	.52	.29	−.05	—	
5. Diminished socioeconomic status	.30	.34	.23	.09	—

Note. From "A Structural Analysis of Health-Related Quality of Life Dimensions" by D. M. Romney, C. D. Jenkins, and J. M. Bynner, 1992, *Human Relations, 45,* p. 170. Copyright 1992 by Tavistock Institute of Human Relations. Adapted by permission.

how the factors affect each other. They then determined how well each model fit the observed data.

The two models are displayed in Figures 6 and 7. The model in Figure 6 is a conventional medical model: Symptoms of illness and neurological dysfunction adversely affect psychosocial and economic factors. The model in Figure 7 is a psychosomatic model: Neurological dysfunction and diminished SES lower morale, which in turn aggravates symptoms and causes poor relationships.

The parameter estimates shown in Figures 6 and 7 were calculated with multiple regression.[8] Three multiple regressions were run for the medical model, using the matrix of correlations as input. Similarly, three regressions were run for the psychosomatic model.

The standardized regression coefficients displayed in Figure 6 were copied from the computer printout. The correlation between symptoms of illness and neurological dysfunction was taken from the input correlation matrix. The three variances of the residuals were calculated by subtracting each of the three R^2s from 1.0. Figure 7 was assembled in a similar manner.

Table 6 contains the direct and indirect effects of the causal variables in each model. The effects were calculated by hand, using the path coefficients displayed in Figures 6 and 7. In keeping with the most common way of reporting indirect effects, the indirect effects shown include only substantively meaningful indirect effects—not spurious and unanalyzable

8. Romney et al. (1992) used LISREL to estimate the parameters of the models.

Figure 6

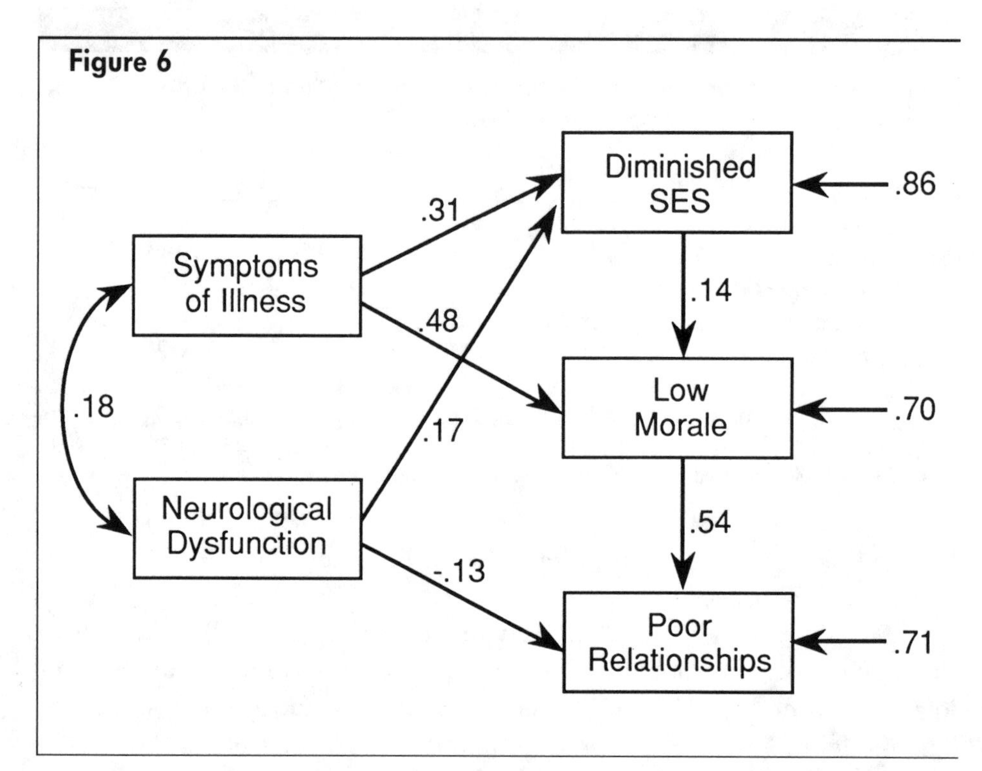

Path diagram depicting medical model of quality-of-life model for cardiac patients. Shown are a correlation (.18), standardized path coefficients (.31, .48, .17, −.13, .14, .54), and variances of residuals (.86, .70, .71). The parameter estimates were calculated from the correlations in Table 5. Due to rounding, one variance differs from Romney et al. (1992) by .01. SES = socioeconomic status. For all values, $p < .05$, two-tailed. From "A Structural Analysis of Health-Related Quality of Life Dimensions" by D. M. Romney, C. D. Jenkins, and J. M. Bynner, 1992, *Human Relations, 45,* p. 172. Copyright 1992 by Tavistock Institute of Human Relations. Adapted by permission.

effects. For example, in Table 6, for the medical model, the indirect effect of symptoms of illness on poor relationships, .28, is calculated (.31 × .14 × .54) + (.48 × .54). Paths passing through the .18 correlation are not included (see Figure 6).

On the whole, the coefficients for these models seem sensible. However, in the medical model, the negative effect of neurological dysfunction on poor relationships seems a little surprising. Also noteworthy are the large error variances. These indicate that there are omitted variables that affect the endogenous variables.

To determine the fit of the models, implied correlations can be calculated for all pairs of variables in each model. These implied correlations are presented in the lower-left triangle of Table 7. These implied correlations by themselves are not very informative. However, comparing

Figure 7

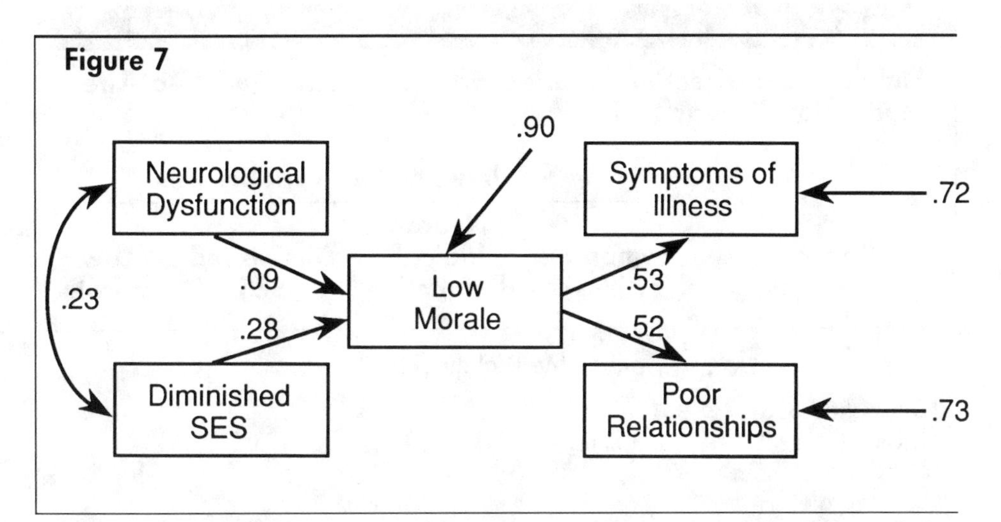

Path diagram depicting a psychosomatic model of quality of life for cardiac patients. Shown are a correlation (.23), the standardized coefficients (.09, .28, .53, .52), and the variances of residuals (.90, .72, .73). SES = socioeconomic status. For all values except the .09, $p < .05$, two-tailed. From "A Structural Analysis of Health-Related Quality of Life Dimensions" by D. M. Romney, C. D. Jenkins, and J. M. Bynner, 1992, *Human Relations, 45*, p. 172. Copyright 1992 by Tavistock Institute of Human Relations. Adapted by permission.

them to the observed correlations in Table 5 allows one to determine how well each model reproduces the original correlations. The absolute values of the differences between the observed and implied correlations are also shown in Table 7, in the upper-right triangle. The average of the absolute differences between the observed and implied correlations, that is, the average absolute residual, for the medical model is .02. The average absolute residual for the psychosomatic model is .05. Thus, the medical model reproduces the original correlations, which ranged from −.05 to .53, quite well, whereas the psychosomatic model does not fit the data as well.[9] The difference in average absolute residual, though not dramatic, favors the medical model.

Competing Hypotheses

When considering—and probably admiring—a model that fits the data, one needs to wonder if there are other theoretically sensible models that might also fit the data.

9. The residuals discussed here represent errors in reproducing the original correlations. They are different from the residuals, or errors, which are the differences between the observed and predicted values for observations.

Table 6

Standardized Direct and Indirect Effects for Two Quality-of-Life Models for Cardiac Patients

Dependent variable	Independent variable			
	Symptoms of illness	Neuro-logical dysfunction	Diminished SES	Low morale
		Medical model		
Diminished SES				
Direct	.31	.17		
Indirect	—	—		
Low morale				
Direct	.48	—	.14	
Indirect	.04	.02	—	
Poor relationships				
Direct	—	−.13	—	.54
Indirect	.28	.01	.08	—
		Psychosomatic model		
Low morale				
Direct		.09	.28	
Indirect		—	—	
Symptoms of illness				
Direct		—	—	.53
Indirect		.05	.15	—
Poor relationships				
Direct		—	—	.52
Indirect		.05	.15	—

Note: SES = socioeconomic status. Dashes indicate that the effect does not exist in the model.

For example, although the medical model fits the quality-of-life data for cardiac patients quite well, it is possible to think of variations on that model that might also fit. (In fact, I was tempted to tinker with the model because of the unexpected negative effect of neurological dysfunction on poor relationships. I tried a model that posited a direct effect of neurological dysfunction on low morale instead of on poor relationships. That is, I moved one arrow. The results of my model were mixed: The coefficients were all positive, which seemed reasonable, but the fit was considerably worse than for the model in Figure 6.)

Table 7

Implied Correlations (Below the Diagonals) and Absolute Values of Residuals (Above the Diagonals) for Two Quality-of-Life Models for Cardiac Patients

Variable	1	2	3	4	5
Medical model					
1. Low morale	—	.01	.03	.01	.01
2. Symptoms of illness	.52	—	.00	.03	.00
3. Neurological dysfunction	.12	.18	—	.02	.00
4. Poor interpersonal relationships	.53	.26	−.07	—	.06
5. Diminished socioeconomic status	.29	.34	.23	.15	—
Psychosomatic model					
1. Low morale	—	.00	.00	.00	.00
2. Symptoms of illness	.53	—	.10	.01	.18
3. Neurological dysfunction	.15	.08	—	.13	.00
4. Poor interpersonal relationships	.52	.28	.08	—	.07
5. Diminished socioeconomic status	.30	.16	.23	.16	—

When evaluating a model, in addition to thinking about alternative ways of arranging the variables in the model, one needs to think about variables that possibly should be in the model but have been omitted. For example, one might try to think of a variable, not in the model, that affected both low morale and poor relationships. A variable such as self-esteem might be a possibility. If there were such a variable, and it were included in the model, it would lower the big direct effect of morale on relationships.

Thinking about alternatives to a model, whether about a rearrangement of the causal ordering of variables already in the model or about the addition or subtraction of relevant variables, amounts to examining some of the assumptions of a path analysis. I now discuss those assumptions more fully.

Assumptions and Issues

Assumptions of Path Analysis

So far this chapter has discussed the nature of path analysis models, how to estimate their parameters, and how to use the estimates to evaluate models. The research value of these models and results is contingent on a set of conditions. The requirements of a path analysis can be grouped into three categories: the assumptions of multiple regression, the assumptions of causal modeling, and general data analysis considerations.

Because path analysis is based on the multiple regression technique, the assumptions of multiple regression, which are discussed in chapter 2, apply here as well. The two most important of these assumptions, because multiple regression is not robust with respect to them, are the assumption of no measurement error and the assumption of no specification error. *Measurement error* refers to inaccuracy in measurement of the observed variables. Multiple regression, and therefore path analysis, assumes that all the variables are accurately measured. *Specification error* refers to inaccuracy in specifying the regression model. The two most worrisome specification possibilities are (a) that a variable is included in the model that does not belong in the model and (b) that a variable that belongs in the model has been omitted (which is even worse). The consequences of violating these assumptions, as well as those stated below, are discussed in the next section.

A second category of path analysis assumptions stems from a consideration of the causal model as a whole. This has to do with whether the causal ordering in the model is correct and whether the correct variables are in the model. (Recall that the results from path analysis are conditional on the proper specification of the model.) Here the notion of specification goes beyond a single multiple regression to encompass the model. For example, an omitted variable that affects two or more of the endogenous variables will cause residuals from different regressions to be correlated, a violation of an assumption. (This violation can be a problem with longitudinal data, where the same variable is measured at different times. Consider the remarriage and well-being example displayed in Figure 2; could there have been, for example, an omitted variable that affected well-being each time it was measured?)

In addition to the statistical assumptions necessary for unbiased estimation of parameters, there are further general data analysis considerations. Multiple regression, and therefore path analysis, is an additive technique. That is, it is assumed that the effect of one predictor on the dependent variable does not depend on the level of another. If interaction among two or more predictors and the dependent variable is suspected, the possibility should be investigated before path analysis is undertaken. If interaction is detected, an interaction term can be incorporated in the regression analysis. For information on constructing interaction terms, see Baron and Kenny (1986) or Jaccard, Turrisi, and Wan (1990). (The detection and handling of interaction are classified here as a general analysis concern: The *use* of an additive model when an interactive model is appropriate is a specification error.)

A second general analysis concern is multicollinearity; if predictor variables are highly intercorrelated, then the parameter estimates are unreliable. On the other hand, modest amounts of multicollinearity are tolerated. Multicollinearity is a problem for path analysis in the same way that it is for multiple regression (see chapter 2).

Finally, there is the question of how big the sample should be to have confidence in the results. The answer depends on how many parameters are being estimated. A general rule of thumb is to have 5–10 as many observations as estimated parameters.[10] (Each straight arrow in a path diagram counts as a parameter, including an arrow for the residuals.)

Effects of Violating Assumptions

The effects of violating the assumptions of multiple regression are stated in chapter 2 of this volume and in many texts. Among the texts, a short, assumption-by-assumption discussion in Lewis-Beck (1980), the longer explanations in Pedhazur (1982), and the monograph by Berry (1993) are particularly useful.

As explained in the sources listed above, in multiple regression, failure to measure variables perfectly has serious consequences, namely, that the estimates of the standardized regression coefficients will be wrong. Baron and Kenny (1986) extended the discussion of this problem to the wider context of causal modeling. In particular, they detailed the effects of measurement error in a mediator. Briefly, the effect of the mediator is underestimated, and the effect of the direct path of the independent variable is overestimated (assuming all coefficients are positive). This means, for example, that if, in the recalled-symptoms model displayed in Figure 1, concurrent symptoms were measured with much error, one would need to recognize that the direct effect of neuroticism on recalled GI symptoms in Model B was inflated.

Asher (1983) also considered the problem of measurement error in the context of path analysis; he discussed techniques (none easy) for

10. This rule of thumb is based on a common recommendation for covariance structure analysis (e.g., Bentler & Chou, 1988). Path analysis is one type of covariance structure analysis.

assessing its consequences. As is noted later in this chapter, an advantage of using a covariance analysis program to perform path analysis is that the assumption that variables are measured without error is not necessary.

Misspecification of a path analysis model as a whole can lead to incorrect conclusions. Furthermore, as has been shown in this chapter, satisfactory results from a path analysis do not prove that it is correct: A model can have a good fit, explain a lot of variance, have significant parameters, and still be totally wrong. Sound theory is the best safeguard against misspecification.

The consideration of assumptions and consequences of violating them is a dismal business. (I don't think I've ever seen a model and data that met the assumptions for a path analysis.) However, that does not mean that assumptions should be ignored. It is often helpful to think of meeting assumptions as one of degree: There are analyses that one can have a lot of confidence in and analyses that are worthless. (Are the variables, if not perfectly measured, measured well? Is the model, particularly with respect to the variables included and omitted, close enough to the probably correct model to merit attention?)

Standardized Versus Unstandardized Path Coefficients

Just as there are standardized and unstandardized regression coefficients, there are standardized and unstandardized path coefficients. The examples in this chapter all use standardized coefficients. To report unstandardized coefficients, simply copy unstandardized rather than standardized coefficients from a multiple regression computer program printout.

The advantage of standardized coefficients is that the units are comparable. For example, it is easy to see in Figure 2 that prior wealth has a greater effect on remarriage than does prior health, even though the original variables were measured on very different scales. Standardized coefficients are useful when one is interested in the relative importance of the predictors. In practice, researchers almost always present standardized path coefficients.

The advantage of unstandardized coefficients is that one can see the impact on the dependent variable of one unit change on the predictor. Suppose that wealth was measured as assets in units of $10,000; then the unstandardized coefficient for wealth would be the effect of an additional $10,000 on the probability of remarriage, a number that might be of interest. As a simple rule of thumb, there are three situations in which one should consider reporting unstandardized coefficients: (a) if one has performed path analyses on two or more different samples of individuals, for example, if one did separate path analyses for citizens of different countries; (b) if the path analyses involved the same variables measured at different times and if the variances of the variables differed at the different times;[11] (c) if the unit of the variables has meaning, for example, dollars, years, calories.

Kim and Ferree (1981) discuss the complexities of the standardization issue and propose a solution that retains some of the virtues of both standardized and unstandardized coefficients. Briefly, they propose standardizing some or all of the variables before doing the analysis and using unstandardized coefficients for interpretation.

Concluding Comments

The Best Way to Estimate Path Coefficients

So far, I have discussed using multiple regression to estimate path coefficients. The same coefficients can be obtained by means of software to estimate the parameters of covariance structure models. Path analysis is a special case of the general covariance structure model. LISREL (Jöreskog & Sörbom, 1989) and EQS (Bentler, 1989) are the two most popular programs for estimating the parameters for these models. The path coefficients from such software, whether standardized or unstandardized, may be interpreted exactly as the corresponding coefficients from multiple regression. Additionally, there are two big bonuses from using such software.

The first advantage of using LISREL (or some similar program) to do a path analysis is that the program prints additional results, such as all of the implied correlations, all total effects, and the standard errors for indirect effects. Most important, the program calculates and prints several measures of overall fit of the model. It is becoming routine to report these measures. (For example, in the quality of life for cardiac patients models in Figures 6 and 7, the authors report the fit measures

11. Following this guideline, unstandardized coefficients might have been reported for the remarriage and well-being model (Figure 2). However, in general, the variances of health, wealth, and well-being did not vary much between the two time periods.

provided by LISREL; the fit of the medical model is excellent whereas the fit of the psychosomatic model is poor.)

The second advantage of using software for covariance structure models, mentioned earlier, is that it provides the opportunity to accommodate to an assumption. One assumption of a conventional path analysis (hiding in the assumptions of multiple regression) is that the variables are measured without error. In a LISREL path analysis, the researcher can make allowance for the measurement error, which in many cases is probably a more realistic model.

Although complete mastery of covariance structure modeling takes time, and may seem daunting, using general covariance software for the type of path analyses discussed in this chapter (recursive models with all variables measured) is quite straightforward. With practice, models are easy to set up and interpret.

On the other hand, using the power of the general covariance structure analysis programs, it is possible to do complicated path analyses that cannot be tackled by regression. Problems that involve comparing path analyses on multiple samples (e.g., citizens of different nations), that are nonrecursive, or that involve unmeasured variables are candidates for general covariance analysis.

Suggestions for Further Reading

For a broad perspective on the nature of causation and of causal inference—a topic not discussed in the present chapter—see the first chapters of textbooks by Kenny (1979) or Heise (1975). For a cautionary discussion of multiple regression and path analysis, see Bibby (1977). Bibby writes in the concluding section of his chapter that "the general linear model is a trap, a snare, and a delusion—although these evil effects may be partially mitigated by adhering to the cautionary word of advice given in previous sections" (Bibby, 1977, p. 76). Also cautionary, but more modern and with more of an emphasis on causal modeling, is an article by Cliff (1983).

Two general, chapter-length discussions of simple path analysis are those by Pedhazur (1982) and Knoke (1985). The Pedhazur chapter includes clear instructions, with many examples, for calculating the various kinds of effects. A somewhat longer general treatment is a monograph on causal modeling by Asher (1983); in addition to simple path analysis, Asher discusses nonrecursive models and the problem of identification.

(Identification is not an issue for the models described in the present chapter, but the problem looms immediately if one is dealing with a nonrecursive model.) Another monograph (Berry, 1984) is totally devoted to nonrecursive models.

If the role of a variable as a mediator is of special interest, see Baron and Kenny's (1986) influential article on moderators and mediators. The authors lay out three conditions that must hold to established mediation. They also give a formula that can be used to hand calculate the standard error of an indirect effect.

If you plan to do a path analysis, you may want to consider using software for covariance structure models to estimate the parameters. Both the *LISREL 7 User's Reference Guide* (Jöreskog & Sörbom, 1989) and *EQS Structural Equations Program Manual* (Bentler, 1989) give an example of a path analysis; each shows a path diagram, the corresponding setup, and the results.

Hayduk (1987) and Bollen (1989) provided in-depth discussions of introductory and advanced topics in covariance structural modeling.

Glossary

COMPOUND PATH A path between two variables that involves two or more links; compound paths may represent indirect, spurious, or unanalyzable effects.

COVARIANCE STRUCTURE MODELS A broad class of statistical models, of which path analysis is one. Covariance structure models in general accommodate latent (unmeasured) variables. However, they can be used when all variables are observed, as they are in path analysis. Covariance structure models are often called *structural equation models* (SEMs).

EFFECT COEFFICIENT The sum of the direct effect and the indirect effects of one variable on another; sometimes called the *total effect*.

ENDOGENOUS VARIABLE A variable that is affected by one or more other variables in the model; in a path diagram, an endogenous variable has one or more straight arrows pointing at it.

ERROR VARIANCE The variance in an endogenous variable that is unexplained by the model; it is equal to $1 - R^2$, where R^2 is the squared multiple correlation when the endogenous variable is regressed on the variables that directly affect it.

EXOGENOUS VARIABLE A variable that has its values determined by a variable or variables not in the model. In a model, an exogenous variable always appears as a cause and never as an effect. In a path diagram, an exogenous variable is linked to other exogenous variables by two-headed, curved arrows and to endogenous variables that it affects by straight arrows.

FULLY RECURSIVE MODEL A model that includes a direct link between each variable and all variables further down the causal chain.

IMPLIED CORRELATION A correlation between two variables that is implied by the model. The results of a path analysis provide parameters that can be used to calculate the implied correlation between any pair of variables, and this correlation can then be compared with the actual (observed) correlation. An implied correlation is sometimes called a *model-implied correlation* or a *reproduced correlation*.

PATH COEFFICIENT A parameter that indicates the magnitude of the direct effect of one variable on another.

RECURSIVE MODEL A model in which the causal linkages run one way; specifically, a model that contains no reciprocal paths, no feedback loops, and no correlations between unmeasured variables.

RESIDUAL PATH COEFFICIENT A path coefficient (a parameter) that indicates the direct effect of unmeasured variables on an endogenous variable.

TOTAL EFFECT The sum of the direct effect and the indirect effects of one variable on another; sometimes called an *effect coefficient*.

References

Asher, H. B. (1983). *Quantitative applications in the social sciences: Vol. 3. Causal modeling* (2nd ed.). Beverly Hills, CA: Sage.

Bachman, J. G., & O'Malley, P. M. (1977). Self-esteem in young men: A longitudinal analysis of the impact of educational and occupational attainment. *Journal of Personality and Social Psychology, 35*, 365–380.

Baron, R. M., & Kenny, D. A. (1986). The moderator–mediator variable distinction in social psychological research: Conceptual, strategic, and statistical considerations. *Journal of Personality and Social Psychology, 51*, 1173–1182.

Bentler, P. M. (1989). *EQS Structural Equations Program manual*. Los Angeles: BMDP Statistical Software.

Bentler, P. M., & Chou, C. (1988). Practical issues in structural modeling. In J. S. Long (Ed.), *Common problems/proper solutions* (pp. 161–192). Newbury Park, CA: Sage.

Berry, W. D. (1984). *Quantitative applications in the social sciences: Vol. 37. Nonrecursive causal models*. Beverly Hills, CA: Sage.

Berry, W. D. (1993). *Quantitative applications in the social sciences: Vol. 92. Understanding regression assumptions*. Newbury Park, CA: Sage.

Bibby, J. (1977). The general linear model—a cautionary tale. In C. A. O'Muircheartaigh & C. Payne (Eds.), *The analysis of survey data* (Vol. 1). New York: Wiley.

Blalock, H. M., Jr. (Ed.). (1971). *Causal models in the social sciences*. Chicago: Aldine.

Boldizar, J. P., Wilson, K. L., & Deemer, D. K. (1989). Gender, life experiences, and moral development: A process-oriented approach. *Journal of Personality and Social Psychology, 57*, 229–238.

Bollen, K. A. (1989). *Structural equations with latent variables*. New York: Wiley.

Cliff, N. (1983). Some cautions concerning the application of causal modeling methods. *Multivariate Behavioral Research, 18*, 115–126.

Curry, R. H., Yarnold, P. R., Bryant, F. B. Martin, G. J., & Hughes, R. L. (1988). A path analysis of medical school and residency performance: Implications for housestaff selection. *Evaluation and the Health Professions, 11*, 113–129.

Decotiis, T. A., & Summers, T. P. (1987). A path analysis of a model of the antecedents and consequences of organizational commitment. *Human Relations, 40*, 445–470.

Greene, R. (1990). The positive effect of remarriage on older widowers' well-being: An integration of selectivity and social network explanations (Doctoral dissertation, University of Michigan, 1990). *Dissertation Abstracts International, 52*, 297A.

Hayduk, L. A. (1987). *Structural equation modeling with LISREL: Essentials and advances*. Baltimore: John Hopkins University Press.

Heise, D. R. (1975). *Causal analysis*. New York: Wiley-Interscience.

Igbaria, M., & Parasuraman, S. (1989). A path analytic study of individual characteristics, computer anxiety and attitudes toward microcomputers. *Journal of Management, 15*, 373–388.

Jaccard, J., Turrisi, R., & Wan, C. K. (1990). *Quantitative applications in the social sciences: Vol. 72. Interaction effects in multiple regression*, Newbury Park, CA: Sage.

Jenkins, C. D., Jono, R. T., Stanton, B. A., & Stroup, C. A. (1990). The measurement of health-related quality of life: Major dimensions identified by factor analysis. *Social Science and Medicine, 31*, 925–931.

Jenkins, C. D., & Stanton, B. A. (1984). Quality of life as assessed in the Recovery Study. In N. K. Wenger, M. E. Mattson, C. D. Furberg, & J. Elinson (Eds.), *Assessment of quality of life in clinical trials of cardiovascular therapies* (pp. 266–280). New York: Le Jacq.

Jöreskog, K. G., & Sörbom, D. G. (1989). *LISREL 7 user's reference guide*. Mooresville, IN: Scientific Software.

Jussim, L. (1989). Teacher expectations: Self-fulfilling prophecies, perceptual biases, and accuracy. *Journal of Personality and Social Psychology, 57*, 469–480.

Kenny, D. A. (1979). *Correlation and causality*. New York: Wiley-Interscience.

Kim, J. O., & Ferree, G. D. (1981). Standardization in causal analysis. *Sociological methods and research, 10*, 187–210.

Knoke, D. (1985). A path analysis primer. In S. B. Smith (Ed.), *A handbook of social science methods* (Vol. 3, pp. 390–407). New York: Praeger.

Larsen, R. J. (1992). Neuroticism and selective encoding and recall of symptoms: Evidence from a combined concurrent–retrospective study. *Journal of Personality and Social Psychology, 62*, 480–488.

Lewis-Beck, M. S. (1980). *Quantitative applications in the social sciences: Vol. 22. Applied regression: An introduction*. Beverly Hills, CA: Sage.

Obeng, K. (1989). Applications of path analysis to transit system maintenance performance. *Transportation*, *15*, 297–316.
Pedhazur, E. J. (1982). *Multiple regression in behavioral research* (2nd ed.) New York: Holt, Rinehart & Winston.
Romney, D. M., Jenkins, C. D., & Bynner, J. M. (1992). A structural analysis of health-related quality of life dimensions. *Human Relations*, *45*, 165–176.
Roth, D. L., Wiebe, D. J., Fillingim, R. B., & Shay, K. A. (1989). Life events, fitness, hardiness, and health: A simultaneous analysis of proposed stress resistance effects. *Journal of Personality and Social Psychology*, *57*, 136–142.
Trevino, L. K., & Youngblood, S. A. (1990). *Journal of Applied Psychology*, *75*, 378–385.
Wright, S. (1920). The relative importance of heredity and environment in determining the piebald pattern of guinea pigs. *Proceedings of the National Academy of Sciences*, *6*, 320–332.
Wright, S. (1921). Correlation and causation. *Journal of Agricultural Research*, *20*, 557–585.
Wright, S. (1925). Corn and hog correlations. *U.S. Department of Agriculture Bulletin* (No. 1300). Washington, DC: U.S. Government Printing Office.
Wright, S. (1960). Path coefficients and path regressions: Alternative or complementary concepts? *Biometrics*, *16*, 189–202.

Appendix

Calculation of the correlation between remarriage and well-being implied by the expanded selectivity model shown in Figure 2. For this particular model, all of the effects are spurious.

Direct effects		None
Indirect effects		None
Spurious effects	$.11 \times -0.2$	
	$.11 \times .57 \times .27$	
	$.11 \times .35 \times .23$	
	$.04 \times .07$	
	$.04 \times .60 \times .02$	
	$.04 \times .08 \times .23$	
	$.00 \times .23$	
	$.08 \times .26 \times .08 \times .23$	
	$.08 \times .26 \times .60 \times .02$	
	$.08 \times .26 \times .07$	
	$.11 \times .40 \times .08 \times .23$	
	$.11 \times .40 \times .60 \times .02$	
	$.11 \times .40 \times .07$	
	$.08 \times .35 \times -.02$	
	$.08 \times .35 \times .57 \times .27$	
	$.08 \times .35 \times .35 \times .23$	
	$.04 \times .40 \times -.02$	
	$.04 \times .40 \times .35 \times .23$	
	$.04 \times .40 \times .57 \times .27$	
	$.00 \times .08 \times .07$	
	$.00 \times .35 \times -.02$	
	$.00 \times .35 \times .57 \times .27$	
	$.00 \times .08 \times .60 \times .02$	
	$.00 \times .35 \times .40 \times .07$	
	$.00 \times .35 \times .40 \times .60 \times .02$	
	$.00 \times .08 \times .40 \times .57 \times .27$	
	$.00 \times .08 \times .40 \times -.02$	
Sum of spurious effects		.0447
Unanalyzed effects		None
Implied correlation	$.0 + .0 + .04 + .0$	.04

Principal-Components Analysis and Exploratory and Confirmatory Factor Analysis

Fred B. Bryant and Paul R. Yarnold

A clinical psychologist wants to develop a questionnaire to measure the desirability of different types of positive life events. He begins by asking a random sample of 100 adults to write brief descriptions of events that make them happy (e.g., "party," "good grade," or "talking to friends"). Those items (events) mentioned by at least 10 respondents are retained for the revised questionnaire. A second adult sample is then given this new questionnaire and asked to rate how happy each event would make them feel, on a 7-point Likert-type scale ranging from *does not make me happy* (1) to *makes me very happy* (7). Principal-components analysis (PCA) could be used to identify the underlying dimensions that explain responses to the questionnaire.

An investigator of leisure studies administers a questionnaire designed to measure the perceived quality of summer vacations among a sample of grade school children. The investigator wants to determine the following: (a) the number of dimensions along which children evaluate the quality of their vacations, (b) the meaning of each of these dimensions of subjective quality, and (c) how these various dimensions interrelate with each other. This is a typical research scenario in which exploratory factor analysis (EFA) is appropriate.

A health psychologist administers a 12-item questionnaire, which measures fear of AIDS, to a sample of 200 high school seniors. The questionnaire is designed to assess fear along three specific dimensions:

We wish to thank Evelyn Perloff for her invaluable pedagogical advice concerning factor analysis, Linda Perloff for help in preparing the figure that accompanies this chapter, and Jennifer Brockway for assistance in locating reference materials.

fear of social stigma, fear of physical suffering, and fear of death. Inspecting the questionnaire items, the researcher hypothesizes that the first four items tap fear of social stigma, the second four tap fear of physical suffering, and the final four tap fear of death. This represents a typical application of confirmatory factor analysis (CFA). As an aside, the investigator also wishes to determine whether this three-factor model replicates across gender and ethnicity. This could be accomplished by means of a multisample CFA.

This chapter discusses in turn each of these three procedures: PCA, EFA, and CFA.

Principal-Components Analysis (PCA)

Imagine that a sample of observations has been obtained on a set of variables measured on interval or ratio scales. In using PCA, there must be at least two variables. In deciding how many observations (subjects) should be used, researchers rely on a rule of thumb, called the *subjects-to-variables* (STV) ratio. For the results of one's analysis to be reliable, that is, for the results to replicate if the analysis is repeated using an independent sample, the minimum number of (randomly selected) observations in one's sample should be at least five times the number of variables: The STV ratio should be 5 or greater. Furthermore, every analysis should be based on a minimum of 100 observations, regardless of the STV ratio (Gorsuch, 1983). For example, imagine a hypothetical application in which one's data consisted of scores for 200 observations each measured on 10 variables. Because the STV ratio is 200/10, or 20, the sample size is sufficiently large by the reliability criterion. The *maximum* number of variables that should be used with a sample of 200 observations is 200/5, or 40 variables. As a check on this, note that 40 variables multiplied by an STV ratio of 5 equals 200 observations. Using more than 40 variables, with 200 observations, would reduce the STV ratio to less than 5.

For the hypothetical example, imagine that the observation scores on each of the 10 variables have been standardized into *z*-score form (thus, for each variable, $M = 0$ and variance $= 1$). When the variables are standardized into *z*-score form, the total variance is equal to the number of variables. Thus, for the hypothetical example, the total variance is equal to 10. The goal of PCA is to identify a new set of a few variables, called *principal components*, that explains all (or nearly all) of this total variance. The fundamental assumption of PCA is that the total variance of a variable reflects the sum of *explained* and *error* variance (Hotelling, 1933; Pearson, 1901).

Consider the first principal component. This component is a linear function of the original variables. In appearance, this linear function is similar to a multiple regression equation except that there is no intercept term. For example, the first principal component might be of the form, $Y = .1$ (Variable 1) $-.3$ (Variable 2) $+.45$ (Variable 3) and so forth. The first principal component has the property that it maximizes the amount of the total variance that is explained, that is, no other linear function can explain more of the total variance than is explained by the first principal component. Geometrically, think of the principal component as a line that passes through a swarm of data points (the sample of observations) that have been plotted or graphed in a multidimensional space: The number of dimensions of this space is equal to the number of variables (each variable is a dimension). In that space, each observation lies at some distance—measured perpendicularly—from the line reflecting the first principal component. Refer to the perpendicular distance of the observation from the first principal component (line) as the *error score* for the observation. The first principal component (line) is associated with the minimum sum of squared error terms for the sample of observations.

Imagine that in the hypothetical example, the first principal component explained a total amount of variance that was equal to 6. Thus, the first principal component explained (6/10) × 100%, or 60% of the variance of the original set of 10 variables. This would reflect tremendous measurement efficiency, or *parsimony*, because this new single variable (i.e., the first principal component) would account for the same overall amount of the total variance as did the 6 original variables. Formally, the linear function, or principal component, is referred to as an *eigenvector*: in this case, the first eigenvector. Also, the amount of the total variance that is explained by an eigenvector is known as the *eigenvalue* (notated with the symbol λ). A subject's score on the eigenvector is computed by entering the subject's scores on the original variables into the linear function (equation).

In the hypothetical example, the first eigenvector explained 60% of the total variance. Thus, 100% − 60%, or 40%, of the total variance remained unexplained. Accordingly, a second principal component (eigenvector) would be computed. Like the first eigenvector, the second eigenvector would be a linear function of the original variables. Also, the second eigenvector would maximize the amount of (remaining) variance that was explained. In the hypothetical example, imagine that the second eigenvector had an eigenvalue of 3. Thus, the second eigenvector explained (3/10) × 100%, or 30%, of the total variance. Taken together, the first two eigenvectors explained 60% + 30%, or 90%, of the total variance of the set of 10 original variables. Note that the first and second (and, indeed, all) eigenvectors were independent, or uncorrelated (geometrically, this means that the principal components are perpendicular to each other). That is, none of the variance explained by the second eigenvector could be explained by the first eigenvector.

Because the first two eigenvectors left 10% of the total variance unexplained, a third eigenvector could be computed. Like the other eigenvectors, the third eigenvector is a linear function of the original variables. Imagine that in the hypothetical example, the eigenvalue of the third eigenvector was equal to 1. The third eigenvector thus explained 10% of the total variance (specifically, the variance that could not be explained by the first two eigenvectors). Taken together, these three eigenvectors explained 60% + 30% + 10%, or 100%, of the total variance of the original set of 10 variables. The number of eigenvectors required to explain 100% of the total variance is known as the *rank* (or *true dimensionality*) of the correlation matrix, which is used to summarize the interrelationships among the original variables. In the present case, the rank of the correlation matrix for all 10 variables is three. And rather than requiring 10 dimensions to describe the unique features of an observation, 3 dimensions—the principal components—could instead be used without losing any of the information (i.e., variance) that was present in the original set of 10 variables.

In actual practice, researchers rarely continue to extract eigenvectors until they have explained 100% of the total variance. Rather, it is a common practice to stop the analysis before all of the total variance has been explained. The following discussion concerns how to determine the appropriate number of eigenvectors and how to interpret each eigenvector, that is, how to identify the theoretical dimension that is implied by the pattern of the variables that are the most important constituents of each eigenvector.

Determining the Number of Eigenvectors

The objective of PCA is to identify the smallest number of factors that together account for all of the total variance in the correlation matrix of the original variables. How does one determine the number of factors to extract (i.e., to retain) in a given analysis? Several different types of *stopping rules* have been developed as an aid in answering this question.

When conducting PCA, researchers sometimes specify, a priori, that successive factors will be extracted until some absolute percentage of the total variance has been explained: This rule is known as the *percentage of variance* criterion. For example, one might specify that factors are to be extracted until 75% of the total variance has been explained. In studies that involve a few strongly correlated variables and a small sample size, it often requires only one or two factors to meet this criterion. However, in studies that involve many moderately correlated variables and a large sample, a relatively large number of factors may be retained by this criterion. In the latter case, it is not uncommon to identify factors that correlate only with a single variable: Such single-variable factors are sometimes called *bloated specifics*.

Another stopping rule is known as the *a priori criterion* (Hair, Anderson, Tatham, & Black, 1992). In some cases, such as when one is attempting to replicate another researcher's findings, one knows in advance how many eigenvectors to extract. This approach is also useful when one has a theoretically motivated idea about the appropriate number of eigenvectors to extract.

For example, Yarnold (1984) hypothesized that three different questionnaires—drawn from different areas of psychology that rarely cross-referenced each other's work—provided measures of two underlying, basic dimensions. These two dimensions included task-focus (concern with "getting the job done") and other-focus (concern for the well-being of others). In this study, a sample of subjects each completed the three questionnaires. Because each questionnaire provided two scores for every subject, six scores were analyzed using PCA. As directed by theory, two eigenvectors were extracted. Together these two eigenvectors explained a modest 56% of the total variance of the six scores. In keeping with the a priori hypothesis, the coefficients (i.e., the numbers in the eigenvector—or linear function—by which the variables were multiplied) of the first eigenvector were large for variables that measured task-focus, and the coefficients of the second eigenvector were large for variables that measured other-focus.

Typically, however, one or both of the following two less ambiguous stopping rules are used to determine the appropriate number of eigenvectors. First, *Kaiser's* (1960) *stopping rule* extracts (i.e., retains) only eigenvectors with eigenvalues of at least 1, that is, the equivalent of the variance of a single standardized variable. Second, Cattell (1966) proposed a graphical procedure, known as the *scree test*, for determining the appropriate number of eigenvectors to extract. To conduct the scree test, the eigenvalues are plotted (on the Y axis) successively by factor (X axis): First plot the eigenvalue for the first eigenvector, then plot the eigenvalue for the second eigenvector, and so forth. Usually, the plot of the eigenvalues drops quickly over the first few eigenvectors (i.e., there is a steep visual descent) and then levels off into a slowly but steadily decreasing pattern of eigenvalues. The eigenvalues (and corresponding eigenvectors) in the steep descent are retained, and the eigenvalues in the gradual descent (including the eigenvalue occurring in the transition from steep to gradual descent) are dropped.

Summarizing research concerning the accuracy of these procedures, Stevens (1986) concluded that Kaiser's (1960) stopping rule should be used for applications in which there are fewer than 30 variables and the communalities (see Exploratory Factor Analysis) are greater than .70 or in which there are at least 250 observations and the communalities are at least .60. Otherwise, the scree test should be used in applications for which there are at least 200 observations and the communalities are reasonably large (Stevens, 1986).

Interpreting the Eigenvectors

Once an appropriate number of eigenvectors has been retained, how does one identify what each eigenvector measures? Intuitively, this question has a relatively straightforward answer. Recall that eigenvectors are linear functions of the original variables and that the coefficients of the eigenvector are the numbers by which the variables are multiplied. Imagine a situation (present in what is known as *simple structure*) in which, for a given eigenvector, some of the coefficients are high (near 1) for a few variables and the remaining coefficients are low (near 0) for the rest of the variables. This eigenvector clearly assesses the attribute that is reflected by the variables for which the coefficients are high. Those variables that have low coefficients are not used in interpreting the eigenvector.

When PCA is conducted through various computer programs, part of the output is known as the *factor loading coefficients*: These are almost always presented in journal articles. These coefficients define the eigenvectors and are typically presented in a table. The columns of the table give the eigenvectors (e.g., 1, 2, 3), and the rows of the table give the variables. The entries in the table indicate the correlation between the given eigenvector and the given variable. For example, if the loading for the first variable on the first eigenvector is .73, then this means that scores on the first variable are correlated at $r = .73$ with scores on the first eigenvector.

Unfortunately, in most cases, the eigenvectors identified in the PCA will be very difficult to interpret. It is possible, however, to rotate the eigenvectors in any arbitrary direction to simplify the task of interpretation (in fact, a common criticism of PCA and related techniques is that there is no unique location for the eigenvectors: the so-called *factor indeterminancy* problem; Steiger, 1979). Specifically, it is desirable—for the sake of interpretation—that a condition known as *simple structure* is obtained when an appropriate rotation is conducted. Thurstone (1947) described simple structure in terms of the following properties. First, each variable should have at least one loading that is near zero on at least one of the eigenvectors, and for applications involving four or more eigenvectors, most of the variables should have loadings that are near zero on most of the eigenvectors. Second, for each eigenvector, there should be at least as many variables with loadings that are near zero as there are eigenvectors. Finally, for every pair of eigenvectors, there should be several variables that load on only one eigenvector: In general, variables should only have a large loading on one eigenvector. When simple structure has been achieved, interpretation of the eigenvectors is often a straightforward procedure.

Several types of rotations are frequently performed in an attempt to achieve simple structure. The types of rotations are best distinguished in terms of whether they are *orthogonal* (uncorrelated) or *oblique* (correlated). The most frequently used orthogonal rotations include *varimax* (which focuses on making as many values in each column of the factor loading coefficient table be as close to zero as is possible) and *quartimax* (which focuses on making as many values in each row of the factor loading coefficient table be as close to zero as is possible). Both of these procedures attempt to produce simple structure while retaining the independence between eigenvectors, that is, scores on the rotated eigenvectors remain uncorrelated. In contrast, in an oblique rotation, scores on different eigenvectors are allowed to be correlated.

Once the eigenvectors have been rotated to achieve simple structure, the coefficients may be examined to determine the central dimension that is being tapped by the eigenvector. For example, in Yarnold's (1984) study of three questionnaires, variables measuring task-focus had high factor loading coefficients on the first eigenvector, and variables measuring other-focus had low factor loading coefficients on this eigenvector. Clearly, the first eigenvector assessed task-focus. At this point, it is appropriate to ask what value of a factor loading coefficient is required for a variable to be considered a constituent of a given eigenvector. Typically, researchers consider variables with factor loading coefficients of at least .30 in absolute value as "loading on the eigenvector" and thus as worthy of consideration in the interpretation of the meaning of the eigenvector. Variables with negative factor loading coefficients are negatively correlated with the eigenvector; eigenvectors that have variables with positive factor loadings as well as variables with negative factor loadings are called *bipolar eigenvectors*. Note that a factor loading coefficient of .30 implies that the variable and the eigenvector share $(.30)^2 \times 100\%$, or 9%, of their variance.

As Stevens (1986) and others have argued, this practice of only considering factor loadings of greater than .30 ignores the number of observations in the sample. That is, because the statistical significance of the correlation between a variable and an eigenvector depends on the sample size, the criterion for classifying a variable as a constituent of an eigenvector should be based on the value of the correlation that is needed to achieve a Type I error rate of $p < .05$, given the sample size. In addition, however, because a substantial number of factor loading coefficients need to be evaluated in a typical PCA, it is also important to adopt a conservative criterion to avoid capitalizing on chance, that is, to ensure that the overall Type I error rate for the entire analysis is equal to $p < .05$ (e.g., see the discussion on the Bonferroni procedure in chapter 8, Multivariate Analysis of Variance). Such precautions are rarely seen in applied research, but they are increasing in popularity as researchers become more familiar with the importance of maintaining control over *experimentwise Type I error* rates.

Exploratory Factor Analysis (EFA)

EFA is closely related to PCA. For example, as in PCA, in using EFA one seeks a small set of easily interpretable eigenvectors, which in EFA are called *factors*. These factors may be rotated by means of either orthogonal (uncorrelated) or oblique (correlated) procedures in an effort to achieve simple structure; and in both PCA and EFA, there is the question of how large a variable's factor loading coefficient must be to use the variable as a constituent in defining the given factor.

The primary difference between EFA and PCA lies in their assumptions. Recall the fundamental assumption of PCA: The total variance of a variable reflects the sum of two (explained and error) components. In contrast, the fundamental assumption of EFA is that the total variance of a variable reflects the sum of three different types of variances: The common and the specific variance are reliable or stable, and the error variance is unreliable. *Common variance* refers to the portion of the total variance that correlates (is shared) with other variables in the analysis. *Specific variance* refers to the portion of the total variance that does not correlate with the other variables (the reliability of a variable reflects the sum of the common and specific variance). *Error variance*, in contrast, reflects inherently unreliable random variation. Whereas PCA finds eigenvectors that maximize the amount of the total variance that is explained, EFA finds factors that maximize the amount of the common variance that is explained.

Two concepts about variables, *uniqueness* and *communality*, are crucial in understanding EFA. The uniqueness of a variable is that portion of the total variance that is unrelated to other variables (i.e., uniqueness = specific variance + error variance), and the communality of a variable is equal to 1 − the uniqueness. Computationally, the communality of a variable is equal to the sum, over all of the factors, of the squared factor loadings for the variable. For example, imagine a two-factor model in which the factor loading of variable *A* on Factor 1 is .80 and the factor loading of variable *A* on Factor 2 is .10. Here, the communality of variable *A* is equal to $(.80)^2 + (.10)^2$, or .65. The communality is an index of the portion of the variance of the variable that is accounted for by the set of factors.

Similarly to PCA, EFA essentially decomposes a correlation matrix into its constituent factors, that is, *dimensions* or *sources of influence*. In PCA, the major diagonal of the correlation matrix, which runs from the upper-left-hand corner of the matrix (reflecting the correlation between the first variable and itself) to the lower-right-hand corner (reflecting the correlation between the last variable and itself), consists of a series of 1s. That is, in PCA the correlation between each variable and itself is assumed to be unity. The resulting factors (eigenvectors) best reproduce the actual data (i.e., measured data values) because the factors reflect the error variance as well as the common and specific variances.

The most commonly reported type of EFA is called *principal-components factor analysis* (not to be confused with PCA). Here, the 1s in the diagonal of the correlation matrix are replaced by different numbers. Specifically, researchers sometimes place the communality of the variables in the diagonal or, instead, sometimes place the reliability (an index reflecting the magnitude of the stable common and specific variance) of the variables in the diagonal. Communalities are used when the objective is to study the nature of the theoretical factors that lead to the observed interitem correlations: The resulting factors best reproduce the observed interitem correlations. In contrast, reliabilities are used when the objective is to study the nature of the theoretical factors that best account for the stable variance of the variables.

Once the correlation matrix has been prepared, principal-components EFA proceeds in much the same manner as PCA. However, the EFA procedure is iterative: It requires more than one pass through the data. In the first step, the initially adjusted correlation matrix (i.e., the correlation matrix with reliabilities, or communalities estimated using PCA, placed in the diagonal) is subjected to PCA. In the second step, the communalities are estimated on the basis of the factors identified in the first step, these reestimated communalities are placed in the diagonal of the correlation matrix, and PCA is conducted again. This procedure is continued (iterated) until there is only negligible change in the communality estimates derived from two successive steps. The resulting solution is typically then rotated to simple structure and interpreted.

PCA is frequently used in a wide array of scientific disciplines; PCA has the ability to summarize a great deal of the total variability in a sample of data points in a parsimonious manner. EFA, in contrast, is seen in the literature with decreasing frequency because of the emergence of the much more powerful confirmatory factor analysis (CFA) procedures, to which we now turn. The reader will no doubt notice how the flexibility of CFA is associated with a large number and variety of analytic options from which one must choose.

Confirmatory Factor Analysis (CFA)

Overview

After first differentiating between exploratory and confirmatory approaches, we introduce the basic rules and building blocks of CFA: what it is, how it works, and how researchers use it to test hypotheses about factor structure. We explain how CFA is used to evaluate the explanatory power of a factor model, to decide which model or models best fit the data. We highlight the information that must be specified in the analysis to estimate a CFA model, we describe the three most commonly used methods of estimation, and we note their statistical assumptions and limitations. We then explain, using actual research examples, how CFA is used to test hypotheses about (a) factor loadings and factor interrelationships within a single group and (b) the equivalence of factor structures across multiple groups. Finally, we conclude by deciphering the various Greek and Roman letters often used to symbolize the observed and estimated values in CFA.

The Logic of CFA

Exploring Versus Confirming Factor Structure

There is one major difference between exploratory and confirmatory factor analysis. The former finds the one underlying factor model that best fits the data; the latter, in contrast, allows the researcher to impose a particular factor model on the data and then see how well that model explains responses to the set of measures. With EFA, one lets the observed data determine the underlying factor model *a posteriori* (i.e., reasoning inductively to infer a model from observed data). With CFA, in contrast, one derives a factor model or models *a priori* (i.e., reasoning deductively to hypothesize a structure beforehand) and then evaluates its goodness of fit to the data. Thus, EFA primarily represents a tool for theory building, whereas CFA represents a tool for theory testing (see Bollen, 1989; Hayduk, 1987; Long, 1983).

Although exploratory and confirmatory procedures are treated separately here for purposes of presentation, in practice the two techniques are often used in tandem. With multiple samples or samples large enough to split randomly in half, EFA can be used to discover a feasible factor structure using one sample's data, and CFA can be used to simplify, refine, and confirm this basic model using the other sample's data (e.g., Moore & Neimeyer, 1991). Thus, the two techniques represent opposite, though complementary, sides of the same research coin. Whereas EFA involves hindsight, CFA requires foresight.

The Basic CFA Model

What follows is a condensed guide for anyone who might encounter a research report using CFA. The interested reader is encouraged to consult the references providing more detailed and expanded coverage of structural modeling in general and of CFA in particular (e.g., Bollen, 1989; Hayduk, 1987; Long, 1983).

To begin with, CFA builds on classical measurement theory. As with

EFA, each measure in one's data set is considered to be an observed indicator of one or more underlying latent constructs or factors. Multiple factors may be hypothesized to underlie a single set of measures, and individual items may load on one or more factors. For example, CFA could be used to test two competing models proposed to underlie a nine-item scale measuring perceived competence: (a) a one-factor model consisting of nine observed indicators of a single, latent construct (self-efficacy) and (b) an oblique, three-factor model consisting of separate but interrelated factors (Academic, Social, and Physical Self-Efficacy), each defined by a different set of three observed indicators. In this way, CFA enables researchers to determine which of several alternative factor models best fits the data.

The CFA model assumes that there are two main sources of variation in responses to observed indicators. Specifically, subjects' scores on observed indicators (or measured variables) are assumed to be influenced by latent constructs (or underlying factors) and by unique-measurement error (i.e., the influence of unmeasured variables and random error). In the case of the hypothetical self-efficacy scale, for example, a researcher would use CFA to model each of the nine measured variables as being influenced by one or more latent factors (e.g., academic, social, or physical self-efficacy) and by unexplained unique error (e.g., variance due to unmeasured latent factors, such as self-esteem, mood, or social desirability, and variance due to random error).

Whereas EFA assumes that the unique errors in the observed indicators are independent (i.e., uncorrelated with one another), CFA, in contrast, allows these measurement errors to be either independent or correlated. With CFA, one can partial out the error variance that variables share as a result of common methods of assessment to examine relationships between variables independent of both unique and correlated measurement error. For example, if some of the nine items comprising the self-efficacy scale were worded positively (e.g., "I feel competent in social situations") and some were worded negatively (e.g., "I don't consider myself to be a good student"), a researcher could use CFA to estimate factor loadings and factor interrelationships, partialing out this shared measurement error. *Common methods of assessment* can refer not only to positively or negatively worded items, as in the above example, but also to assessment approaches, such as self-report and behavioral observations.

Like EFA, CFA examines the relationships among a set of measures. CFA can be used to analyze the structure underlying either item intercorrelations or item covariances. The *covariance* between two measures is their correlation multiplied by both their standard deviations. Thus, Pearson correlation coefficients represent standardized covariances, for which the variance of each item is fixed to one. Covariances contain important information about group differences in variability that can be obscured by correlations.

Model Fitting

Unlike exploratory analyses, which extract factors from the data in the one way that maximizes the common variance (principal-components factor analysis) or total variance (PCA) explained, CFA uses whatever model the user specifies to generate a predicted set of item interrelationships (i.e., correlations or covariances). The difference between each of these predicted interrelationships and the actual observed interrelationships is referred to as a *fitted residual*. The *goodness of fit* of a particular model can be assessed by examining the overall size of the fitted residuals that it produces (i.e., the degree of correspondence between the interrelationships predicted by the model and the interrelationships actually observed). The closer these residuals are to zero, the better the model fits the data.

Researchers often examine the standardized residuals, in judging how well a CFA model fits the data. A *standardized residual* is a fitted residual divided by its estimated standard error. Standardized residuals are independent of the units of measurement in the observed indicators and are therefore more clearly interpretable (Jöreskog & Sörbom, 1989). Also, for each estimated *parameter* in the model (parameters are any of the terms in a CFA model that take on numerical values in the factor solution), CFA also furnishes a probability level indicating the likelihood that the given parameter is different from zero. This information can be used to decide which observed indicators can be eliminated from the model, without sacrificing reliability.

Each CFA yields an overall maximum likelihood chi-square and an associated p value, which indicates the probability that the matrix of fitted residuals generated by the model is different from zero. Contrary to other inferential statistical tests for which significant p values represent greater accuracy of prediction, with CFA, a statistically significant chi-square denotes a model that fails to reproduce the observed data accurately (i.e., the residuals it generates are significantly different from zero). With CFA, one seeks a model that produces a *non*significant p value, thereby striving to confirm the null hypothesis. Setting aside issues concerning philosophy of science and the appropriateness of accepting the null hypothesis (cf. Popper, 1968), it is important to remember that when it comes to evaluating CFA, models that produce nonsignificant p values fit the data well whereas models with significant p values fit the data poorly.

Incremental Fit

In many applications, however, one is not so much interested in finding a model that produces a nonsignificant chi-square, but rather in determining whether one particular model fits the data better than another. When these competing models are *nested*, then their chi-squares can be directly contrasted to test the hypothesis that one model fits the data better than the other. Factor models are considered to be nested when the one that is more restrictive can be obtained by imposing constraints on the one that is more general (Bentler, 1990). In the self-efficacy example, an orthogonal (restrictive) version of the three-factor model is obtained when the covariances among the academic, social, and physical self-efficacy factors are fixed at zero. The orthogonal model has all of the same parameters as the more general oblique model, except that some parameters (factor covariances) have been fixed at set values, in contrast to the oblique model, in which the factor covariances are free to vary. Thus, the orthogonal model is nested within the oblique model.

To compare the fit of two nested models, one simply computes the difference in the chi-square statistics associated with each model (see Bentler & Bonett, 1980; Long, 1983; Mulaik et al., 1989). In other words, subtract the chi-square for the general model from the chi-square for the restrictive model: This is $\Delta\chi^2$. Next subtract the degrees of freedom for the general model from the degrees of freedom for the restrictive model: This is Δdf. Finally, evaluate the $\Delta\chi^2$ as if it were an ordinary chi-square, using Δdf as the degrees of freedom for the significance test. If the test is significant, then the model with the smaller individual chi-square is considered to provide a relative improvement in fit over the other. This strategy allows the user to systematically test hypotheses about which particular factor models best fit the data. Note that when one evaluates the fit of an individual model, a statistically nonsignificant chi-square reflects a good representation of the data; but that when one compares the goodness of fit of two nested models, on the other hand, a significant difference in chi-squares ($\Delta\chi^2$) means that one model provides an improvement (or deterioration) in fit as compared with the other.

Evaluating Goodness of Fit

Because the chi-square statistic is extremely sensitive to sample size, however, it may be of little value in evaluating overall goodness of fit when large samples are used. With large samples, even reasonable models are likely to produce statistically significant chi-square values (see Alwin & Jackson, 1980; Bentler, 1990). For this reason, differences in chi-square values may be more informative than the chi-square values themselves (Jöreskog, 1971b, 1978). Accordingly, Jöreskog (1978) offered the following advice:

> If a value of χ^2 is obtained that is large compared to the number of degrees of freedom, the fit may be examined by an inspection of the residuals, i.e., the discrepancies between the observed and reproduced values. Often the results of an analysis, the inspection of residuals or other considerations will suggest ways to relax the model somewhat by introducing more parameters. The new model usually yields a smaller χ^2. If the drop in χ^2 is large compared to the difference in degrees of freedom, this is an indication that the change made in the model represents a real improvement. If, on the other hand, the drop in χ^2 is close to the difference in number of degrees of freedom, this is an indication that the improvement in fit is obtained by "capitalizing on chance" and the added parameters may not have real significance and meaning. (p. 448)

The Ratio of Chi-Square to Degrees of Freedom (χ^2/df)

In keeping with this advice, one useful heuristic for comparing the relative fit of various factor models is χ^2/df: the ratio of the chi-square to the degrees of freedom (Hoelter, 1983). As this ratio decreases and approaches zero, the fit of the given model improves. Use of this ratio allows one to compare the fit of alternative models, controlling for differences in their complexity. The more parameters that one's model contains, the greater the model's complexity. Thus, a model that contained more estimated parameters (i.e., fewer degrees of freedom) might yield a smaller chi-square than a simpler model with fewer estimated parameters, but the two models might produce equivalent χ^2/df ratios. In this case, the simpler model might be preferred, because the more complex model would achieve a lower chi-square without adding meaningful information.

Goodness-of-Fit Indexes

Another way to gauge how well a CFA model fits the data is to compute a *goodness-of-fit index*. A variety of different indexes of relative fit have been developed, including the Tucker-Lewis coefficient (TLC; Tucker & Lewis, 1973); Jöreskog and Sörbom's (1989) goodness-of-fit (GFI) and adjusted-goodness-of-fit (AGFI) indexes; Bollen's (1989) incremental fit index (IFI); Bentler and Bonett's (1980) normed fit (NFI) and nonnormed fit (NNFI) indexes; Bentler's (1990) normed comparative fit (CFI) and noncentralized nonnormed fit (NCNFI) indexes; and Maiti and Mukherjee's (1991) indexes of structural closeness (ISC). Despite variations in their formulas, most of these comparative fit indexes basically reflect how much better the given factor model fits the data—relative to the most restrictive model, typically a *null model*, which specifies there are no common factors, and that sampling error alone explains the item covariances (Tanaka, 1993). These various fit indexes share the common feature of ranging between zero and one, with higher values indicating better fit. Bentler and Bonett (1980) have suggested that fit indexes have a minimum acceptable level of .90, as a rule of thumb in using CFA to evaluate the adequacy of factor models.

Root Mean Square Residual (RMSR)

RMSR is a measure of the average size of the residuals generated by a particular model, and it can be used to compare the fit of two or more different models for the same data. RMSR represents the absolute value of the average fitted residuals for a given CFA model. Fitted residuals are the difference between (a) the actual correlations (or covariances) among the observed indicators and (b) the correlations (or covariances) predicted by a particular model. The closer RMSR is to zero, the better the fit of the model.

Modification Index (MI)

The MI can be estimated for each constrained (or fixed) parameter in a CFA model. The MI is a measure of the predicted decrease in chi-square if the particular parameter were freed (relaxed) and the model was re-estimated. The meaningfulness of a constrained parameter's MI may be judged by means of a chi-square distribution with one degree of freedom. The constrained parameter corresponding to the largest such index is the one that if estimated rather than fixed, will maximally improve fit. Although researchers sometimes use these MI values to decide how to revise a model to make it fit better, this practice is prone to capitalize on chance, and it is recommended only when relaxing a parameter makes sense from a substantive viewpoint (Jöreskog & Sörbom, 1989).

The Mechanics of CFA

Knowing a bit about the mechanics of CFA helps in understanding how researchers use it to test hypotheses about factor structure. Two FORTRAN computer programs are most commonly used to perform CFA:

LISREL (Jöreskog & Sörbom, 1989), which stands for *LI*near *S*tructural *REL*ationships, and EQS (Bentler, 1989). As input for analysis, these software packages use either raw data or a matrix of item correlations (or covariances) provided by the researcher. To estimate a particular factor model, CFA requires the researcher to specify information about three sets of parameters: (a) factor loadings, (b) factor interrelationships, and (c) measurement errors.

Items, Factors, and Factor Loadings

First, the user must specify the number of observed indicators to be analyzed, the number of latent factors presumed to underlie these indicators, and the pattern of loadings for each observed indicator on each factor. In deciding which factor loadings to include in a CFA model, researchers seek to develop parsimonious models in which individual items load on as few factors as necessary to reasonably fit the data. In this way, they balance their desire to explain variance in subjects' responses with their desire for conceptual parsimony.

In searching for an appropriate CFA model, researchers typically consider a range of alternative models. At one end of the continuum is a highly restrictive null model, which contains no latent factors, no factor loadings, and no factor variances or factor covariances. At the other end of the continuum is the *fully saturated* or unrestricted model, in which all factor loadings and factor interrelationships are free to be estimated and for which there is no falsifiable structural hypothesis (i.e., the model fits the data perfectly). CFA is used to develop models that approach the perfect fit of the saturated model but that include fewer estimated parameters (i.e., fewer factor loadings or fewer factor interrelationships that are not fixed to a specified value but rather are left free to be estimated in the factor solution).

Factor Variances and Interrelationships

Having specified the number of underlying factors and the pattern of factor loadings that define each factor, the user must also specify the nature of the relationships among these latent factors. One may specify either an orthogonal model (in which the factors are independent), an oblique model (in which the factors are intercorrelated), or a combination of these two models (in which some factors intercorrelate whereas others do not). By comparing the chi-square values resulting from models with orthogonal versus oblique factors, researchers can test the hypothesis that the factors are interrelated. In addition, the CFA user can constrain two or more factor covariances to be equal within the same model; the difference between the resulting chi-square and the chi-square obtained when estimating these factor covariances individually is used to test the hypothesis that the factor covariances in question are different from one another.

Unique and Correlated Measurement Errors

The final piece of information that the user must provide concerns the unique error in each observed indicator and the relationships among these measurement errors. One either may specify these unique-error terms as being independent of one another or may allow specific-error terms to correlate with one another. Researchers often allow for correlated measurement error in their CFA models, particularly when these shared-error terms improve the fit of models that are well-grounded in a priori theory.

Methods of Estimation

There are three main methods commonly used to obtain estimates of model parameters in CFA: (a) maximum likelihood (ML), (b) generalized least squares (GLS), and (c) unweighted least squares (ULS). Unlike ULS, ML and GLS methods have the advantage of being *scale invariant*, or *scale free* (Long, 1983). This means that parameter estimates obtained through ML or GLS will not necessarily change if different units of measurement are used. ML and GLS methods also offer the additional advantage of providing the user with an overall chi-square statistic for testing the adequacy of a given model. These benefits make ML and GLS the methods of choice in estimating CFA models. The most popular method of estimation is probably ML.

Statistical Assumptions and Limitations

Multivariate Normality

As with EFA, CFA assumes multivariate normality. This means that besides assuming each observed indicator is normally distributed, all linear combinations of these indicators are also assumed to be normally distributed. Violations of multivariate normality can distort goodness-of-fit indexes and invalidate the conclusions drawn from statistical tests (Browne, 1984; Hu, Bentler, & Kano, 1992).

Both *restricted range* (i.e., variables taking on only a small set of values) and *distributional skewness* attenuate factor loadings. When the variables' distributions markedly depart from normality, ML and GLS methods produce unreliable goodness-of-fit statistics and standard errors (Jöreskog & Sörbom, 1989). Special CFA techniques have been developed for analyzing ordinal and other nonnormal variables (see Muthen, 1993).

Sample Size

Sample size is another important concern in CFA. In general, factor-analytic research requires large samples (Guadagnoli & Velicer, 1988). Ironically, when using the ML chi-square value to evaluate CFA models, small samples provide less power to detect a model's true lack of fit, thereby inflating the model's apparent goodness of fit. Thus, residuals that truly are nonzero are not detected as such, because of low statistical power. Whereas inadequate sample size biases one toward finding no difference when analyzing mean differences (i.e., Type II error), having too few subjects actually biases one toward finding good fit when analyzing interrelationships using CFA (i.e., Type I error).

The effect of sample size on goodness-of-fit indexes depends on the specific index used. Some goodness-of-fit indexes (e.g., NFI) are distorted by small samples, whereas others (e.g., NNFI) reflect model fit accurately at all sample sizes (Anderson & Gerbing, 1984; Marsh, Balla, & McDonald, 1988). As with EFA, a rough guideline for the minimum number of subjects required is 5–10 times the number of observed indicators.

Identification

The issue of identification concerns whether it is possible to determine uniquely the values of the structural parameters from the observed interrelationships. CFA involves the mathematical solution of a set of structural equations containing both known (fixed) and unknown (free) parameters. A CFA model is considered to be identified when only one estimated value can be obtained for each of its free parameters. If there are too many unknown (free) parameters to solve the model's parameter estimates uniquely, then the model is said to be *underidentified*, and CFA estimates cannot be trusted. Parameter estimates derived from underidentified models are arbitrary and are not uniquely deduced from the observed data. To correct this problem of model underidentification (or indeterminacy), the user must specify (fix) the values of some parameters to reduce the number of unknowns.

Exact identification is said to occur when a CFA model contains precisely the right number and combination of known and unknown parameters to yield unique estimates. For a model to be identified, a minimum of k^2 elements must be fixed in the matrix of factor loadings, in which k is the number of latent factors in the model (Alwin & Jackson, 1979, 1980). If a CFA model contains more than enough known parameters for exact identification, then the model is said to be *overidentified.* In this case, multiple estimates can be derived for free parameters, and the exact solution obtained is essentially arbitrary. To correct this problem of overidentification, the user must free some parameters so as to increase the number of unknowns. In general, researchers strive for CFA models that contain neither too many (i.e., underidentification) nor too few (i.e., overidentification) unknown parameters.

A related problem is that to estimate the parameters in a CFA model, the user must fix the measurement scale for each latent factor. This is typically done by constraining either the factor variances to one (thereby yielding a standardized solution) or by fixing one loading on each factor (usually that of the highest loading indicator) to one. The latter procedure is used when the user specifically wishes to examine the factor variances or to explore differences in factor structures between groups (Jöreskog & Sörbom, 1989).

Model Misspecification

Whereas identification concerns whether a given model has the right mix of known and unknown elements to provide a unique mathematical solution, *model misspecification* concerns whether the given model is reasonable and complete in the face of the data. A CFA model can be properly identified but still be misspecified. When a particular CFA model is different from the structures that the observed data would support, then the given model is said to be misspecified. Trivial misspecification (e.g., including a factor loading when the correct model has none) appears to have little effect on goodness-of-fit indexes. More substantive misspecification (e.g., omitting an important factor loading), on the other hand, dramatically lowers the value of goodness-of-fit indexes (La Du & Tanaka, 1989). Model misspecification results in improper CFA solutions, in which unique errors are negative, factor intercorrelations exceed one, or parameter estimates are exceptionally large (Bagozzi & Yi, 1988).

Standardization

Researchers sometimes standardize their observed indicators before conducting CFA. This practice is reasonable when the data set includes variables measured in different metrics, such as response scales of different sizes or composite total scores with different ranges. In such situations, measures with greater variance will have stronger correlations and factor loadings, making estimates of latent factor less reliable than they would be if all indicators were measured in the same metric. To correct this problem, researchers standardize each of their observed indicators by transforming raw scores to z scores before conducting CFA.

Standardization of observed indicators is also important when researchers combine the data from separate groups to obtain a larger pooled sample for CFA. If the separate groups show significant mean differences across the indicators, then simply combining the raw data from these different groups can produce spurious correlations for the pooled group (cf. Blyth, 1972). Before pooling the data from separate groups, researchers should first standardize the raw data separately within each group and then pool these data.

However, it is never proper to standardize data separately within groups when one wishes to examine structural differences between groups. To do so would effectively eliminate the common metric of measurement between groups and would make comparisons of factor loadings and factor variances meaningless. It is thus appropriate for researchers to analyze unstandardized variance–covariance matrices when comparing factor structure across groups (Cunningham, 1978; Jöreskog & Sörbom, 1989).

Another type of standardization commonly used is a standardized factor solution. A standardized solution represents a CFA model in which the variances of the latent factors have been fixed at one. This type of solution changes the relationships among the latent factors from covariances into correlations, thereby making them more easily interpretable.

Structural Hypothesis Testing Using a Single Sample

Having provided a general orientation to the logic, mechanics, and limitations of CFA, we now explain and give empirical examples of how researchers typically use it to test structural hypotheses. We first consider single-group hypothesis testing and then between-groups hypothesis testing.

Testing Competing Factor Models

Perhaps the most common use of CFA is to see whether one particular factor model fits the data better than others. Here CFA is used to evaluate the fit of alternative configurations of factor loadings, so as to determine the most appropriate measurement model. For example, McKennell and Andrews (1980) used CFA to test hypotheses about the structure underlying a set of measures of subjective well-being. Models that contained separate factors representing affective and cognitive self-evaluations were found to fit the data significantly better than nested models, which did not distinguish between affect and cognition. This evidence confirms theoretical speculation regarding the relative independence of affective and cognitive experience (Campbell, 1980).

More recently, Weinfurt, Bryant, and Yarnold (1994) used CFA to compare two competing measurement models for the Affect Intensity Measure (AIM; Larsen, 1984), a self-report measure of the characteristic strength of individuals' positive and negative emotions: (a) a unidimensional model that constrains all 40 items to load on a single underlying factor and (b) a multidimensional model, based on EFA, consisting of two positive affect and two negative affect factors that are interrelated. Imposed on the AIM data of 673 college undergraduates, this latter, oblique four-factor model provided a significant improvement in fit over the one-factor model, $\Delta\chi^2 = 2313.08$, $\Delta df = 5$, $p < .0001$. This finding casts doubt on a major theoretical assumption underlying the AIM (Larsen, 1984)—that affect intensity is unidimensional—and it supports the notion that people make separate self-evaluations of positive and negative affective experience (Bradburn, 1969; Bryant & Veroff, 1982, 1984).

Testing Factor Interrelationships

Another popular application of CFA is to determine whether oblique models that allow for correlated factors fit the data better than orthogonal models with independent factors. An illustration of this type of CFA is provided by Bryant and Yarnold (1989), who analyzed the responses of 1,203 college undergraduates to the Student Jenkins Activity Survey (SJAS), a popular measure of Type A behavior. Although an orthogonal, two-factor model is assumed to underlie responses to the SJAS (Glass, 1977), an oblique model that allowed these Hard Driving/Competitive and Speed/Impatience factors to correlate provided a significant improvement in fit, $\Delta\chi^2 = 316.00$, $\Delta df = 1$, $p < .0001$. This evidence suggests that it is more accurate to conceptualize competitiveness and impatience as being interrelated, as opposed to independent, dimensions of Type A behavior.

Another use of CFA is to test hypotheses about differences in the strength of the relationship between pairs of latent factors. For example, Bryant (1989) used CFA to test hypotheses about the relationships among factors underlying the responses of 524 college students to a set of 15 measures of perceived control. CFA was used first to confirm the fit of an oblique, four-factor model (perceived control over good and bad events and over good and bad feelings), relative to alternative factor models. A modified version of this four-factor model was then estimated, in which the covariance between the factors representing perceived control over good and bad events was constrained to equal the covariance between the factors representing perceived control over good and bad feelings. The model with this *equality constraint* fit the data significantly worse than did the model without this constraint, $\Delta\chi^2 = 7.81$, $\Delta df = 1$, $p < .001$. Confirming the a priori hypothesis, inspection of the estimated correlations between factors indicated that this was because beliefs about personal control over good and bad feelings are more independent than are beliefs about personal control over good and bad events.

Multisample Hypothesis Testing

Testing Hypotheses About Factorial Invariance

Besides evaluating how well a given factor model fits the data of a single sample, CFA also allows researchers to determine whether the same factor structure holds across multiple groups. This approach, known as *simultaneous CFA*, enables one to systematically test hypotheses about the invariance of factor loadings, factor variances–covariances, and unique error terms for a given model across independent samples (see Alwin & Jackson, 1979; Jöreskog, 1971a; Jöreskog & Sörbom, 1989).

Three sets of hypothesis tests are involved in using CFA to examine a model's factorial invariance. The first hypothesis tested is that the multiple groups share the same factor loadings. This is tested by performing two types of simultaneous CFAs and comparing the chi-square values obtained from each. The first analysis specifies that all groups have identical factor loadings, and the second analysis allows the factor loadings to be computed individually. A statistically significant difference in chi-squares ($\Delta\chi^2$) between these two analyses indicates that the given model generates different factor loadings across groups. A nonsignificant difference in chi-squares indicates that the factor loadings are equivalent across groups.

Assuming equivalent factor loadings, one then uses the same procedure to test the hypothesis that the factor variances–covariances are equivalent across groups. Here, the first analysis specifies equal factor loadings, but not equal factor variances–covariances, across groups. The chi-square from this analysis is compared with that obtained when the factor loadings and the factor variances–covariances are constrained to be invariant across groups. A significant difference in chi-squares ($\Delta\chi^2$) indicates that the given model yields different factor variances–

covariances across groups, whereas a nonsignificant difference indicates that the factor variances–covariances are invariant.

Finally, assuming equivalent factor variances–covariances, one uses the same procedure to test the hypothesis that the unique error terms are the same for each group. This is tested by comparing (a) the chi-square obtained when constraining factor loadings and factor variances–covariances (but not the error terms) to be identical across groups to (b) the chi-square obtained when factor loadings, factor variances–covariances, and error terms are estimated separately for each group. A significant difference in chi-squares ($\Delta\chi^2$) indicates that the given model yields different unique errors across groups, whereas a nonsignificant difference indicates that the model produces equivalent error terms across groups. Failure to reject all three hypotheses concerning factor loadings, variance–covariances, and error terms is strong evidence that the factor model is equivalent across groups.

Comparing Factor Loadings Between Groups: An Empirical Example

An example of the use of simultaneous CFA to test hypotheses about equality of factor loadings among groups is provided by Bryant and Veroff (1982). Using both EFA and CFA, these researchers investigated the dimensions underlying people's self-evaluations of psychological well-being, using data from two nationwide representative-sample, cross-sectional surveys: one conducted in 1957 (n = 2,460) and the other in 1976 (n = 2,264). Eighteen observed indicators of well-being were constructed from items common to both surveys—assessing feelings of well-being, self-perceptions, symptoms of distress, and various aspects of adjustment in marriage, parenthood, and work. EFA was first conducted separately for men and women within each year to determine the appropriate number of factors and to delineate the underlying structural models for each group. Scree plots of latent roots revealed that for both men and women in 1957 and 1976, three basic factors (labeled *Unhappiness*, *Strain*, and *Personal Inadequacy*) underlay self-evaluations of well-being.

CFA was first used to refine the models developed from EFA for each separate group, to derive more parsimonious models that both provided interpretable factor structures and reasonably fit the data. These separate-group models were then imposed on the data of each of the other groups, to gauge the relative strengths of year and sex differences. The results of these separate-group CFAs (taken from Bryant & Veroff, 1982, p. 663) are reported in Table 1.

To interpret the findings in Table 1, examine the data set of the

Table 1

Goodness-of-Fit Statistics for Selected Three-Factor Models Using the Data of Separate Groups

Data set	Model	χ^2	*df*	χ^2/df	TLC
1957 men	1957 men	326.03	131	2.49	.86
	1957 women	347.56	133	2.61	.84
	1976 men	412.91	134	3.08	.80
1957 women	1957 women	330.67	133	2.49	.87
	1957 men	403.62	131	3.08	.82
	1976 women	462.95	132	3.51	.81
1976 men	1976 men	262.63	134	1.96	.85
	1976 women	280.45	132	2.12	.83
	1957 men	297.82	131	2.27	.81
1976 women	1976 women	311.87	132	2.36	.83
	1976 men	366.03	134	2.73	.78
	1957 women	388.93	133	2.92	.76

Note. From "The Structure of Psychological Well-Being: A Sociohistorical Analysis," by F. B. Bryant and J. Veroff, 1982, *Journal of Personality and Social Psychology, 43,* p. 663. Copyright 1982 by the American Psychological Association. Adapted with permission of the author. Entries are the results of single-group confirmatory factor analyses performed by means of COFAMM (Sörbom & Jöreskog, 1976), a forerunner of LISREL. The χ^2/df ratio reflects a model's goodness of fit to the data, relative to the number of estimated parameters. As this ratio decreases and approaches zero, the fit of the given model improves (Hoelter, 1983). TLC represents the Tucker-Lewis coefficient (Tucker & Lewis, 1973), a goodness-of-fit index. This coefficient reflects the amount of common variance accounted for by the model, relative to the amount of total variance. As the TLC increases and approaches 1.0, the fit of the given model improves.

men in 1957. Three different CFA models were applied to this data set: the CFA model developed using the sample of 1957 men, which logically should provide the best fitting model; the CFA model developed using the 1957 data for women; and the CFA model developed using the data set of 1976 men. Consider first the Tucker–Lewis coefficient (TLC; Tucker & Lewis, 1973) measure of fit associated with each model applied to the data set of 1957 men. As expected, the highest TLC (.86) was achieved when the CFA model for 1957 men was applied to the data set for 1957 men. The worst TLC (.80) was achieved when using the CFA model based on the 1976 data for men applied to the data set of 1957 men. And the TLC for the CFA model for women in 1957 applied to the data set of 1957 men was intermediate (.84). Why should the 1957 CFA model for women applied to the 1957 data set for men provide a better fit than when the model of 1976 men was applied to the data set of 1957 men?

This finding suggests that men and women were more alike in 1957 than men were alike across decades. Note that the same pattern of findings is shown by examining the χ^2/df ratios. Continuing to examine the fit between different models and data sets reveals the following pattern of results: The patterns of relative fit suggest that men's and women's models fit each other better within years than men's models fit men across years or women's models fit women across years. In other words, historical differences were relatively stronger than gender differences in the structuring of psychological well-being.

In the final step of their data analysis, Bryant and Veroff (1982) used simultaneous CFA to test hypotheses about between-groups differences in factor loadings. To evaluate historical change in the structure of well-being, they tested, separately for men and women, whether the model for either year fit the data of both years equally well. The only comparison that yielded equivalent factor loadings across years was that for the 1957 women's model applied to the women's data in both years, $\chi^2(14, N = 1{,}465) = 23.63, p < .10$. These findings indicate that historical change in the structure of well-being has been greater for men than for women.

To evaluate gender differences in factor structure, Bryant and Veroff (1982) tested, separately within each year, whether the model for either gender fit the data of men and women equally well. Again, only one comparison yielded equivalent factor loadings for men and women: The 1976 men's model applied to men's and women's data in 1976, $\chi^2(13, N = 1{,}258) = 20.09, p < .10$. These findings suggest that men and women were more alike in 1976 than they were in 1957, with this historical convergence in structure due mostly to shifts for men. This research demonstrates that CFA can be used to answer questions that have both theoretical and practical importance.

LISREL Notation

In reporting CFA results, researchers often use Roman and Greek letters to represent the observed indicators, latent factors, and estimated parameters in their factor models. For this reason, it may be useful to briefly discuss the letters and symbols most commonly used. In LISREL (Jöreskog & Sörbom, 1989), arguably the most popular computer program for conducting CFA, each required parameter of the CFA model is designated by a Greek or Roman letter: (a) Each observed indicator is designated by the letter x. (b) Each factor or latent construct is denoted by the Greek letter xi (ξ), pronounced "zī". (c) Each factor loading is designated by the

Greek letter lambda (λ), and the pattern of loadings (λs) for the observed (x) indicators on the factors is referred to as the lambda-x matrix (λ_x), with λ_xs representing the loadings of the observed indicators on the ξs. (d) The variance and covariances of each factor are designated by the Greek letter phi (ϕ), and the matrix of variances–covariances among the latent constructs is referred to as the *phi matrix*. (e) The unique-error term for each observed indicator is designated by the Greek letter delta (δ), and the matrix consisting of the variance of the unique-error terms and their covariances is referred to as the *theta-delta matrix* (θ_δ).

Figure 1 presents a schematic diagram of a hypothetical CFA model in LISREL. In diagramming a CFA model, each observed indicator is enclosed in a square and designated by a Roman letter x. In this hypothetical model, there are nine observed indicators, labeled x_1–x_9. The effect of measurement error on each observed indicator is marked by a small straight line to the indicator, and each unique error term is designated by a Greek letter delta (δ). In this hypothetical model, the unique errors for the nine observed indicators are designated δ_1–δ_9. Each latent construct (or factor) is enclosed in a circle and designated by a Greek letter xi (ξ). This hypothetical model has three latent factors, labeled ξ_1–ξ_3. The effect of a latent factor on an observed indicator (i.e., a factor loading) is marked by a straight line from the factor to that indicator and is designated by a Greek letter lambda (λ_x). (This λ_x should not be confused with λ, the symbol for eigenvalue.) In this model, there are nine estimated factor loadings: The first three indicators (x_1–x_3) have loadings (λ_1–λ_3) on the first latent factor (ξ_1); the middle three indicators (x_4–x_6) have loadings (λ_4–λ_6) on the second latent factor (ξ_2); and the last three indicators (x_7–x_9) have loadings (λ_7–λ_9) on the third latent factor (ξ_3). All other factor loadings (i.e., for x_1–x_3 on ξ_2 and on ξ_3; for x_4–x_6 on ξ_1 and on ξ_3; and for x_7–x_9 on ξ_1 and on ξ_2) have been fixed at zero. The covariances among the latent factors are marked with curved paths, and each is designated by a Greek letter phi (ϕ). In this oblique, three-factor model, three factor covariances are estimated ($\phi_{2,1}$, $\phi_{3,1}$, and $\phi_{3,2}$). The variances of the three latent factors are omitted from the diagram because this is a standardized CFA solution: Each factor's variance is constrained to equal 1.0. If this were an unstandardized CFA solution, then the factor variances would be reported in parentheses within the circle of each latent factor and the user would also be required to fix one factor loading on each factor (usually that of the highest loading item) to a value of 1.0, to define the measurement scale for each latent factor.

Figure 1

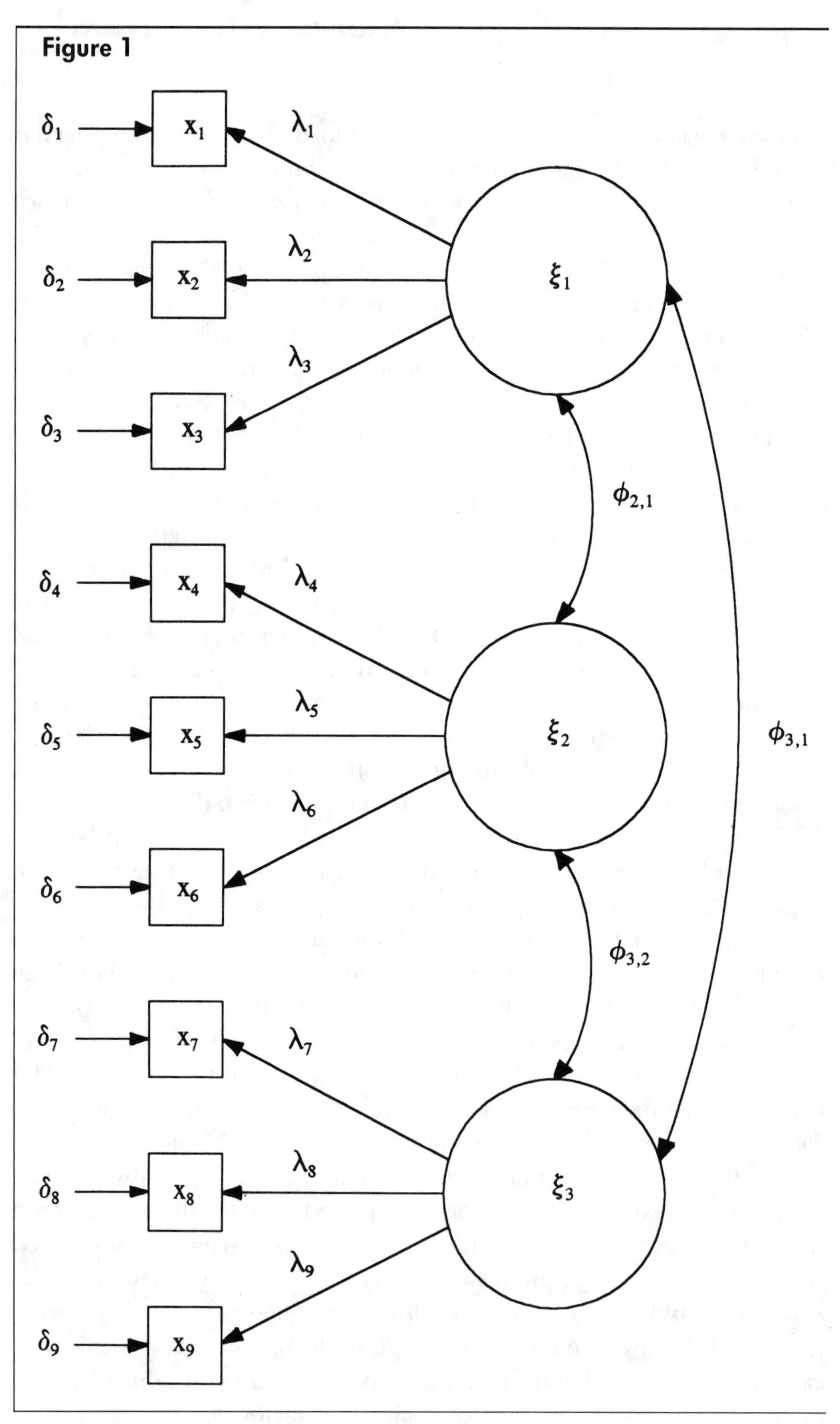

Schematic diagram of a hypothetical-confirmatory factor analysis model, containing three factors, for a hypothetical nine-item measure.

Conclusion

PCA and EFA are largely used as dimension-reducing procedures. For a collection of continuous variables, these techniques can identify a small set of synthetic variables, called *eigenvectors* or *factors*, that explain most of the total (PCA) or common (EFA) variation present in the original variables. CFA is typically used for purposes of theory testing. CFA requires that the researcher develop precise structural hypotheses. Structural features (*parameters*) that are usually modeled include factor loadings, factor variance–covariances, factor intercorrelations, and so forth. The goodness of fit of different models to the data can be evaluated to assess relative plausibility of alternative hypothesized structural models.

Suggested for Further Reading

Thurstone (1947) and Fruchter (1954) provided excellent, though dated, introductions to PCA and EFA. Relatively modern, detailed, and readable discussions of PCA were presented by Dunteman (1989) and Stevens (1986), and of PCA and EFA by Amick and Walberg (1975), Cattell (1966), Dillion and Goldstein (1984), Green (1978), Hair et al. (1992), Kleinbaum, Kupper, and Muller (1988), and Tabachnick and Fidell (1983). Relatively comprehensive discussions were presented by Gorsuch (1983), Harman (1976), Jolliffe (1986), Lawley and Maxwell (1971), and Morrison (1976).

The reader interested in learning more about CFA is referred to Bollen (1989) and Hayduk (1987). These books provide a thorough and comprehensive overview of the logic, application, and interpretation of both single-group and multisample CFA. For technical details regarding the CFA procedure, see Jöreskog and Sörbom (1989).

Glossary

A PRIORI CRITERION A "stopping rule" for determining the appropriate number of eigenvectors (factors) to extract, in which one knows in advance how many factors should be extracted.

BIPOLAR EIGENVECTOR (OR BIPOLAR FACTOR) An eigenvector (factor) on which at least one variable has a positive factor loading coefficient and at least one variable has a negative factor loading coefficient.

COMMON VARIANCE The portion of the total variance that a set of variables has in common.

COMMUNALITY Equal to the sum over all factors of the squared factor loadings for a variable, communality indicates the variance which a variable has in common with the other variables in the analysis.

CONFIRMATORY FACTOR ANALYSIS (CFA) A multivariate statistical technique for testing hypotheses about the dimensions underlying a set of measured variables. With CFA, the user specifies a particular factor model (i.e., a configuration of factor loadings, factor variances–covariances, and unique errors in the measured variables), and its relative fit to the data is evaluated.

CORRELATION MATRIX A tabular procedure for summarizing all possible correlations between a set of variables. Both the rows and the columns of the table consist of the variables (e.g., Variable 1, 2, 3). Because the correlation (*r*) between a variable and itself (e.g., Variable 1 and Variable 1) is always equal to unity, the diagonal of the correlation matrix consists of 1s. Off-diagonal entries are the *r* between the indicated variables. For example, the entry in column 4, row 5 of a correlation matrix would contain the *r* between Variable 4 and Variable 5. Note that this entry is the same as that for column 5, row 4 of the correlation matrix (entries of which are symmetric about the diagonal).

COVARIANCE The covariance between two measures is their correlation multiplied by both their standard deviations. A Pearson correlation coefficient represents a standardized covariance, for which the variance of each item is fixed to 1.

DIAGONAL OF THE CORRELATION MATRIX The cells in a correlation matrix, running from the upper-left-hand corner to the lower-right-hand corner, which indicate the *r* between a given variable and itself.

EIGENVALUE An index, denoted using the symbol λ, that indicates the portion of the total variance of a correlation matrix that is explained by an eigenvector.

EIGENVECTOR The linear functions of the variables, identified using principal-components analysis, that maximize the amount of the total variance in a correlation matrix that is explained.

EQUALITY CONSTRAINT Parameters in a confirmatory factor analysis (CFA) model that have been forced to have identical estimated values in the factor solution. An equality constraint specifies that the particular parameters involved be of equal value, but it does not specify beforehand what that value must be. Instead, the CFA program is allowed to estimate the exact value for equality constrained parameters.

EQS A popular computer program for performing confirmatory factor analysis (Bentler, 1989).

ERROR VARIANCE In principal-components analysis, the portion of the total variance that is not explained by the principal component(s). In exploratory factor analysis, the portion of the total variance that is inherently unreliable random variation.

EXACT IDENTIFICATION A condition that occurs when a confirmatory factor analysis model contains precisely the right number and combination of known and unknown parameters to yield unique estimates from the observed data.

EXPERIMENTWISE TYPE I ERROR The Type I error rate for the experiment taken as a whole. In studies involving more than one test of a statistical hypothesis, the Type I error rate (p) for the experiment as a whole (the experimentwise Type I error rate) is much higher than the nominal Type I error rate (p) used to evaluate each of the individual hypotheses.

EXPLAINED VARIANCE In principal-components analysis, the portion of the total variance that can be explained by the principal component(s).

EXPLORATORY FACTOR ANALYSIS (EFA) A methodology that is used to identify linear functions or factors, which explain the theoretical maximum amount of (remaining) common variance in a correlation matrix.

FACTOR LOADING COEFFICIENT The correlation between a variable and an eigenvector (factor).

FACTORS The eigenvectors that are extracted in a principal-components analysis. Also, dimensions that are extracted in an exploratory factor analysis or posited to exist in confirmatory factor analysis.

FITTED RESIDUALS The difference between the correlations or covariances predicted by a particular factor model and the actual observed values. The closer these residuals are to zero, the better the model fits the data.

FIXED PARAMETER A factor loading, factor variance, factor covariance, or unique-error term in a CFA model that has been assigned a specific value by the user.

FREE PARAMETER A factor loading, factor variance, factor covariance, or unique-error term that has been allowed to vary in a CFA model, to be estimated by the computer program in the analysis.

GOODNESS-OF-FIT INDEX (GFI) A coefficient, ranging from zero to one, that reflects the improvement in fit gained by a given CFA model over the most restrictive factor model, which is typically a null model. As a rule of thumb, a minimum GFI value of .90 has been recommended in judging the acceptability of CFA models (Bentler & Bonett, 1980).

IDENTIFICATION A mathematical issue concerning whether a CFA model contains a sufficient combination of known (fixed) and unknown (free) parameters to yield unique estimates from the observed data.

KAISER'S STOPPING RULE A stopping rule for determining the appropriate number of eigenvectors (factors) to extract, in which one retains all factors with associated eigenvalues of 1 or greater.

LAMBDA (λ) In CFA through LISREL, the pattern of factor loadings is referred to as the *lambda matrix*, with λ_xs representing the loadings of the observed indicators on the latent factors. In PCA, each eigenvalue is denoted by the symbol λ.

LINEAR FUNCTION A model of the form: Y = X1*Variable 1 + X2 *Variable 2, and so forth. Here, the Xs are the function coefficients.

LISREL An acronym for *LI*near *S*tructural *REL*ationships (Jöreskog & Sörbom, 1989), perhaps the most popular computer program for performing CFA in the social sciences.

MODEL MISSPECIFICATION A condition that occurs when a specified CFA model is different from models that the data would support. Misspecified models yield improper solutions, in which unique errors are negative, factor intercorrelations exceed 1, or parameter estimates are massive.

MODIFICATION INDEX (MI) For each fixed parameter in a CFA model, a measure of the predicted decrease in chi-square if the particular parameter is freed and the model is reestimated. The constrained parameter corresponding to the largest such index is the one that, if estimated rather than fixed, will maximally improve the model's fit.

NESTED MODELS Models are considered to be nested when one, which is more restrictive, can be obtained by imposing constraints on the other, which is more general. In such cases, one can compute the difference in the chi-squares (and degrees of freedom) associated with each model, to determine whether one model fits the data better than the other.

NULL MODEL A baseline model for use in judging the relative fit of common-factor models in CFA. It specifies that the sole source of variance in subjects' responses is sampling error and that no common variance exists. This is typically the most restrictive CFA model considered, and it is the conceptual opposite of the unrestricted, saturated model, in which all parameters are estimated.

OBLIQUE ROTATION Used in an attempt to achieve simple structure, this factor rotation method allows eigenvectors (factors) to be correlated.

ORTHOGONAL ROTATION Used in an attempt to achieve simple structure, this factor rotation method forces eigenvectors (factors) to remain uncorrelated ($r = 0$).

OVERIDENTIFICATION A condition that occurs when a model contains too few unknown parameters to allow unique estimates from the data. CFA estimates of underidentified parameters are arbitrary and cannot be trusted.

PARAMETERS Any one of the terms in a CFA model that takes on a numerical value in the factor solution. These consist of the factor loadings, factor variances, factor covariances, and unique- and shared-error variances that constitute the CFA model. Each parameter is either fixed at a specified value by the user or left free to be estimated in the factor solution.

PERCENTAGE OF VARIANCE CRITERION A stopping rule for determining the appropriate number of eigenvectors (factors) to extract, in which one retains factors until some portion of the total (common) variance has been explained. This is an a priori method: The desired (target) variance is specified before the analysis is conducted.

PRINCIPAL COMPONENT A linear function of a set of variables that explains the theoretical maximum amount of (remaining) total variance in a correlation matrix.

PRINCIPAL-COMPONENTS ANALYSIS (PCA) A methodology that is used to identify linear functions or factors, which explain the theoretical maximum amount of (remaining) total variance in a correlation matrix.

QUARTIMAX ROTATION Used in an attempt to achieve simple structure, this factor rotation method forces eigenvectors (factors) to be uncorrelated ($r = 0$). This rotation method focuses on making as many values in each row of the factor loading coefficient table to be as close to zero as is possible.

RANK OF A CORRELATION MATRIX The number of principal components needed to explain 100% of the total variance of a correlation matrix.

RATIO OF CHI-SQUARE TO DEGREES OF FREEFOM (χ^2/df) Used to compare the relative fit of alternative factor models for the same data. As this ratio decreases and approaches zero, the fit of the given model improves. Use of this ratio allows one to compare the fit of alternative models, controlling for differences in their complexity.

RELIABILITY In exploratory factor analysis, the sum of the common and specific variance.

ROOT MEAN SQUARE RESIDUAL (RMSR) A statistic used to judge the goodness of fit of a CFA model. RMSR represents the absolute value of the average fitted residuals for a given CFA model. Fitted residuals are the difference between (a) the actual correlations (or covariances) among the observed indicators and (b) the correlations (or covariances) predicted by a particular model. The closer RMSR is to zero, the better the fit of the model.

ROTATION OF FACTORS A procedure in which the eigenvectors (factors) are rotated in an attempt to achieve simple structure.

SCREE TEST A stopping rule for determining the appropriate number of eigenvectors (factors) to extract, in which one graphs the eigenvalues of successive eigenvectors (factors) and then draws a line of best fit, which indicates the change in eigenvalues over successive factors. The factors with eigenvalues that lie on the path of steep descent in this plot are retained. The factors with eigenvalues that come later lie on the scree slope and are not extracted.

SIMPLE STRUCTURE A condition in which variables load at near 1 (in absolute value) or at near 0 on an eigenvector (factor). Variables that load near 1 are clearly important in the interpretation of the factor, and variables that load near 0 are clearly unimportant. Simple structure thus simplifies the task of interpreting the factors.

SIMULTANEOUS CONFIRMATORY FACTOR ANALYSIS Using CFA to estimate a given factor model, using the data of multiple groups in the same analysis. Researchers use this procedure to test hypotheses about the equivalence of factor loadings, factor variances–covariances, and unique errors across groups.

SPECIFIC VARIANCE The portion of the total variance that is reliable but that does not correlate with the other variables in the analysis.

STANDARDIZED RESIDUAL A fitted residual that has been divided by its estimated standard error. These are independent of the units of measurement in the observed indicators and are therefore more clearly interpretable than fitted residuals.

STANDARDIZED SOLUTION An estimated factor model in which the variances of the latent factors have been fixed at 1. This type of solution makes it easier for the user to interpret the relationships among the factors.

STOPPING RULES Procedures used to determine the appropriate number of eigenvectors (factors) to extract.

SUBJECTS-TO-VARIABLES RATIO The ratio of the number of subjects (or observations) in one's sample divided by the number of variables used in the analysis.

THETA DELTA (θ_δ) In LISREL, the matrix of unique errors for observed variables and the interrelationships among these error terms.

TOTAL VARIANCE In PCA, the variance of the set of variables that are analyzed. If the variables have been standardized into *z*-score form, then the total variance of the set of variables is equal to the number of variables. In EFA, the total variance is equal to the sum of three components: common, specific, and error variance.

UNDERIDENTIFICATION A condition that occurs when a model contains too many unknown parameters to allow unique estimates from the data. CFA estimates of underidentified parameters are arbitrary and cannot be trusted.

UNIQUENESS In EFA and CFA, the portion of the total variance that is unrelated to other variables. Thus, uniqueness = specific variance + error variance. Also, uniqueness is equal to 1 − the communality.

Unrestricted (or Saturated) Model A baseline model for use in judging the relative fit of common-factor models in CFA. In the fully saturated model, all parameters are free and there is no falsifiable structural hypothesis, only the equivalent of an exploratory solution.

Varimax Rotation Used in an attempt to achieve simple structure, this factor rotation method forces eigenvectors (factors) to be uncorrelated ($r = 0$). This rotation method focuses on making as many values in each column of the factor loading coefficient table to be as close to zero as is possible.

Xi (ξ) In LISREL, latent constructs or factors are referred to as *xis*.

z-Score (or Standardized Score) A score transformed using the formula: *z* score = (original score − *M*)/*SD*. A standardized variable has a mean of 0 and a variance of 1.

References

Alwin, D. F., & Jackson, D. J. (1979). Applications of simultaneous factor analysis to issues of factorial invariance. In D. J. Jackson (Ed.), *Factor analysis and measurement in sociological research* (pp. 249–279). London: Sage.

Alwin, D. F., & Jackson, D. J. (1980). Measurement models for response errors in surveys: Issues and applications. In K. F. Schuessler (Ed.), *Sociological methodology, 1980* (pp. 68–119). San Francisco: Jossey-Bass.

Amick, D. J., & Walberg, H. J. (1975). *Introductory multivariate analysis for educational, psychological and social research*. Chicago: University of Illinois at Chicago Press.

Anderson, J. C., & Gerbing, A. W. (1984). The effect of sampling error on convergence, improper solutions, and goodness-of-fit indices for maximum-likelihood confirmatory factor analysis. *Psychometrika, 49*, 155–173.

Bagozzi, R. P., & Yi, Y. (1988). On the evaluation of structural equation models. *Journal of the Academy of Marketing Science, 16*, 74–94.

Bentler, P. M. (1989). *EQS structural equations program manual*. Los Angeles: BMDP Statistical Software.

Bentler, P. M. (1990). Comparative fit indexes in structural models. *Psychological Bulletin, 107*, 238–246.

Bentler, P. M., & Bonett, D. G. (1980). Significance tests and goodness of fit in the analysis of covariance structures. *Psychological Bulletin, 88*, 588–606.

Blyth, C. R. (1972). On Simpson's paradox and the sure-thing principle. *Journal of the American Statistical Association, 67*, 364–366.

Bollen, K. A. (1989). *Structural equations with latent variables*. New York: Wiley.

Bradburn, N. M. (1969). *The structure of psychological well-being*. Chicago: Aldine.

Browne, M. W. (1984). Asymptotically distribution-free methods for the analysis of covariance structures. *British Journal of Mathematical and Statistical Psychology*, *37*, 62–83.

Bryant, F. B. (1989). A four-factor model of perceived control: Avoiding, coping, obtaining, and savoring. *Journal of Personality*, *57*, 773–797.

Bryant, F. B., & Veroff, J. (1982). The structure of psychological well-being: A sociohistorical analysis. *Journal of Personality and Social Psychology*, *43*, 653–673.

Bryant, F. B., & Veroff, J. (1984). Dimensions of subjective mental health in American men and women. *Journal of Health and Social Behavior*, *25*, 116–135.

Bryant, F. B., & Yarnold, P. R. (1989). A measurement model for the short form of the Student Jenkins Activity Survey. *Journal of Personality Assessment*, *53*, 188–191.

Campbell, A. (1980). *The sense of well-being in America*. New York: McGraw-Hill.

Cattell, R. (1966). The meaning and strategic use of factor analysis. In R. B. Cattell (Ed.), *Handbook of multivariate experimental psychology* (pp. 174–243). Chicago: Rand McNally.

Cunningham, W. R. (1978). Principles for the identification of structural differences. *Journal of Gerontology*, *33*, 82–86.

Dillion, W. R., & Goldstein, M. (1984). *Multivariate analysis: Methods and applications*. Berkeley: University of California Press.

Dunteman, G. H. (1989). *Principal components analysis*. Newbury Park, CA: Sage.

Fruchter, B. (1954). *Introduction to factor analysis*. New York: Van Nostrand.

Glass, D. C. (1977). *Behavior patterns, stress, and coronary disease*. Hillsdale, NJ: Erlbaum.

Gorsuch, R. L. (1983). *Factor analysis*. Hillsdale, NJ: Erlbaum.

Green, P. E. (1978). *Analyzing multivariate data*. Hinsdale, IL: Dryden Press.

Guadagnoli, E., & Velicer, W. F. (1988). Relation of sample size to the stability of component patterns. *Psychological Bulletin*, *103*, 265–275.

Hair, J. F., Anderson, R. E., Tatham, R. L., & Black, W. C. (1992). *Multivariate data analysis with readings* (3rd ed.). New York: Macmillan.

Harman, H. H. (1976). *Modern factor analysis*. Chicago: University of Chicago Press.

Hayduk, L. A. (1987). *Structural equation modeling with LISREL*. Baltimore: Johns Hopkins University Press.

Hoelter, J. W. (1983). The analysis of covariance structures: Goodness of fit indices. *Sociological Methods and Research*, *11*, 325–344.

Hotelling, H. (1933). Analysis of a complex of statistical variables into principal components. *Journal of Educational Psychology*, *24*, 417–441, 498–520.

Hu, L., Bentler, P. M., & Kano, Y. (1992). Can test statistics in covariance structure analysis be trusted? *Psychological Bulletin*, *112*, 351–362.

Jolliffe, I. T. (1986). *Principal component analysis*. New York: Springer-Verlag.

Jöreskog, K. G. (1969). A general approach to confirmatory maximum likelihood factor analysis. *Psychometrika*, *34*, 183–202.

Jöreskog, K. G. (1971a). Simultaneous factor analysis in several populations. *Psychometrika*, *36*, 409–426.

Jöreskog, K. G. (1971b). Statistical analysis of sets of congeneric tests. *Psychometrika*, *36*, 109–133.

Jöreskog, K. G. (1978). Structural analysis of covariance and correlation matrices. *Psychometrika*, *43*, 443–477.

Jöreskog, K. G., & Sörbom, D. G. (1989). *LISREL 7 user's reference guide*. Chicago: Scientific Software.

Kaiser, H. F. (1960). The application of electronic computers to factor analysis. *Educational* and Psychological Measurement, *20*, 141–151.

Kleinbaum, D. G., Kupper, L. L., & Muller, K. E. (1988). *Applied regression analysis and other multivariable methods* (2nd ed.). Boston: PWS-Kent.

La Du, T. J., & Tanaka, J. S. (1989). Influence of sample size, estimation method, and model specification on goodness-of-fit assessment in structural equation models. *Journal of Applied Psychology, 74,* 625–635.

Larsen, R. J. (1984). Theory and measurement of affect intensity as an individual difference characteristic. *Dissertation Abstracts International, 85,* 2297B. (University Microfilms No. 84-22112).

Lawley, D. N., & Maxwell, A. E. (1971). *Factor analysis as a statistical method.* New York: Elsevier Science.

Long, J. S. (1983). *Confirmatory factor analysis.* Beverly Hills, CA: Sage.

Maiti, S. S., & Mukherjee, B. N. (1991). Two new goodness-of-fit indices for covariance matrices with linear structures. *British Journal of Mathematical and Statistical Psychology, 44,* 153–180.

Marsh, H. W., Balla, J. R., & McDonald, R. P. (1988). Goodness-of-fit indexes in confirmatory factor analysis: The effect of sample size. *Psychological Bulletin, 103,* 391–410.

McKennell, A. C., & Andrews, F. M. (1980). Measures of cognition and affect in perceptions of well-being. *Social Indicators Research, 8,* 257–298.

Moore, M. K., & Neimeyer, R. A. (1991). A confirmatory factor analysis of the Threat Index. *Journal of Personality and Social Psychology, 60,* 122–129.

Morrison, D. F. (1976). *Multivariate statistical methods* (2nd ed.). New York: MeGraw-Hill.

Mulaik, S. A., James, L. R., Van Alstine, J., Bennett, N., Lind, S., & Stilwell, C. D. (1989). Evaluation of goodness-of-fit indices for structural equation models. *Psychological Bulletin, 105,* 430–445.

Muthen, B. O. (1993). Goodness of fit with categorical and other nonnormal variables. In K. A. Bollen & J. S. Long (Eds.), *Testing structural equation models* (pp. 205–234). London: Sage.

Pearson, K. (1901). On lines and planes of closest fit to systems of points in space. *Philosophical Magazine,* May 2, 559–572.

Popper, K. R. (1968). *The logic of scientific discovery.* New York: Harper Torchbooks.

Sörbom, D. G., & Jöreskog, K. G. (1976). *COFAMM: Confirmatory factor analysis with model modification, a FORTRAN IV program.* Chicago: National Educational Resources.

Steiger, J. H. (1979). Factor indeterminancy in the 1930's and the 1970's: Some interesting parallels. *Psychometrika, 44,* 157–167.

Stevens, J. P. (1986). *Applied multivariate statistics for the social sciences.* Hillsdale, NJ: Erlbaum.

Tabachnick, B. G., & Fidell, L. S. (1983). *Using multivariate statistics.* New York: Harper & Row.

Tanaka, J. S. (1993). Multifaceted conception of fit in structural equation models. In K. A. Bollen & J. S. Long (Eds.), *Testing structural equation models* (pp. 10–39). London: Sage.

Thurstone, L. L. (1947). *Multiple-factor analysis: A development and expansion of the vectors of mind.* Chicago: University of Chicago Press.

Tucker, L. R., & Lewis, C. (1973). A reliability coefficient for maximum likelihood factor analysis. *Psychometrika, 38,* 1–10.

Weinfurt, K. P., Bryant, F. B., & Yarnold, P. R. (1994). The factor structure of the Affect Intensity Measure: In search of a measurement model. *Journal of Research in Personality, 28,* 314–331.

Yarnold, P. R. (1984). Note on the multidisciplinary scope of psychological androgyny theory. *Psychological Reports,* 55, 936–938.

Multidimensional Scaling

Loretta J. Stalans

Imagine that the Criminal Justice Commission has hired you as a consultant. The commission plans to create criminal sentencing legislation that is based on what the public believes is fair punishment for burglary (breaking and entering a place when a person is not present) and robbery (taking property from someone with the threat of physical harm). Your task is to determine why some crimes are seen as similar (e.g., burglary of a business dwelling and purse snatching) and other crimes are seen as different (e.g., armed robbery and unarmed robbery). You speculate about the critical features defining people's perceptions of the appropriate punishment for these crimes and think of two features: (a) the threat of physical harm and (b) the worth of the property that was stolen. The commission also suggests other potential characteristics of crimes: (a) the offender's prior criminal history, (b) the offender's age, and (c) the presence of extenuating circumstances. You need a procedure that will discover which of these potential characteristics underlies the public's perception of which criminal acts should receive similar punishment.

In another example, suppose you are a clinical psychologist. You want to know how George, a client, perceives his relationships with family members, acquaintances, and friends. You have a hunch that the primary identifying characteristic underlying his interactions with others is control and that the secondary characteristic is self-interest. How can you test

I thank Larry G. Grimm, Harry S. Upshaw, Paul R. Yarnold and two anonymous reviewers for many helpful suggestions, which improved this chapter. I am currently an assistant professor of criminal justice at Loyola University and was formerly an assistant professor at Georgia State University.

this clinical hypothesis? Simply asking George why he interacts more with some people and less with others may not provide accurate data. You need an objective way to test whether control and self-interest are the defining characteristics of George's interactions with other people. You decide to use an overt, objective assessment tool. You ask George to list all of the people that he knows on a personal basis. You then instruct him to sort these people by how often he interacts with each person. When sorting these persons, George judges how similar each person is to the other persons in terms of how often he interacts with them. After he completes this task, you ask him to rate each person on several dimensions, such as allowing him to control them, promoting his self-interest, emotional support, similar hobbies, and similar values. These ratings can be used to determine why George interacts with some people more than with other people. From these data, you are prepared to discover the characteristics defining George's social world.

The issues addressed in these examples appear different on the surface, but they actually are addressing the same question. In both of these examples, the question is, What features underlie judgments about the similarity among members or items of a category? (Throughout this chapter, the term *members* refer to people or animals in a category. The term *items* is used when the members of a category refer to events [e.g., crimes], inanimate objects [e.g., chairs], or plants [e.g., trees] in a category.) The first example explores lay views about the similarity among crimes based on their perception of appropriate punishment. The second example examines the pattern in George's similarity judgments of how often he interacts with each person. The following are other examples of topics that can be addressed using similarity data: the kinship structure of a tribe, the characteristics associated with smell, the characteristics associated with who talks to whom in an organization, and the perceived quality of consumer products or politicians.

What statistical tool can be used to understand the systematic patterns in similarity data? *Multidimensional scaling* (MDS) procedures are designed to detect the hidden structure of similarity judgments. That is, MDS is a statistical tool to determine which particular characteristics are most important in discerning the pattern in similarity judgments out of a set of plausible defining features. Because MDS analyzes pairwise comparison data (perceived relatedness between two items of a category), it has a wide array of applications in business, psychology, political science, health fields, communication, advertising, and criminal justice.

Constructing a map (not simply reading an already constructed map) is a useful and common analogy for what metric MDS attempts to accomplish. Metric MDS scaling assumes that dissimilarities are measured at an interval or ratio level: The difference between the numbers 3 and 4 is the same as the difference between the numbers 4 and 5. This assumption is true when the dissimilarities are derived by means of interval or ratio measures such as amount of time interacting with different toys or the airline miles between two cities.

Creating a map of the location of several key cities in the United States (Atlanta, Boston, Chicago, Denver, Houston, Miami, New York, New Orleans, Los Angeles, San Francisco and Seattle) would involve measuring airline mileage between these cities. With MDS, one would then use this airline mileage to compute *distance*, which refers to the space between two objects (cities) as measured with a straight line. One would then spatially represent the location of each city in relation to the others, on the basis of the computed distances. The computed distance is not actual distance because airline mileage has measurement error.

The constructed map provides both direction and distance. For example, Los Angeles and San Francisco may be very close to each other, and both may be very far away from Boston and Miami. The coordinated axes obtained from the MDS solution have no special significance; the MDS solution reveals the number of meaningful directions in the visual plot but does not indicate their exact location. The researcher must find and interpret the meaningful directions. That is, the MDS solution may not produce a map with North directly in the center of the top half and South in the center of the bottom half; instead, the coordinated axes may have South on top and North on the bottom or Northeast in the top center and Southwest in the bottom center. Thus, the researcher rather than MDS decides and labels the meaningful directions. For example, geographers at some point in history decided what directions on a visual map were important; they labeled one dimension *East to West* and the other dimension *North to South*. These interpreted dimensions on the map are often used to find specific cities, such as Atlanta. The directions that researchers interpret and label from the MDS-created spatial plot are called *dimensions*.

Imagine someone gave you the flying mileage between each pair of these cities. You can arrange the cities based on the pairwise data of flying mileage (similar to how you arrange pieces in a jigsaw puzzle until they all fit in the picture) and create a visual representation of the cities of the United States. Your arrangement of the cities would be an easy task if you already knew the geography of the United States and had error-free data.

Similar to a jigsaw puzzle, the appropriate location of each item (e.g., city) in the completed picture and the most important features determining the pattern among the items are often unknown! Moreover, a person may not know in advance whether a one-dimensional picture (a line), two-dimensional picture (a plane), three-dimensional picture (a sphere), or higher dimensional picture is needed to capture the relatedness among the items.

MDS is an especially useful tool when the appropriate visual representation is unknown. MDS mathematically transforms the perceived relatedness among items into a visual representation of distance. That is, distance (the space between two items) in a spatial plot is an analogy for the perceived relatedness among the items. This spatial representation, which shows the location of each item of a category, is called a *configuration*, and is the most important result produced by MDS. By using MDS to analyze the distance data of the cities, one can learn about the geography of the United States. The visual map of the United States produced by MDS generally will be an approximate picture because the dissimilarity data contain measurement error.

The configuration provides a visual picture of the pattern among the perceived relatedness of items. Researchers attempt to discover the dimensions that can explain the pattern of relatedness among the items presented in a configuration. Researchers, however, must use both their intuition and their statistical results and tools to aid their interpretation of a configuration. The interpretation of a configuration consists of identifying relevant dimensions. These discovered dimensions represent continuous, directional features in the pattern of items (e.g., East to West, passive to active, accidental to intentional). The discovered dimensions allow researchers to locate specific items in a configuration. More important, the discovered dimensions can be used to predict the relatedness among a new sample of items. For illustration, consider the two discovered dimensions of cities in the United States. These two dimensions, again, are East to West and North to South. If Memphis is in the new sample of cities and is identified as being in the Southeastern region, one would predict that it was closer to Atlanta than to Boston.

Metric MDS has some features in common with other scaling procedures, such as factor analysis. Both metric MDS and factor analysis assume at least interval measurement, and both attempt to discover the hidden dimensions in the data. They both assign numbers to items so that these numbers are connected to some relationship in the real world.

Types of Multidimensional Scaling Procedures

Metric scaling was first invented but is rarely used because of the stringent assumptions about data measurement. *Nonmetric MDS scaling* is the most frequent type used in published articles. In comparison with metric scaling, nonmetric scaling makes a less stringent assumption about how the data were measured, because it assumes only that the data values are measured at an ordinal level. An ordinal level of measurement assesses only the rank position of similarity; for example, the finishing place of three horses in a race. The winner was the fastest, the second-place finisher was the next fastest, and the third-place finisher was the slowest. The finishing time between the winner and second-place finisher may be 1 s whereas the distance between the second-place finisher and third-place finisher may be 2 min. Ordinal measurement thus assesses the relative similarity (highest to lowest) but does not measure the absolute difference among positions. Nonmetric scaling is generally more appropriate when the perceived relatedness among items is measured with rating scales; the difference between numbers in rating scales are rarely perceived as equal. Thus, the interval/ratio assumption generally is not true when the perceived-relatedness data are derived by means of rating scales or ranks.

Another basic distinction among MDS procedures is whether a procedure is *unweighted* or *weighted*. Unweighted MDS procedures assume all subjects place the same importance on the features used to make comparisons among items of a category; procedures with this assumption also are known as *classical*. In contrast, weighted MDS assumes individuals may differ on both the characteristics they use to define a category and the importance they place on each characteristic. Because of this assumption, weighted MDS procedures also are called *individual-differences scaling*.

Although a detailed discussion of how to interpret individual-differences MDS is not presented in this chapter, it is important to understand the kinds of problems suitable for individual-differences scaling (for a detailed discussion of three-way MDS, see Rabinowitz, 1986; Schiffman, Reynolds, & Young, 1981). These weighted procedures are capable of answering more complex questions than are unweighted MDS procedures. Individual-differences scaling can discover differences between individuals, contexts, or occasions. For example, political scientists can address whether conservatives place greater importance on the amount of harm done in distinguishing among criminal acts, whereas liberals place greater importance on the offender's background and drug use history. Market researchers can determine whether males and females differ in their opinion about the most important characteristics defining a quality automobile. Communication researchers can test the hypothesis that self-interest and power determine who talks to whom in an organization during the 3 months before raise negotiations, whereas shared hobbies and political attitudes determine who talks to whom in an organization during the other 9 months. Clinical psychologists can evaluate whether abused children differ from nonabused children in how they perceive social relationships. Legal scholars can address whether judges and citizens use the same characteristics of robberies in determining the appropriate punishment for each robber. Policymakers can obtain an empirical answer to the question, Do taxpayers at all levels of social status use the same dimensions to define the preferred way to reduce the budget deficit? Sociolegal scholars can address whether people have the same dimensions in their expectations about how they will be treated by different legal institutions (e.g., police, trial courts, Internal Revenue Service, and appellate courts). Weighted or individual-differences MDS scaling, thus, allows for both individual differences among how respondents define the relatedness for a given category and how the same respondent defines a category on different occasions or across situations.

Purpose and Outline of Chapter

The purpose of this chapter is to assist readers in learning how to interpret nonmetric two-way MDS results reported in published articles. Nonmetric, two-way MDS assumes the data are measured at an ordinal level and that all respondents use the same features of the items to form their similarity judgments. The chapter is intended for readers who are unfamiliar with MDS procedures and wish to acquaint themselves with problems best suited for MDS procedures and with how to interpret statistical results. This chapter is not intended either as a guide on how to perform MDS or as a guide on the differences between MDS procedures; several articles and books provide detailed discussions on these topics (Kruskal & Wish, 1978; Rabinowitz, 1986; Schiffman et al., 1981).

The next two sections of this chapter describe data collection for MDS and discuss the nature of the results produced by MDS. After this are a survey of the statistical tools for selecting the best configuration of results and a summary of intuitive and statistical methods to interpret a selected MDS configuration. Instructions on how to detect violations of MDS assumptions precede a comprehensive example in which various phases of the MDS procedure are illustrated. Suggestions for Further Reading and a Glossary complete the chapter.

Data Needed for Nonmetric MDS

In describing the data collection methods and how to interpret the results from MDS, the example about the public's perception of crime at the beginning of the chapter will be used. Pretend you are the consultant for the crime commission and are preparing to determine the underlying characteristics of the public's perception of crime. You want to answer the question, In comparing crimes on the basis of similar or different required punishment, do people use the offenders' criminal histories, offenders' age, amount of goods or money stolen, the amount of physical harm, or the threat of physical harm in forming these similarity judgments? The answer to this question can take several forms: (a) Respondents may use only one of these features (e.g., threat of physical harm); (b) respondents may use two features (e.g., threat of physical harm and offenders' criminal history); (c) respondents may use three or four features. You have decided to use MDS as the statistical tool because it is designed to indicate which particular features are most important in defining people's perceptions of the relatedness among crimes out of a set of features that are plausibly important.

How many people do you need to interview to use MDS? One great advantage of MDS is that it requires very few respondents: The number of respondents can be as low as one and as high as the researcher's resources and stamina allow. When the purpose is to generalize to a population, a representative sample randomly selected from the population increases the possibility that the results apply to the population. For data collected from more than one respondent, two-way MDS procedures use the perceived similarity for each pair of crimes averaged across respondents.

Assume that you have decided to interview 5 people as a preliminary study. What type of data do you need to collect to use MDS? Data representing the similarity or difference between each pair of crime stories must be collected. For each possible pair of crimes in the set, MDS requires a number reflecting how similar or different each pair is judged to be. This number represents the perceived relatedness between a pair of items (e.g., crimes) and is called a *proximity*. The relatedness among items can be measured as the perceived difference among items, in which higher numbers represent greater dissimilarity. The relatedness among items also can be measured as the perceived similarity among items, in which higher numbers represent greater similarity. Although similarity measures often are the inverse of dissimilarity measures, analyses of similarity measures sometimes produce MDS solutions that are different from analyses of dissimilarity measures. (For a detailed discussion of this issue, see Schiffman et al., 1981).

There are a number of ways to obtain proximity data. Assume you have selected nine stories of burglaries, nine stories of unarmed robberies, and nine stories of armed robberies (each story is a single, doubled-spaced page). In these stories, you have varied four characteristics: (a) the offenders' criminal histories, (b) the offenders' age, (c) the amount of goods or money stolen, and (d) the amount of physical harm. See Table 1 for a description of 27 hypothetical stories of robbery that manipulate these four characteristics. One direct way to collect proximity data for this set of crime stories is to ask respondents to rate the similarity between each pair of crimes. Respondents, for example, are instructed to compare two crimes on the basis of the punishment that the offender should receive. Instructions for comparison judgments often suggest how the similarity judgments should be made. For example, respondents may be told that crimes should be compared on whether they require different punishment. To maximize the spread of the similarity judgments (i.e., variance), the instructions can leave out how the similarity judgments should be made.

Respondents can represent their perception of similarity between two crimes by marking a 5-in. line with the anchor of *exactly the same* at one end and the anchor of *completely different* at the other end. Scales such as this one, in which the opposing ends are complete opposites, are called *bipolar scales*. For this set of 27 crimes, respondents will make 351 ratings (assuming each respondent compares all possible combinations of crimes). Clearly, this method can be very laborious for the respondents.

To make the task less laborious, there are alternative data collection methods that will answer your research questions. For example, you can ask respondents to indicate their preferred punishment for each crime from among various sentencing options (e.g., probation, probation with community service, probation with jail, or some years in prison). Alternatively, you can ask respondents to sort the crimes into piles on the basis of similarity (crimes perceived to be similar would be placed in the same pile and crimes perceived as different would be placed in separate piles). For these two data collection techniques, you must transform the collected data into measures of proximities. One such frequently used transformation is to compute correlations. Correlations represent the association between two items and thus are useful measures of perceived similarity. For example, correlations between the sentencing preferences for each pair of crimes can be computed. These computed correlations represent the similarity among crime stories. A derived measure of relatedness, such as a correlation, is called a *profile proximity measure*. (See Kruskal & Wish, 1978, for a detailed, technical discussion of the biases associated with each data collection method.)

The Nature of the Results Produced by Nonmetric MDS

This section first describes the results produced by MDS when the data are perceived relatedness rather than actual distances. It then illustrates how to interpret configurations using a subjective method and prepares the reader to understand the statistical method used to interpret configurations. In a later section, a statistical method for interpreting configurations is discussed.

Before learning how to interpret a configuration obtained from nonmetric MDS using the subjective method, an example may clarify what MDS produces when the data are perceived relatedness among items rather than actual distances. Recall the example about the public's perception of crimes. In this example, MDS then provides a visual representation of the set of crimes so that two crimes perceived to be similar are closer to each other and are far away from two crimes perceived to be different from them. For example, a residential burglary resulting in a loss of $100 might be seen as only slightly different from a residential burglary resulting in a loss of $50. These two burglaries, however, are seen as very different from any armed robbery, regardless of the amount stolen. MDS provides a visual display of these perceptions by spatially placing the two burglaries very close together and farther away from any of the armed robberies. MDS then creates a map in which the perceived

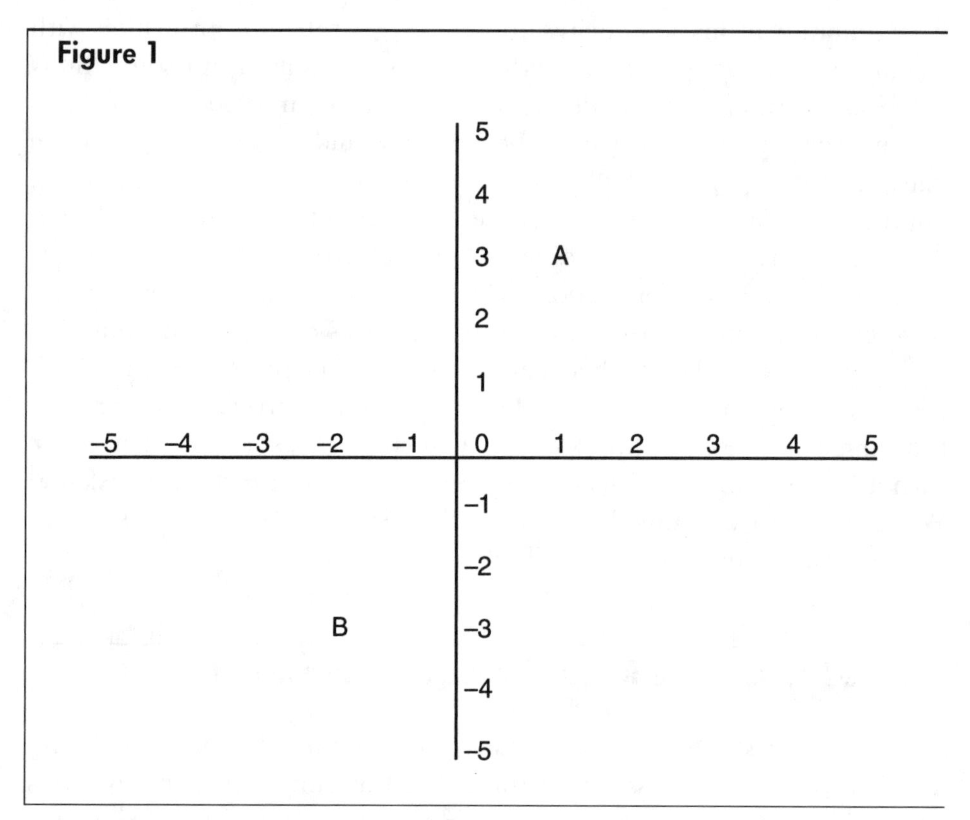

Configuration plot, which shows how coordinate numbers are used to locate specific instances of a category.

difference between pairs of crimes is represented as being "far away from" and perceived likeness is represented as being "close to."

Each of these crimes is represented as a point in the configuration. The *coordinate axes* of the configuration provide the directions of the points and have measured interval spaces called *coordinate numbers*. Coordinate numbers can be negative, zero, or positive and are used to locate each point (e.g., crime story) in the configuration.

Figure 1 contains a visual illustration of an MDS-derived configuration. The configuration contains only two points (i.e., crimes) to illustrate how the use of coordinate numbers can locate a specific point in an MDS-derived configuration. For the horizontal axis, a positive number indicates that the point is located on the right side of the axis, and a negative number indicates that a point is located on the left side of the axis. For the vertical axis, a positive number indicates the point is located in the top half of the plot and a negative number indicates that the point is located in the bottom half of the plot. A zero indicates that the point is located on an axis. For example, if the coordinates for a robbery story are (1, 3), that means the robbery story can be located by moving one space to the right on the horizontal axis and three spaces up on the vertical axis.

You may be wondering why are the coordinate numbers so important. There are two reasons to understand the concept of a coordinate number. First, the coordinate numbers can locate each item in the configuration as illustrated in Figure 1. Each item has its own coordinate numbers. Each dimension provides one coordinate to assist in locating an item in the configuration; thus, there are two coordinate numbers for a two-dimensional solution, there are three coordinate numbers for a three-dimensional solution, and so forth. Second and more important, the coordinate numbers are used in the statistical method for interpreting the configuration. The role of the coordinate numbers in the statistical method is covered in more detail later.

Having learned about the basic structure and purpose of an MDS-derived configuration, it is time to learn the subjective method for interpreting a configuration. The subjective approach involves examining the configuration to discover underlying patterns that are based on what one knows about the items. The fact that the coordinate axes in two-way MDS have no special significance should be remembered when using the subjective method. That is, the configuration should be examined from all directions because the coordinate axes are no more meaningful than lines drawn in any other direction.

The easiest way to learn the subjective method is by attempting it. Consider this problem, which is an example of how MDS can be used in research on consumer preferences. Consumers were asked to rate the similarity of 15 fruits on the basis of how much they enjoyed the taste of the fruit. Researchers asked the question, What characteristics of fruit determine satisfaction? You can discover the answer using the subjective approach. Examine the MDS two-dimensional configuration of hypothetical data presented in Figure 2. Take as much time as you need and look for something systematic in the pattern of fruits. It is important that you try to discover the underlying meaning of this pattern.

In case you are having a difficult time discovering an appropriate interpretation, here is a hint. Ask yourself what is common about coconuts and pineapples (which are on the far-left side of the plot) and how are these fruits different from peaches and strawberries (which are on the

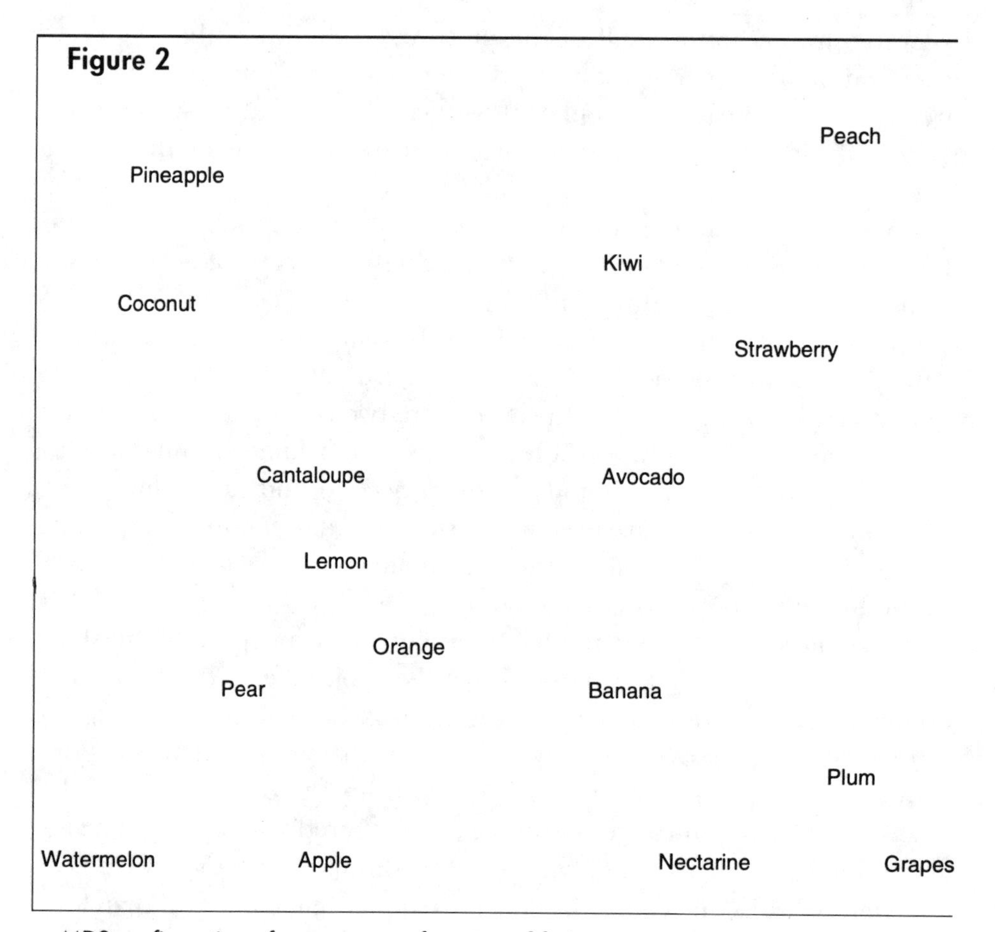

MDS configuration of consumer preferences of fruit.

far right-side of the plot). Then repeat this procedure using other fruits on the perimeter of the configuration.

There are at least two interpretations of the MDS configuration of the perceived relatedness among fruits. First, think about how the fruits in the top half of the configuration differ from the fruits in the bottom half. The fruits in the top half have fuzzy skins and the fruits in the bottom half have smooth skins. Thus, one interpretation of the vertical axis is *fuzzy* to *smooth*. Now let's consider how the fruits in the left side of the diagram differ from the fruits in the right side of the diagram. What do pineapples and coconuts have in common? They both have hard outer skins. What do grapes and plums have in common? They both have soft outer skins. Note that fruits on the left side have hard skins whereas the fruits on the right side have soft skins. Thus, one interpretation of the horizontal axis is *hard skins* to *soft skins*. You are now on your way to becoming an expert at interpreting MDS solutions using the subjective approach. The subjective approach is very useful in reading published work because the authors of the research may have overlooked an alternative or additional interpretation of their MDS configuration. Using the subjective approach, readers can formulate questions that can be examined in their own future research.

There, however, are several problems with the use of the subjective method as the only approach to interpreting a configuration. First, researchers often forget to look in all directions. Perhaps there are additional undiscovered dimensions in the hypothetical solution of relatedness among fruits; I did not notice these dimensions because I carefully examined only the top versus the bottom half and the right versus the left half. I committed the first error by not examining the diagonals of the configuration! A second problem is the difficulty of examining and interpreting three or higher dimensional configurations. A third problem is the difficulty of guarding against the human tendency to find patterns where they do not exist (Kruskal & Wish, 1978).

Statistical Tools Used to Select the Best Configuration

MDS produces several configurations that vary in the number of dimensions used to locate items in the spatial plot. In one-dimensional solutions, the data fall along a straight line. In two-dimensional solutions, there are two lines that define the location of items (a plane). In three dimensional solutions, the configuration can be imagined as a sphere (three dimensions: width, height, and depth). Although configurations with more than three dimensions can be derived and interpreted, the shape of these configurations is difficult to visualize. MDS does not automatically select the most appropriate dimensional solution; researchers must select which solution best fits the similarity data.

One method used to select among configurations with different numbers of dimensions is to examine how well each configuration fits the data. One statistical concept used to determine how well a configuration fits the data is called *stress*. The values of stress range from 0 to 1. Small stress values (e.g., 0 to .15) indicate a good fit, with 0 indicating the best possible fit (assuming other problems are not present; I describe these problems a little later). Larger stress values, thus, indicate a worse fit, with 1 indicating the worst possible fit. Stress often is called a *badness of-fit-measure* because the larger values indicate a worse fit.

There are several assumptions and considerations associated with the use of stress as an indicator of badness of fit. An important consideration in interpreting the value of stress is that metric scaling almost always results in higher values of stress than does nonmetric scaling. The interpretation of stress also assumes that the lowest possible stress has been reached, that is, that MDS has found the solution that best fits the data. The interpretation of stress as a badness-of-fit indicator also assumes that the influence of the ratio of items to dimensions on the value of stress is slight. When the number of items is four times greater than the number of dimensions, the number of items and dimensions is not an important consideration in interpreting the stress value. The crime example can illustrate this point. For the 27 crime stories (i.e., items), stress is interpretable for configurations of seven dimensions or less (7 dimensions × 4 = 28 crime stories or items) without substantial influence from the number of items or dimensions.

As the number of items becomes closer to the number of dimensions, the interpretation of stress as an indicator of badness of fit becomes less appropriate. For example, a stress value of zero (the best possible value) will occur well over 50% of the time for random data of seven items in four dimensions (Kruskal & Wish, 1978). Stress then is a very biased indicator of how well the configuration fits the data for solutions in which the number of items approaches the number of dimensions.

When the assumptions of MDS are met, stress assists in selecting the appropriate dimensionality. *Dimensionality* refers to the number of directions (i.e., coordinate axes) needed to locate a particular point (e.g., one crime story) in the configuration. The use of stress as a tool for choosing among configurations with different numbers of dimensions is based on the notion that for any given proximity data, there exists some ideal dimensionality that represents the *true* pattern of relatedness among items in the world. To determine whether the stress value for a higher dimensional solution is significantly lower, the stress values are plotted across the number of dimensions. Figure 3 illustrates a hypothetical plot of stress values for different dimensional solutions. This plot is similar to the scree plot used in factor analysis.

An intuitive approach in selecting the appropriate dimensionality is to look for an *elbow* (i.e., a bend in the line of points) in the plot of points. As shown in Figure 3, a slight elbow occurs at the third dimension, which has a stress value of .05. The stress for a two-dimensional solution is .12, and stress for the three-dimensional solution is .05. Given the presence of the bend and .07 improvement in stress, three dimensions seems ap-

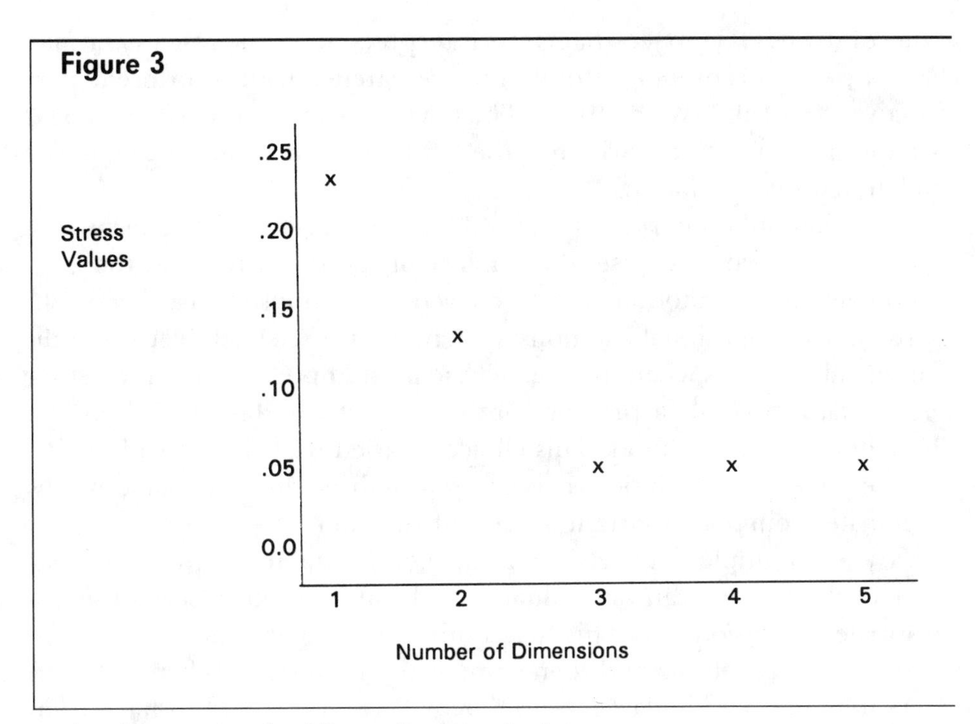

Plot of stress values for different dimensional solutions.

propriate for these hypothetical data. Although this example portrays the detection of elbows as an easy task, it is often relatively difficult with real data because measurement error can reduce the sharpness of the elbow.

Kruskal and Wish (1978) describe several rules of thumb in using stress as an indicator of the *true* number of dimensions: (a) An elbow seldom should be used if stress is above .10; (b) because a very large stress for a one-dimensional solution has limited accuracy, an apparent elbow at two dimensions is less informative and should be interpreted with caution; and (c) a stress value of .15 or less for a one-dimensional solution suggests that the true configuration is one dimensional.

Whether all dimensions are interpretable is another important consideration in the selection of the best MDS-derived configuration. Configurations with fewer dimensions sometimes are less interpretable than are configurations with a greater number of dimensions (Kruskal & Wish, 1978). For example, if the choice is between a two-dimensional configuration in which one of the dimensions has no clear interpretation and a three-dimensional configuration in which all dimensions have clear interpretations, a three-dimensional configuration should be chosen even if three-dimensional and two-dimensional configurations have the same value of stress. Of course, what is less interpretable to one researcher may have a clear interpretation to another researcher with a different perspective (Kruskal & Wish, 1978). Thus, when reading journal articles, it is important to evaluate whether the configuration provides additional undetected interpretations.

In addition to interpretability and fit, the selection of dimensionality depends on the ease of use, the stability of the structure, and the generalizability across alternative data collection methods (Kruskal & Wish, 1978). Two-dimensional solutions sometimes are selected over three dimensional solutions when they reflect the most important and interesting theoretical aspects of a problem but do not fit the data as well as the three-dimensional solution. This choice is based on the ease with which the two-dimensional solutions compared with three-dimensional solutions illuminate the most important aspects of the data.

When multiple data sets are available, the stability and generalizability of dimensions can be evaluated and can be used to select the appropriate dimensionality. This is accomplished by performing MDS separately for each sample and then comparing the results. When different MDS solutions are similar, the original MDS study is said to have been replicated (or confirmed) by the MDS results of the second study. If two dimensions are replicated across different samples and different data collection methods and a third dimension appears during only one of the data collection methods, researchers should choose the two-dimensional solution because it is robust (i.e., replicates and generalizes) across samples and methods.

A Statistical Method Used to Interpret a Selected MDS Configuration

In this section, I briefly describe the statistical method commonly used to interpret an MDS-derived configuration that a researcher has selected as the solution that best fits the data. In the last section, I illustrate how to interpret the statistical results using hypothetical data.

Because of the limitations of the subjective method of interpretation covered earlier, researchers also rely on a statistical method to guide their interpretation. The most common statistical method used is based on linear regression (see chapter 2). With this method, researchers collect data on additional variables that they believe will have a systematic relationship to the position of stimuli in a configuration. For example, respondents in the study on public views of crime may be asked to rate each crime on several dimensions. These ratings can be used to interpret the visual map of the perceived relatedness among crimes obtained from the MDS solution. Examples of dimensions that should be rated include (a) the extent to which the crime was planned, (b) the amount of harm done to property, (c) the extent to which the offender is a drug addict, (d) the likelihood of physical harm, (e) the likelihood the offender can become a productive member of society, (f) how worried the respondent is about becoming a victim of a similar crime, and (g) the severity of physical injuries. For each of these dimensions, you can ask respondents to indicate their opinions by means of bipolar scales ranging from 1 to 7. For example, the scale of the likelihood of physical harm can be anchored with *not at all* likely (1) and *extremely likely* (7). For each scale, the responses are averaged across respondents.

These averages then are used as dependent measures in regression equations to assist in the interpretation of the visual map of crimes obtained from MDS. The obtained coordinates of the selected MDS configuration are used as the predictor variables. Recall that the coordinate numbers locate each crime story in the configuration and thus capture the pattern of perceived relatedness among the crimes. The objective of the regression analysis is to obtain some weighted combination of the coordinates that explains each dependant variable as well as possible. An example can highlight the objective of the regression analysis. Assume the coordinates from an MDS configuration of public views of the differences among crimes predict very well the perceived threat of physical harm, which is the dependent variable in the regression equation. This finding from a regression equation indicates that the perceived threat of physical harm is an important feature that underlies the public's perception of the relatedness among crimes.

Researchers use statistical measures of goodness of fit from the regression analysis to determine the dependent variables that are best explained by the weighted combination of coordinate numbers. Two such goodness-of-fit measures are most often found in published articles: multiple R correlation and R^2 (also called the *coefficient of determination*). The multiple R correlation reveals the strength of the relationship between the weighted combination of coordinates and the dependent variable; it ranges from 0 to 1, with higher numbers indicating better fit. R^2 derives from the multiple R correlation; it ranges from 0 to 1, with higher numbers indicating better fit. R^2 provides the amount of variance in the dependent variable accounted for by the weighted combination of coordinates. With either the multiple R correlation or R^2 as an indicator of goodness of fit, the variables that best explain the coordinates can be determined.

In addition to the multiple R correlation and R^2 approaches, another statistical concept examined in the linear regression is the *regression weight*. The regression weight is a regression coefficient normalized so that the sum of squares equals one (see chapter 8 for the lay meaning of sum of squares) and indicates the angle and direction of a dimension in relation to the derived-MDS coordinate axis. Because it indicates the angle and direction of a dimension, it is often called a *direction cosine* in MDS articles. The regression weight then provides the necessary information to determine where the dimension is located in MDS configuration plot; researchers, using the information gleaned from the regression weight, generally highlight the appropriate interpretation by drawing the dimension on the configuration plot. (This is illustrated in Figure 7.)

This overview of how linear regression is used to interpret an MDS configuration highlights the two most important statistical concepts, multiple R correlations and regression weights, that assist the researcher in interpreting the dimensions. Similar to stress, the multiple R correlation is a measure of how well the MDS coordinates of a dimension fit a possible interpretation. The regression weight indicates whether the discovered interpretation lies close enough to the MDS coordinate axis to be a reasonable interpretation of the axis. Kruskal and Wish (1978) provide some rules of thumb in using these two statistics from linear regression to interpret an MDS configuration of the pattern of relatedness among items. Kruskal and Wish (1978) recommended the following:

> In order for a rating scale (or any other variable) to provide a satisfactory interpretation of a dimension, two conditions are necessary: (1) the multiple correlation for the scale must be high (indicating that the scale can be well fitted by the coordinates of the configuration), and (2) the scale must have a high regression weight on that dimension (indicating that the angle between the dimension and the direction of the associated scale is small). Although it is desirable to have multiple correlations in the .90s for a good interpretation of a dimension, correlations in the .80s and upper .70s have to suffice in many instances. (pp. 37, 39)

Detecting Violations of MDS Assumptions

A central concept of MDS is that the distance between points should correspond to the proximity data. Recall that MDS transforms the prox-

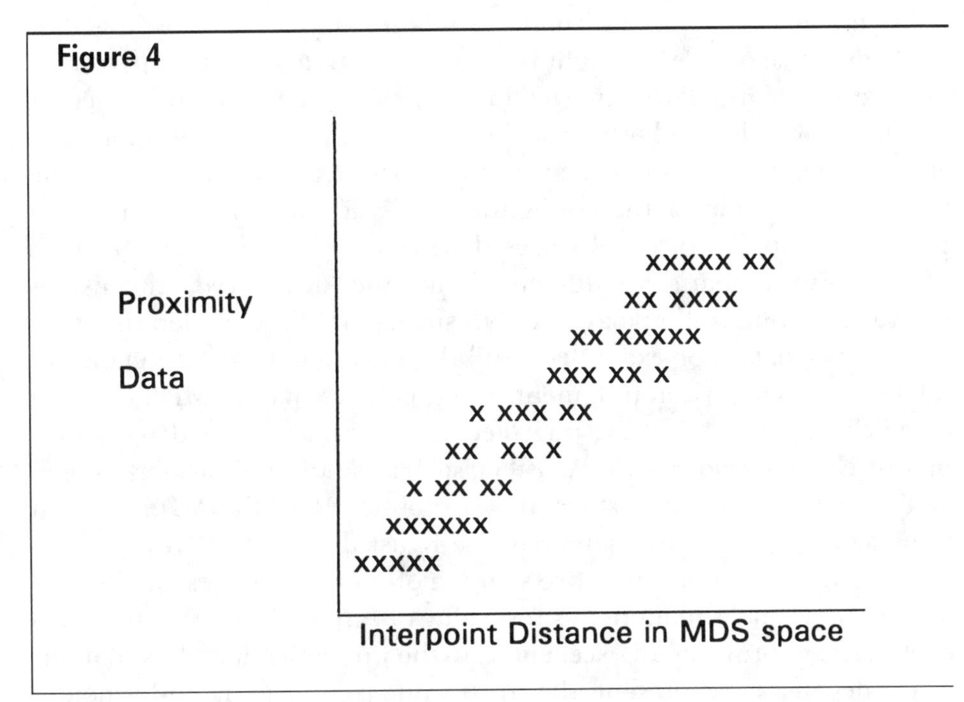

Scatter diagram of a normal relationship between distance and dissimilarity data.

imity data (i.e., perceived relatedness among items) into distance data. A *scatter diagram* provides a visual picture of the relationship between proximity data and MDS-computed distance data. A scatter diagram shows distances on the horizontal axis and proximities on the vertical axis (in other presentations, distances may be on the vertical axis and proximities may be on the horizontal axis). Scatter diagrams are very useful tools for detecting problems with an MDS configuration.

MDS assumes the relationship between distance and proximity data is smooth. *Smooth* refers to the continuous relationship between distance and proximity data. Figure 4 contains a scatter diagram showing a normal, smooth relationship between distance and proximity data, measured as dissimilarities. Because the data are measured as dissimilarities, larger distances correspond to larger differences, and smaller distances correspond to smaller dissimilarity (i.e., items are very similar). Notice how the data points are close together and large gaps are absent; these features define a smooth relationship between distance and proximity data.

One common problem that violates the assumption of a smooth relationship between distance and proximity data is called *degeneracy*. Degeneracy means that the points of the configuration are located in a few tight clusters. For example, all of the burglaries may be very close together in a tight circle on the top right side of the configuration (are perceived to be very similar), all the unarmed robberies may be very close together in a tight circle in the bottom of the left side of the configuration, and all the armed robberies may be very close together in a tight circle in the top of the left side of the configuration. A degenerative solution is a problem because it often obscures the presence of important features used to form comparison judgments. Thus, the "discovered" dimensions in an MDS configuration may be only a small part of the hidden structure in similarity data. For example, recall the illustration of a degenerative solution for comparison judgments of crimes. A separate MDS for each crime cluster (i.e., burglary, unarmed robberies, and armed robberies) may reveal dimensions within clusters, such as amount of property stolen and offender's criminal history, that are obscured in the MDS solution using all the crimes in one analysis (see Kruskal & Wish, 1978).

In summary, a degenerative solution often yields a stress value close to zero. In normal situations, a stress value of zero indicates that the MDS configuration provides an excellent fit to the proximity data. In situations where degeneracy is present, the stress value is misleading and is not an indicator of good fit because important dimensions underlying similarity judgments may be obscured. When degeneracy occurs, researchers should note the clustering and avoid any other substantive conclusions. Degenerative solutions are most likely to occur when there is a natural clustering to the items and nonmetric scaling is being used.

How does one know if a degenerative solution is present? The derived configuration of a degenerative solution is shown in Figure 5, and the scatter diagram for a degenerative solution is shown in Figure 6. The configuration plot shows three tight clusters of points, and the scatter diagram shows three large steps. The typical scatter diagram of a degenerative solution is a picture resembling a staircase consisting of a few large steps. This staircase picture occurs because all of the distances take on only a small number of unique values. In contrast, a scatter diagram of a normal solution is the absence of large open spaces between points (i.e., a smooth relationship).

Another possible violation of assumptions occurs when MDS stops prematurely at a less-than-optimal visual representation of the similarity data. How does MDS perform the task of providing an optimal representation of the perceived relatedness among crimes? A technical and detailed description of how MDS creates a configuration is beyond the scope of this chapter. A brief summary of the most common method,

Figure 5

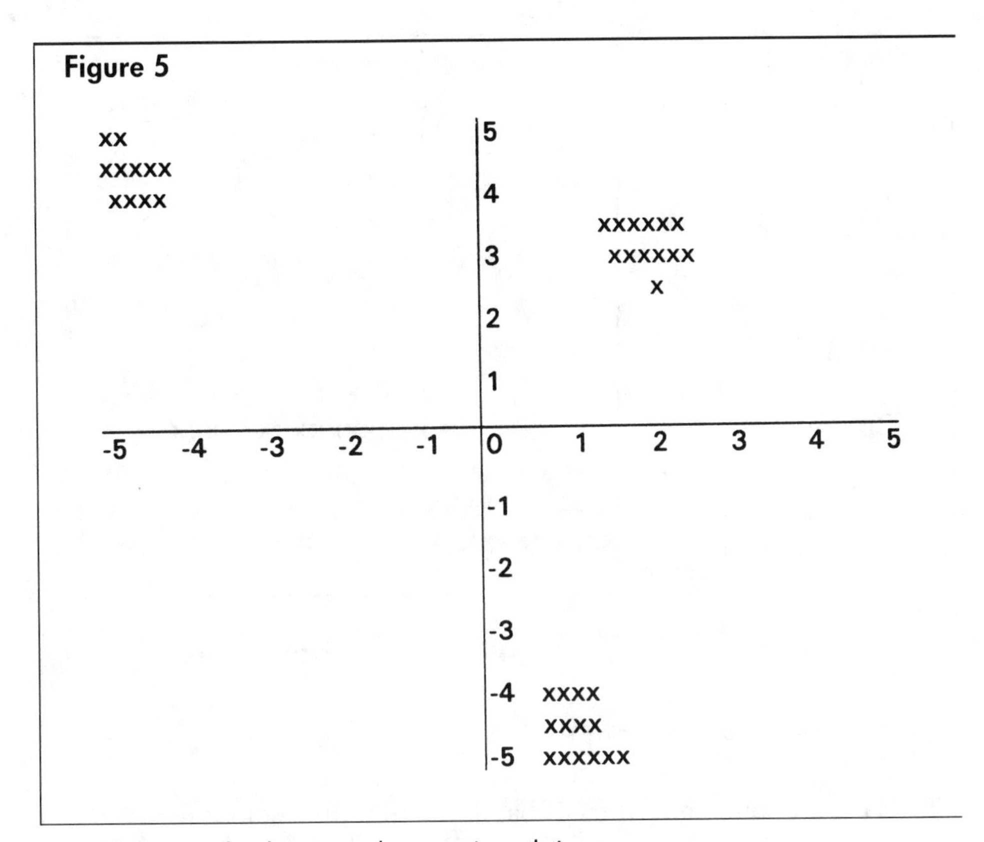

Configuration plot showing a degenerative solution.

called *steepest descent*, can highlight the analytical procedure. Steepest descent is an iterative method. The computer starts with a guess of the pattern of items, moves to a better estimate based on additional computation, and continues to correct the estimate until it reaches the best possible estimate.

One assumption made in the interpretation of MDS configurations is that there is an ideal configuration of the items and that MDS has discovered this ideal pattern; the best possible solution for any proximity data is called a *global minimum solution*. When MDS finishes with a less-than-optimal solution, this suboptimal solution is called a *local minimum solution*. Although local minimum solutions (less-than-optimal solutions) often occur because of the iterative nature of the procedure, they generally are not drastically different from the global minimum solution. Researchers are concerned with detecting *bad* local minimum solutions; a bad local minimum solution is one that is drastically different from the ideal solution that exists in the real world.

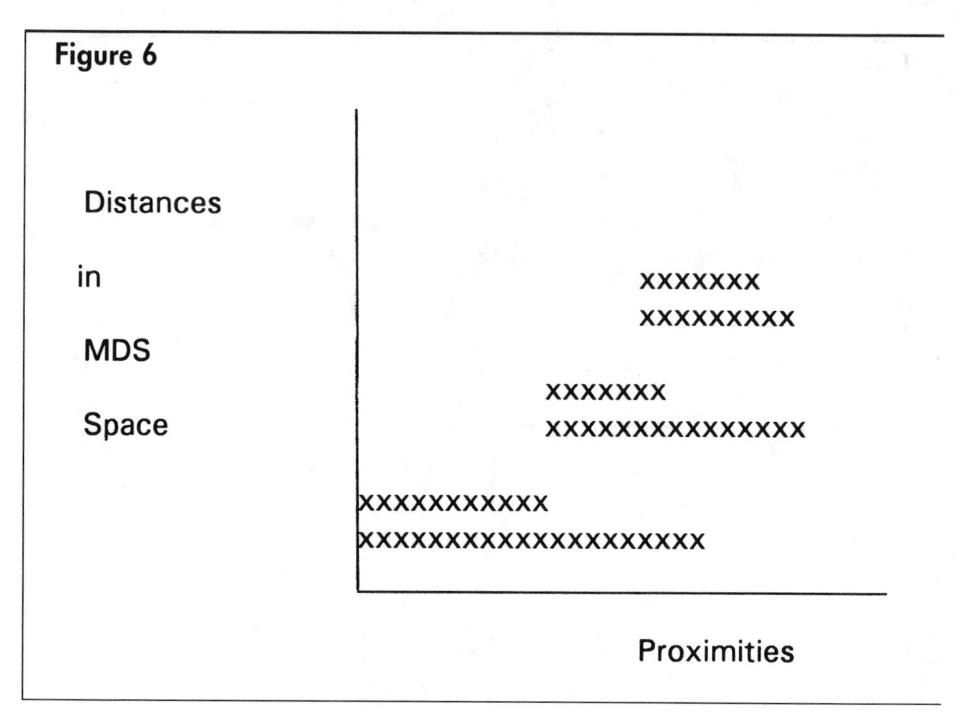

Scatter diagram of a degenerative solution.

What are some indicators of bad local minimum solutions? One way to check on the possibility of local minimum solutions is to compare the stress values across configurations with different numbers of dimensions. For illustrative purposes, consider these hypothetical stress values. The stress value for the one-dimensional solution is .40; for the two-dimensional solution, .23; for the three-dimensional solution, .29. This pattern of stress values is suspicious. First, the stress values are all high, and second, the stress value for the three-dimensional solution is much larger than the stress value for the two-dimensional solution. When the lowest possible stress values have been reached, stress should either decrease or reach the same prior level as the number of dimensions increases. Stress values for higher dimensions often will be approximately the same as the next lowest dimensional solution because stress may not change much from one dimensionality to the other.

A potential problem with local minimum solutions then is indicated when the stress value of a higher dimensional solution (e.g., three dimensional) is much larger than the stress value of a lower dimensional solution (e.g., two dimensional). To examine further the possibility of a local minimum solution, several MDS analyses using different starting configurations (i.e., different first guesses about what the configuration is like) should be performed. If a local minimum solution is present in the original analysis, it may be possible to find a configuration without suspiciously high stress values, using a different starting configuration. To start with a different configuration, a researcher makes a guess about the relationships among the items on the basis of either some theoretical hunches, prior empirical results, or a guess without a supporting rational judgment.

An Illustrated Example of How MDS Results Are Interpreted

This section uses hypothetical results from nonmetric, two-way MDS to demonstrate an appropriate interpretation of an MDS configuration. The hypothetical data is generated for the example on public's perceptions of burglaries and robberies covered throughout this chapter. Imagine once again that you are a consultant for the crime commission to assist in developing criminal sentencing legislation. Similarity data have been collected and are analyzed by means of two-way, nonmetric MDS. The question addressed through MDS is, Which characteristics of crimes (the offenders' prior criminal convictions, offenders' age, amount of goods or money stolen, the amount of physical harm, or threat of physical harm) are important in the public's view in determining punishments for specific criminal acts?

Respondents ($N = 5$) rated the perceived similarity of each pair of crime stories of the 27 stories used. (Refer back to the data collection section on p. 143 for a more detailed description of the collected data.) Table 1 contains a brief description of each of the 27 hypothetical crime stories. The stories varied on five dimensions: (a) type of crime [burglary, unarmed robbery, or armed robbery]; (b) number of prior convictions [none, three, or six]; (c) amount of property stolen; (d) amount of physical harm [none, minor (not requiring medical assistance), or severe (requiring hospitalization)]; and (e) age of the offender [teenager, young adult, or over 35]. It could be useful to pause here to become familiar with the labels of the different crime stories and to note how the characteristics vary across crime stories. For example, in Table 1, note that the labels for all burglaries begin with *BU*, the labels for all unarmed robberies begin with *UR*, and the labels for all armed robberies begin with *AR*. Note that a zero is in the labels of all stories containing no prior convictions, that a three is in the labels of stories containing three prior con-

Table 1

Description of Hypothetical Crime Stories

Story label	Type of robbery	Prior convic-tions	Amount stolen	Physical harm	Age
BU0S	Burglary	None	$50	None	Teenager
BU0L	Burglary	None	$5,000	None	Young adult
BU0M	Burglary	None	$500	None	Over 35
BU3M	Burglary	3	$100	None	Teenager
BU3L	Burglary	3	$10,000	None	Over 35
BU3S	Burglary	3	$10	None	Young adult
BU6L	Burglary	6	$3,000	None	Teenager
BU6S	Burglary	6	$50	None	Over 35
BU6M	Burglary	6	$750	None	Young adult
UR0L	Unarmed	None	$3,000	None	Over 35
UR0S	Unarmed	None	$50	Severe	Young adult
UR0M	Unarmed	None	$750	Minor	Teenager
UR3L	Unarmed	3	$5,000	Severe	Teenager
UR3M	Unarmed	3	$500	None	Young adult
UR3S	Unarmed	3	$10	Minor	Over 35
UR6L	Unarmed	6	$10,000	Minor	Young adult
UR6S	Unarmed	6	$30	None	Teenager
UR6M	Unarmed	6	$400	Severe	Over 35
AR0L	Armed	None	$5,000	Severe	Teenager
AR0S	Armed	None	$10	None	Over 35
AR0M	Armed	None	$500	Minor	Young adult
AR3L	Armed	3	$3,000	Minor	Teenager
AR3S	Armed	3	$40	Severe	Over 35
AR3M	Armed	3	$400	None	Young adult
AR6M	Armed	6	$450	Severe	Young adult
AR6L	Armed	6	$4,000	None	Teenager
AR6S	Armed	6	$20	Minor	Over 35

victions, and that a six is in the labels of stories containing six prior convictions. Note that the labels of all stories resulting in a small amount stolen have as their last letter an *S*, that the labels of all stories resulting in a moderate amount stolen have as their last letter an *M*, and that the labels of all stories resulting in a large amount stolen have as their last letter an *L*. This familiarity with the example can facilitate the ease of interpreting the hypothetical configuration. For instance UR0M indicates a vignette in which an unarmed robbery was committed by a person with no prior convictions and in which a moderate amount of money was stolen.

For this hypothetical data set, the stress value for the one-dimensional solution is .200, for the two-dimensional solution .070, for the three-dimensional solution .055, and for the four-dimensional solution .045. The decline in stress from the one-dimensional to two-dimensional solution is substantial (.200 − .055 = .13), whereas the decline in stress from two to three dimensions is comparably small (.07 − .055 = .015). The small difference in stress between the two-dimensional solution and the three-dimensional solution suggests that an elbow occurs at the two-dimensional solution. This very modest decline in stress from two to three dimensions suggests that the two-dimensional solution best fits the data. If the assumptions of MDS are not violated, the .07 stress level for the two-dimensional solution indicates that this MDS configuration has a good fit with the proximity data.

Neither local minimum nor degenerative solutions are present in these data. The decline in stress as the number of dimensions increases and the relatively low values of stress suggest that a bad local minimum solution is not present. As shown in Figure 7, the crimes are well spread out, suggesting the absence of a degenerative solution.

In interpreting a configuration, an important point to remember is that the amount of stress reduction does not indicate the relative importance of a dimension. The amount of variability a dimension has in the set of crime stories influences the amount of reduction in stress because the formula for computing stress reduction gives greater weight to larger distances. For example, the crime stories varied only slightly on the amount of extenuating circumstances. If there is not much variation on a dimension, it has little chance to make a significant reduction in stress and to appear in the MDS configuration. Thus, one threat to internal validity in MDS research is that a relevant dimension may have very little variation in the set of rated items (Schiffman et al., 1981).

After selecting the two-dimensional configuration as the best fitting configuration, the next step is to identify the dimensions that explain the pattern of crimes displayed in the configuration represented in Figure 7. A dimensional interpretation of MDS configurations assumes that people see relatedness as a matter of degree along a dimension. For example, they may distinguish among crimes by the amount of harm done, so that crimes that are closer in the amount of harm done (e.g., $50 stolen compared with $75 stolen) are perceived to be more similar than crimes that are further apart in the amount of harm done (e.g., $50 stolen compared with $5,000 stolen). In illustrating how to interpret MDS configurations

Figure 7

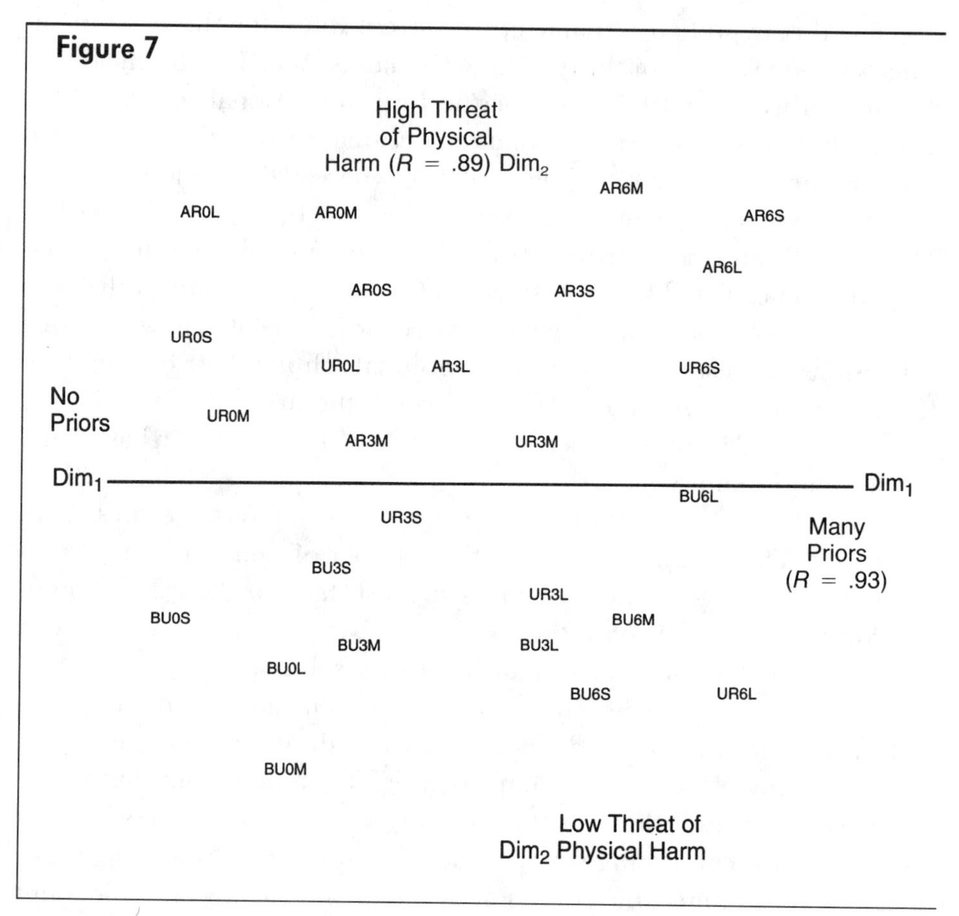

Configuration plot of the two-dimensional solution. See Table 1 for explanation of labels. Dim = dimension.

in published articles, a dimensional approach to interpretation is described.

Figure 7 contains the configuration plot of crime stories using the alphabet labels in Table 1. Linear regression was used to discover the meaning of this configuration. The dependent variables were the bipolar ratings described under the data collection section on p. 153, and the manipulations in the crime stories (prior convictions, amount stolen, physical harm, age, and weapon present). For example, the linear regression equation for ratings on the bipolar scale of perceived threat of physical harm is

$$\text{perceived threat} = a + b_1\text{Dim}_1 + b_2\text{Dim}_2$$

The dependent variable is the average rating of perceived threat for each

Table 2

Multiple Regression of Bipolar Scale Ratings on Dimensions of Relatedness Among Crime Stories

	Regression Weights		
Higher pole of a dimension	Dim_1	Dim_2	*R*
A. Much property stolen	.473	.881	.307
B. Severe physical harm	.450	.893	.362
C. Habitual drug user	.760	.650	.510
D. Offender over 35 years old	.690	.724	.210
E. Remain unproductive citizen	.993	.118	.921
F. Many prior criminal convictions	.992	.130	.930
G. High threat of physical harm	−.253	−.967	.890
H. No extenuating circumstances	−.434	−.901	.639
I. Well planned	.807	.591	.760
J. Type of crime	−.302	−.952	.950

Note. Dim = dimension.

crime story, and the independent variables are the coordinates of the two MDS-derived axes. The symbol *a* is the constant in the regression equation, *b* is the direction cosine (standardized regression weight), Dim_1 refers to the coordinate numbers associated with Dimension 1 in an MDS solution, and Dim_2 refers to the coordinate numbers associated with Dimension 2 in an MDS solution.

Table 2 contains the regression weights and multiple *R* correlations for each measured variable. The dependent variables are identified in the far-left column. The regression weights of the MDS-computed coordinates for each dependent variables are presented in the rows. The regression weights for the horizontal MDS-derived coordinate axis are under the column Dim_1, and the regression weights for the vertical MDS-derived coordinate axis are under the column Dim_2.

To interpret the two-dimensional configuration, the first step is to examine the multiple *R*s. High multiple *R*s (.80 to 1.00) indicate which dependent variables provide a good fit with the pattern of relatedness among crimes in the configuration. Recall that the coordinate numbers indicate the location of each crime and thus define the pattern of relatedness among the crimes; thus, a high multiple *R* indicates that a dependent variable fits the pattern of crimes in the configuration.

Two dimensions provide adequate interpretations of the horizontal MDS axis (Dimension 1 in Figure 7). The multiple *R* correlation for the number of prior criminal convictions is .93 (Table 2, row F, third column of numbers). The multiple *R* correlation for perceptions about whether the offender can become a productive citizen is .92 (Table 2, row E, third column of numbers). Both of these high multiple *R*s indicate that these dimensions are important features affecting the pattern of relatedness among crimes. However, another criterion must be met before these two features (number of prior criminal convictions and likelihood of becoming a productive citizen) can be identified as features that can explain a significant part of the public perception of the differences among crimes. The other criterion is a large regression weight.

The regression weights for these two dimensions are very large and close to each other [(.992 and .993; first column of numbers, rows F and E, respectively)]. A regression weight of .992 for number of prior criminal convictions indicates that this perceived dimension is very close to the first coordinate axis generated by MDS; A .992 regression weight means that the dimension of prior criminal convictions is at a 7.1° angle (a regression weight of .992 corresponds to a cosine of 7.1) to the first (horizontal) MDS-generated dimension. As discussed earlier, a small angle means a dimension provides a reasonable interpretation of the first dimension in the two-dimensional MDS configuration. Because both the number of prior convictions and the likelihood of becoming a productive citizen have high multiple *R* correlations and small angles to the first MDS dimension, there are two reasonable interpretations of the first dimension. Recall that regression weights above .90 represent small angles, but regression weights from .70 to .89 often must suffice.

The second dimension also has two interpretations. Both the perceived threat of physical harm and crime type have high multiple *R* correlations (.89 and .95) and have very similar regression weights (−.967 and −.952). The regression weights indicate the direction of the axis, and the cosines of these weights indicate the angle with the MDS coordinate axis. A −.967 regression weight has a 14.7° angle with the second MDS coordinate axis. A −.952 regression weight has a 17.8° angle with the second MDS coordinate axis. Both angles are sufficiently small to provide a reasonable interpretation of the dimension.

The fact that there are two plausible interpretations for each dimension is not problematic; indeed, the additional interpretations are very informative. For example, the perceived threat of harm is a subjective broader perception, whereas the crime type is an objective feature of the crime. This example highlights that a two-dimensional solution can have more than two underlying interpretations. Additional interpretations can provide supplementary information and should not necessarily be seen as competing explanations. Thus, the four interpretations of the two-dimensional configuration of the relatedness among crime provide a more complete understanding about how the public differentiates among crime in selecting the appropriate punishment.

This hypothetical example addressed which features of crimes are most important in people's judgments about punishment for specific criminal acts. A two-dimensional MDS configuration provided the best fit for the proximity data. On the basis of the results from regression analyses, the first dimension had two interpretations: perceptions about whether an offender could become a productive citizen and the number of prior criminal convictions. The second dimension also had two interpretations: threat of future physical harm and type of crime (burglary, unarmed robbery, or armed robbery).

What are the implications and appropriate uses of these findings? What are the implications of using each of these interpretations for creating policies about how judges can punish certain criminal offenders? Recall that the public judged the similarity of two crimes on whether they should receive similar or different punishments. Are each of the discovered dimensions equally susceptible to discriminatory or unequitable punishments? The answer is no! Broader subjective perceptions such as the likelihood of becoming a productive citizen are more susceptible to discriminatory and disparate treatment on the basis of race, gender, and sexual orientation than are objective characteristics, such as the number of prior criminal convictions. For purposes of policy such as establishing sentencing guidelines, the objective characteristics of crimes are more appropriate to distinguish among crimes than are the subjective perceptions. In contrast, the broader subjective dimensions can generate a broader theoretical model of how people make comparisons among crimes in their judgments about the appropriate punishment. The broader theoretical model can identify other relevant features of the crime story that may be associated with perceptions about the likelihood that an offender will become a productive citizen. For example, a robber with three prior convictions who has successfully completed drug therapy and has been a good employee for the last 6 months may be seen as significantly different from a robber with three prior convictions who has refused drug therapy and who has never been employed. The broader subjective dimensions, thus, apply more to academic-oriented research, whereas the observable characteristics, such as the number of prior convictions, apply more for developing sentencing guidelines and legislation.

Conclusions

In this chapter, I highlighted the usefulness of MDS procedures across a broad range of academic and applied topics. I then discussed two major distinctions among MDS procedures (two-way versus three-way MDS and metric versus nonmetric MDS); described conceptually how two-way, nonmetric MDS discovers the structure of relatedness among items; explained in nontechnical language the statistical concepts and problems associated with this procedure; and illustrated how to interpret MDS results in published articles.

Suggestions for Further Reading

The chapter provides sufficient information to understand published results concerning two-way MDS but provides insufficient information to use MDS as a statistical tool in one's own research. Readers who would like an in-depth knowledge of MDS procedures should read additional sources. Guttman (1968) described an alternative way to interpret MDS configurations. Kruskal and Wish (1978) provided a detailed monograph that described both nonmetric, two-way MDS and individual scaling (three-way) MDS. Schiffman et al. (1981) described the statistical concepts associated with several three-way MDS procedures and also provided information about the computer programs available to conduct each procedure. Rabinowitz (1986) provided a brief introduction to nonmetric, two-way MDS and individual scaling MDS. Shepard (1972) provided an introduction to data collection methods and MDS procedures, and Torgerson (1958) provided a conceptual discussion of scaling in general.

Glossary

CONFIGURATION The spatial representation of the location of items to each other as produced by MDS using proximity data.

COORDINATE AXIS A dimension produced by MDS. It provides systematic directions in the pattern of points.

COORDINATE NUMBERS The numbers produced by MDS that can locate each item in the configuration. These numbers are used as criterion variables in the statistical method of interpreting the pattern in the MDS configuration.

DEGENERACY A problem that can occur in MDS solutions, in which the items in a configuration are located in a few tight clusters.

DIMENSIONALITY The number of coordinate axes needed to locate a particular point in the MDS configuration.

DIMENSIONS The directions of the pattern of points in the configuration. Researchers must discover and interpret the systematic directions in the MDS-derived configuration.

DISTANCE The space between two items as measured using a straight line. MDS computes distance measures.

GLOBAL MINIMUM SOLUTION The best possible MDS configuration for any given proximity data.

LOCAL MINIMUM SOLUTION A less-than-optimal MDS configuration.

MULTIPLE *R* CORRELATION A measure of how weakly or strongly the criterion variables (e.g., MDS coordinates) are related to a dependent variable.

PROFILE PROXIMITY MEASURE Datum that has been transformed into proximity. One common transformation of data that are not collected as similarity data is a correlation.

PROXIMITY A number that indicates the similarity or difference between two items. MDS analyzes proximity data.

R^2 Another measure of how weakly or strongly the criterion variables are related to a dependent variable. R^2 provides the percentage of variance of the dependent variable accounted for by the criterion variables.

REGRESSION WEIGHTS Values that indicate the direction and angle of a discovered dimension (i.e., dependent variable) in the configuration.

SCATTER DIAGRAM A plot of the relationship between MDS-computed distances and the original proximity data.

STEEPEST DESCENT An analytical procedure used in MDS programs for creating the solution. It is an iterative computational method.

STRESS A measure of how well or poorly the configuration fits the data. Stress ranges from 0 to 1, with 0 indicating the best possible fit and 1 indicating the worse possible fit.

References

Guttman, L. (1968). A general nonmetric technique for finding the smallest coordinate space for a configuration of points. *Psychometrika, 33*, 469–506.

Kruskal, J. B., & Wish, M. (1978). *Multidimensional scaling*. Beverly Hills, CA: Sage.

Rabinowitz, G. (1975). An introduction to nonmetric multidimensional scaling. *American Journal of Political Science, 19*, 343–390.

Rabinowitz, G. (1986). Nonmetric Multidimensional Scaling. In W. D. Berry & M. S. Lewis-Beck (Eds.), *New tools for social scientists: Advances and applications in research methods* (pp. 77–107). Newbury Park, CA: Sage.

Schiffman, S. S., Reynolds, M. L., & Young, F. W. (1981). *Introduction to multidimensional scaling: Theory, methods, and applications*. San Diego, CA: Academic Press.

Shepard, R. N. (1972). A taxonomy of some prinicipal of data and of multidimensional methods for their analysis. In R. N. Shepard, A. K. Romney, & S. Nerlove (Eds.), *Multidimensional scaling: Theory and applications in the behavioral sciences* (Vol. 1, pp. 21–47). New York: Seminar Press.

Torgerson, W. S. (1958). *Theory and methods of scaling*. New York: Wiley.

Wish, M., & Carroll, J. D. (1974). Applications of individual differences scaling to studies of human perception and judgment. In E. C. Carterettee & M. P. Friedman (Eds.), *Handbook of perception* (Vol. 2, pp. 449–491). San Diego, CA: Academic Press.

Analysis of Cross-Classified Data

Willard Rodgers

Other chapters in this volume describe procedures that are appropriate for analyzing quantitative data: how much, how many, how often, or how long, for example. But often of interest are phenomena that cannot be described in quantitative terms: what kind, where, or under what circumstances, for example. In this case, an approach to data analysis is needed that can use such qualitative information effectively and meaningfully.

Compare the following examples of the two kinds of information. Suppose, first, that we want to examine the income distribution in the United States. To do so, we might examine data about the total income in a calendar year for a representative sample of all households and then use those data to estimate the average income of all households, or the proportion of households with incomes below a poverty threshold or above an affluence threshold. Now suppose that instead of looking at *amount* of income, we want to examine the primary *sources* of income for households in the United States. We could examine data from the same sample of households to determine the primary source of income for each, perhaps distinguishing the following categories: labor earnings, earnings from investments, private pensions, and government transfers.

Using the quantitative data about income in dollars, we may want to compare the average income of different kinds of households, such as those headed by women relative to those headed by men or the relationship between income and the age of the household head; or to compare the income distribution in 1992 with that in 1980; or to compare income dispersion in the United States with that in other countries. A variety of

Table 1

Cross-Classification of High School Seniors' Use of Illicit Drugs by U.S. Region

	n				
U.S. region	No illicit drugs	Mari-juana only	Few pills	More pills	Any heroin
Northeast	1,750	498	313	304	23
North Central	2,544	641	461	547	53
South	2,851	665	468	503	49
West	1,511	496	294	367	33

Note. Responses were based on past and present usage.

procedures, some of which are described in this book, can be used to answer questions such as these, which involve quantitative variables. For example, multiple regression can be used to predict the annual income of a household, headed by a single woman, with two children. Analysis of variance can be used to determine if there is a significant difference between the incomes of households headed by a single father versus a single mother. In each of these examples, the dependent variable of interest is quantitative: the amount of income.

In this chapter, the focus is on the second type of data, that is, on data that are referred to as *qualitative* rather than *quantitative*, because they do not correspond to dimensions that can be quantified in terms of counts or amounts. Given the data on primary source of household income, for example, we might wish to examine the relationship between this household characteristic and other characteristics of those households, such as presence of a male adult or age and education of the household head. Some of the other household characteristics that may be of interest to us may also be qualitative, for example, race, gender, and marital status.

The most general procedure for analyzing qualitative data consists, at heart, of simply cross-classifying individuals according to each of the characteristics of interest. Table 1 is an example of a cross-classification (or *contingency table*) with respect to two variables. In this table, high school seniors are cross-classified according to information obtained from a 1992 survey. I have more to say about the source of the data later; for now, focus your attention on just this table, which conveys information about two characteristics of the seniors: where they were living and their lifetime use of illicit drugs. Region of residence is a qualitative variable, because there is no quantity that can be measured and associated with region of residence. The other variable, illicit drug use, is one that could be quantified—for example, in terms of how long a student had been using any illicit drugs, or the amount consumed, or the street value of those drugs—but because of the variety of illicit drugs, there is no single dimension that could be quantified to characterize each student. Instead, in the table, the students are simply categorized according to their reports of lifetime use of marijuana, pills (e.g., amphetamines, psychedelics, and barbiturates), and heroin. We learn from the table, for example, that 1,750 students lived in the Northeast and reported that they had used no illicit drugs, 498 lived in the Northeast and reported that they had used marijuana at least once in their lifetime but no other drugs, and so on for the other 18 combinations of the two characteristics.

The categories for each of the variables that define the cross-classification must be *exhaustive* and *mutually exclusive*: Every person must belong in one and only one category for each of the variables. The *mutually exclusive* requirement means that clear criteria must be established for distinguishing the categories. In our example, region of residence is defined by the place and date on which the data were collected, a rule that permits unambiguous assignment to one region even for students who moved from one region to another. Similarly, students are assigned to the last-listed of the categories on the illicit drug use variable for which they meet the criteria, so, for example, any student who reports ever having used heroin is put in the fifth category regardless of use of marijuana or pills.

Some notation will be useful in talking about cross-classification tables such as this one. One of the variables defines the rows of the table, and the number of rows (referred to by the letter I), is equal to the number of categories that we distinguish with respect to the row variable: In our example, the row variable is region of residence, and $I = 4$. The second variable defines the columns of the table, and the number of columns (referred to by the letter J) is equal to the number of categories that we distinguish with respect to the column variable: In our example, the column variable is illicit drug use, and $J = 5$. The cells of a table are defined by combinations of categories for the row and column variables, and the number of cells is simply $I \times J$. In the example, the number of cells is $4 \times 5 = 20$; one of those cells is defined by the combination of living in the South (the third row on the region of residence variable) and having used more pills (the fourth category of the illicit drug use variable), and 503 students are in that cell. Because of the requirement that the categories of each variable be mutually exclusive, those 503 students are uniquely assigned to that cell; and because of the requirement that the categories of each variable be exhaustive, every student falls into one of the 20 cells of the table.

We could cross-classify individuals by more than two variables at a time. For example, we could first look at male students only, and generate a table like the one above, in which those boys were cross-classified by region and lifetime drug use, followed by a second half of the table in which girls were cross-classified by the same characteristics. And then we could repeat the whole set of frequencies for data collected from high school seniors in 1980 to compare with the frequencies from 1992. As is true for cross-classifications with respect to two variables, cross-classifications with respect to more than two variables still require that the categories of each of the variables be exhaustive and mutually exclusive, so that each observation falls into one, and only one, cell.

To extend the notation beyond a two-variable cross-classification, begin by counting the number of categories for each variable. To be concrete, suppose that we were to cross-classify students with respect to four variables: the two that we have already looked at (region of residence and illicit drug use), gender, and year. Represent the number of categories on the first variable by the letter I (e.g., $I = 4$, the number of categories for region of residence); the number of categories on the second variable by the letter J (e.g., $J = 5$, the number of categories for illicit drug use); the number of categories on the third variable by the letter K (e.g., $K = 2$, the number of categories for gender); number of categories on the fourth variable by the letter L (e.g., $L = 17$, the number of categories for year, ranging from 1976 to 1992); and so on for as many variables as are used in the cross-classification. The number of cells can be found as the product of the number of categories on those variables, $I \times J \times K \times L \times \ldots$; for our example, the number of cells is $4 \times 5 \times 2 \times 17 = 680$.

Thus, contingency tables created by cross-classifying individuals with respect to two or more variables include a predetermined number of cells. An experimental or observational study yields counts, numbers of observations, for each of those cells; in general, those frequencies are not the same across cells: Some cells may have large numbers of cases, and others may have relatively few or even none. How do we make sense of such a collection of frequencies? What types of questions can we answer from cross-classifications, and how?

The most basic question that we often wish to address is whether there is any relationship between two variables. That is, does knowledge about one characteristic of an individual (say, region of residence) provide any information about another characteristic (say, lifetime use of illicit drugs)? If there is a relationship between two variables, what is the nature of that relationship, and how strong is it?

With cross-classifications by three or more variables, we can address more complex (and more interesting) questions. For example, if there is indeed a relationship between region of residence and use of illicit drugs, is it stronger for boys than for girls? Or does that relationship disappear when we control for the type of place in which students live (e.g., whether they live in large cities, towns, or rural areas)?

The analysis of cross-classifications (or *contingency table analysis* as it is sometimes called) is a useful technique in part because it is so flexible. We do not need to be able to make quantitative measurements to use this approach. We do not even have to assign a rank order to the categories (e.g., that students who have used a few illicit pills are in some sense intermediate between those who have used many such pills and those who have used only marijuana). All that is required is that we be able to classify individuals into a set of mutually exclusive categories with respect to each of the variables of interest. On the other hand, if we know more, for example, if it is sensible to assign a rank order to a set of categories, we can use that information in the analysis. For example, we might classify high school seniors into five or six categories according to their grade point averages and, for some purposes, treat those simply as unordered categories, but for other purposes, we could use the quantitative dimension that underlies those several categories.

Contingency table analysis does not require that we treat one of the variables as the dependent variable that we are trying to explain in terms of one or more other variables. Because the procedure starts with a simple cross-classification of the individuals with respect to the entire set of variables of interest, the variables are all treated equivalently. On the other hand, if we are trying to address a question about why individuals are in particular categories of one of the variables (e.g., what factors are associated with different levels of drug use by high school students), the procedure is flexible enough to allow us to address that type of question as well.

With that much of an introduction, we are ready to plunge into a more detailed examination of contingency table analysis and of the range of questions that we can address using this procedure. Before we do so, however, let's look at the source of the data that we will use for examples.

Monitoring the Future

Every year, thousands of high school seniors are selected to take part in a study that is designed to assess the frequency of drug use by American youths. In this chapter, we look at the answers given to a few of the many questions asked in that survey, so it will be useful to know a few things about the study as a whole.

The survey is called Monitoring the Future, because it keeps tab on many social indicators with respect to succeeding classes of high school seniors and thereby provides a preview of changes that may take place in the adult population as these students move out of their teenage years and into their 20s and beyond. The first group of students to be asked these questions in the Monitoring the Future study consisted of those who graduated from high school in 1975; thus most members of that class are now (i.e., in 1994) about 37 years old.

The study is sponsored by the National Institute of Drug Abuse, and the primary objective is to provide information about levels and trends in the use of both legal and illegal drugs; however, the study also asks questions about many other types of attitudes and behaviors, including educational and occupational plans, gender role, family values and expectations, religious and political attitudes, views about important social problems, and delinquent behaviors and victimization. The study is conducted by a team of researchers (led by Lloyd Johnston, Jerald Bachman, and Patrick O'Malley) at the University of Michigan. The data have been made available to researchers everywhere, and many articles have been written that are based on analyses of these data. More information about the design and content of the study are available elsewhere (e.g., Bachman, Johnston, & O'Malley, 1989).

Most of the drugs that are asked about in the survey, such as marijuana, cocaine, and LSD, are illegal in this country; others, such as alcohol and nicotine, cannot legally be purchased by teenagers and are frowned on by parents and teachers. Given this, can we believe the answers that are given to these questions? Is it not likely that many high school seniors would hesitate to admit to using an illegal drug? This was an important concern in designing the Monitoring the Future study, and steps were taken to assure the respondents that they could be candid in their answers with no fear of disclosure. The questionnaires are administered in a classroom setting. Each respondent reads the question from a booklet and marks his or her response in that booklet, so that no one else can see the answers. Several different forms of questionnaires are used, so that it would be difficult for anyone else even to have a sense of the question a particular respondent is reading and answering.

The people who administer the study in the schools take considerable care to assure the students that their answers will be kept confidential and that, in particular, no one in their school or home will ever be told how they answered any question. The study administrators also tell the students that if any question is objectionable to them, or would be to their parents, they should simply not answer that question.

It is not possible to verify the answers of individual respondents, but the distributions of answers given by the entire sample have been compared with information from other sources, and these comparisons have been reassuring. Moreover, other indirect methods have been used to evaluate the overall accuracy of the responses and provide reassurance that the data are of good quality. Although there have no doubt been individual students who were less than fully forthright in their answers, from all indications the data provide a picture of American youth that is generally quite accurate.

What Are the Odds?

Some of the questions these high school seniors answer every year are used to obtain such basic information as their gender, their race, the region of the country, and the size of the place (i.e., city, town, and so on) in which they live. These simple categorizations are necessary to examine differences among subgroups: For example, are teenagers in the South more or less likely than those in the West to drink alcohol or to use crack?

One of these background questions asked whether the respondent was male or female. In 1992, 6,767 of the 14,371 respondents were boys compared with 7,604 girls. That is, fewer than half of the respondents that year were male. One way to say this more precisely would be to note that 47.1% of the respondents were boys and that 47.1% is less than 50%. Suppose that each responding high school student wrote the answer to the question about their gender onto a slip of paper and that we collected all of those slips of paper from the 14,371 respondents into a big bag and then pulled out one slip at random. The *probability* that the slip would be marked "male" would be 47.1%, sometimes expressed as $p = .471$. Another way to describe a distribution is in terms of the *odds*: the ratio of the number of cases in one category to the number in the other category. What are the odds of selecting a slip marked "male" rather than "female" from the box with the 14,371 slips? Because there were 6,767 slips marked by boys and 7,604 slips marked by girls, the odds would be 6,767 to 7,604, or 0.890, that a particular slip would be marked "male." Contrariwise, the odds that the slip would be marked "female" would be 7,604 to 6,767, or 1.124.

I can state a simple rule, then: If the chances of an outcome occurring are equal to the chances of it not occurring, the odds are exactly 1 (this corresponds to a probability of .5). If the chances of an event occurring are fewer than the chances of it not occurring, the odds are smaller than 1 (this corresponds to a probability of less than .5); if there is no chance of an event occurring, the odds and the probability are both at their lower limit of 0. If the chances of an event occurring are more than the chances of it not occurring, the odds are greater than 1 (this corresponds to a probability of greater than .5); if an event is certain to occur, the probability is at its upper limit of 1 and the odds are *infinite*, that is, there is no upper limit for the odds. Thus, if the number of boys in the sample is equal to the number of girls, the odds for boys are exactly 1. If, as in our example, the number of boys in the sample is smaller than the number of girls, the odds for boys are less than 1. And if the number of boys in the sample is larger than the number of girls, the odds for boys are greater than 1.

In 1992, the odds for boys (.890) in Monitoring the Future's sample of high school seniors were less than 1, though not by much. The odds give us a way to describe the distribution of boys and girls, just as do our earlier observations that about 47.1% of the respondents were boys, and that the probability was .471 that a randomly selected respondent would be male. In their own way, each of these statements tells us exactly the same thing.

So why, you should be asking yourself, have I bothered to introduce the *odds* notion at all? If this chapter were only concerned with describing the composition of a sample with respect to one characteristic, such as gender, then you would be justifiably miffed with me for wasting your time and a tree or so worth of paper. But we are not going to stop with simple descriptions of samples, one characteristic at a time. We are going

Table 2

Distribution of 1992 "Monitoring the Future" Respondents by Gender

Gender	*n*
Female	7,604
Male	6,767
Odds that male	0.890

to consider the relationships among two, three, and more characteristics, and we will find that the *odds* concept is a very useful one for describing those relationships.

To anticipate later discussion, Table 2 summarizes what we have said about the distribution of the sample of 1992 respondents between boys and girls: The odds of being a boy are simply the ratio of the number of boys in the sample (6,767) to the number of girls (7,604): 6,767/7,604 = 0.890.

What Are the Odds for a Particular Type of Person?

One of the most popular drugs among American youths is marijuana. The Monitoring the Future survey asks high school seniors several questions about their past and current use of marijuana, how easy it is to obtain it, how they regard those who use it, and so on. One of the questions reads as follows: "On how many occasions (if any) have you used marijuana (grass, pot) or hashish (hash, hash oil) in your lifetime?" Of the 14,371 students who answered that question in 1992, 9,722 said that they had never used marijuana or hashish; 1,398 had done so once or twice (they did not say whether or not they inhaled, though); 759 had used it 3 to 5 times; 496, 6 to 9 times; 580, 10 to 19 times; 456, 20 to 39 times; and 960 said that they had used it 40 or more times. Altogether, then, there were 4,649 who said they ever had used marijuana, compared with the 9,722 who professed total abstention. If we were to select at random 1 of the 14,371 students, the odds would be 4,649 to 9,722, or 0.478, that the student would report having used marijuana. Table 3 summarizes this information.

Now let's take this one step further and select a particular type of student, for example, a female student. As we saw earlier, there were 7,604 female students who answered these questions in 1992. Of those female students, 5,391 said that they had never used marijuana, and the

Table 3

Marijuana Use by Sample Respondents

Marijuana use	*n*
Never used	9,722
Ever used	4,649
Odds that have used	0.478

remaining 2,213 said that they had done so at least once (see Table 4). Therefore, if we selected at random a female student from this sample, the odds will be 2,213 to 5,391 or 0.410, that she had reported having used marijuana. These are called the *conditional odds*, because they are conditional on the selected student's being female rather than male. The conditional odds for girls are different from the *unconditional* odds of 0.478, which apply to all students, regardless of gender, who answered the question in 1992. The conditional odds for girls are 0.410, lower than the unconditional odds, which tells us that if we want to guess that a student reported using marijuana, our chances of being correct go down if the student turns out to be female.

We can also calculate the odds for male students, of course: Of the 6,767 boys who answered the question in 1992, 4,331 denied ever using marijuana compared with 2,436 users, so the odds, conditional on being male, are 2,436 to 4,331, or 0.562. The odds of a student being a marijuana user are higher if the student is male than if the student is female.

We have learned something new about marijuana use: The answers to this question are related to another characteristic of the respondents,

Table 4

Use of Marijuana by Gender

Marijuana use	*n*
Girls	
Never used	5,391
Ever used	2,213
Conditional odds that have used	0.410
Boys	
Never used	4,331
Ever used	2,436
Conditional odds that have used	0.562

Table 5

Conditional Odds That Student Has Used Marijuana, on the Basis of Gender

Marijuana use	Girls	Boys
Never used	5,391	4,331
Ever used	2,213	2,436
Conditional odds for use	0.410	0.562

namely, their gender. If we know the gender of a student, we change our assessment of the odds that the student has used marijuana: If the student is female, the odds go down, and if the student is male, the odds go up. I summarize this relationship in Table 5.

We also change the odds that a student is male if we know whether that student reports having used marijuana. Recall that overall, the odds that a randomly selected student is male are 6,767 to 7,604, or about 0.890. If we know that a student reports never having used marijuana, the conditional odds that the student is male are 4,331 to 5,391, or 0.803. On the other hand, the conditional odds that a student is male, conditional on that student reporting having used marijuana, are 2,436 to 2,213, or 1.101. Table 6 summarizes this information.

Now let's cross-classify the respondents with respect to another pair of variables. Recall that in the Introduction to this chapter, we looked at a cross-classification of the respondents in the Monitoring the Future study in 1992, according to their reported use of illicit drugs and the region of the country in which they lived. We now ask whether the odds of having tried marijuana are related to region. The numbers of respondents who lived in each of four regions were as follows: 2,888 lived in the Northeast; 4,246 lived in the North Central United States; 4,536 lived in the South; and the remaining 2,701 lived in the West. For now, I want to focus on a comparison between those living in the South and those

Table 6

Conditional Odds That a Student Is Male, on the Basis of Marijuana Use

Marijuana use	Girls	Boys	Conditional odds that male
Never used	5,391	4,331	0.803
Ever used	2,213	2,436	1.101

Table 7

Conditional Odds That Student Has Used Marijuana, on the Basis of Region of U.S.

	n	
Marijuana use	Non-South	South
Never used	6,528	3,194
Ever used	3,307	1,342
Conditional odds that have used	0.507	0.420

living in the rest of the country. By adding up the numbers living in the three non-South regions, we see that the odds that a randomly selected student is living in the South are 4,536 to 9,835, or 0.461.

Among the 9,835 students not living in the South, 3,307 reported having used marijuana, so the odds of being a marijuana user, conditional on living in the non-South, are 3,307 to 6,528, or 0.507. Among those 4,536 students who lived in the South, 1,342 reported having used marijuana, so the odds of being a marijuana user, conditional on living in the South, are 1,342 to 3,194, or 0.420. I summarize this relationship in Table 7.

Again, we can look at the same relationship from a different perspective: If we know that a student reported never having used marijuana, the conditional odds of living in the South are 3,194 to 6,528, or 0.489; whereas if we know that a student reported having used marijuana, the conditional odds of living in the South are 1,342 to 3,307, or 0.406. I summarize this information in Table 8.

How Different Are the Odds for Different Types of People?

We observed that the odds that a student had used marijuana differed depending on whether the student was male or female. We noted that

Table 8

Conditional Odds That Student Lives in South, on the Basis of Marijuana Use

	n		
Marijuana use	Non-South	South	Conditional odds for South
Never used	6,528	3,194	0.489
Ever used	3,307	1,342	0.406

Table 9

Odds Ratio for Using Marijuana, on the Basis of Gender

Marijuana use	Girls		Boys
Never used	5,391		4,331
Ever used	2,213		2,436
Conditional odds that have used	0.410		0.562
Odds ratio		1.370	

this indicated the nature of a relationship between the two variables: Marijuana use was related to gender. We also observed that the odds differed depending on whether the student lived in the South or in another part of the country. Thus, marijuana use was also related to region of residence. In this section, we go one step further by describing the strength of the relationship between two variables.

It would often be useful to be able to have a single number that conveys information about the relationship between two variables. For example, we may want to be able to say something about the nature and the strength of the relationship between gender and marijuana use. Or we may want to make a comparative statement: is the relationship between region and marijuana use stronger or weaker than the relationship between gender and marijuana use? One such number is called the *odds ratio*, and it is simply the ratio of two *conditional* odds.

Recall that the odds for marijuana use, conditional on a student's being female, are 0.410; and that the odds for marijuana use, conditional on a student being male, are 0.562. If we divide the conditional odds for boys by the conditional odds for girls, we obtain the following odds ratio: 0.562 divided by 0.410 is 1.370. That is, the odds of being a marijuana user are 1.37 times higher for male high school seniors than they are for female high school seniors. I can summarize this by expanding Table 5 to make Table 9.

Remember, though, that we also looked at this relationship from a different perspective, by observing that the conditional odds that a student was male differed depending on whether the student had reported having used marijuana. Among users, the conditional odds of being male are 1.101, compared with conditional odds of just 0.803 among nonusers. What if we were to use the ratio of these two conditional odds to summarize the relationship between gender and marijuana use, instead of the ratio of the two conditional odds given in the Table 9? It makes no difference to the computed value of the odds ratio. That is, 1.101 divided

Table 10

Relationship Between Gender and Marijuana Use

Marijuana use	Girls		Boys	Conditional odds that male
Never used	5,391		4,331	0.803
Ever used	2,213		2,436	1.101
Conditional odds that have used	0.410		0.562	
Odds ratio		1.370		

by 0.803 is 1.370, exactly the same number (except, perhaps, for rounding) we got by dividing 0.562 by 0.410. The odds ratio is the same, regardless of the way in which we look at the two variables. The conditional odds of a student being male are 1.370 times as high among users as among nonusers, and the conditional odds of being a user are 1.370 times as high among boys as among girls. If you think about the calculation of the odds ratio in the two different ways, you will see why they are the same:

$$\frac{2{,}436/4{,}331}{2{,}213/5{,}391} = \frac{2{,}436/2{,}213}{4{,}331/5{,}391} = 1.370.$$

This relationship between the two variables is summarized in Table 10.

The odds ratio tells us the nature of the relationship between the two variables: Because it has a value greater than 1, we know that the odds of being a user are higher for boys than for girls (and that the odds of being a boy are higher among users than among nonusers). A value equal to 1 would have indicated no relationship at all: It would tell us that the conditional odds of being a user are the same regardless of gender. And a value less than 1 would have indicated that the odds of being a user are lower for boys than for girls (and that the odds of being a boy are lower among users than among nonusers).

The odds ratio also tells us something about the strength of the relationship between these two variables. The further the odds ratio is from 1, the more different are the conditional odds, and the stronger the relationship is between the two variables. The lowest possible value of the odds ratio is 0, but there is no upper limit to the odds ratio. This lack of symmetry is an annoying feature of the odds ratio because it makes it awkward to use as a measure of the strength of the relationship, so a somewhat different measure is often used: the *logarithm* of the odds ratio.

In case you have forgotten your high school math, I remind you that the common logarithm of a number is the power to which 10 must be raised to equal that number; for example, 100 is 10^2 (that is, 10 to the power of 2), so 2 is the logarithm of 100. The logarithm of 1 is 0, because 10 (or any other number) raised to the 0 power is defined as equal to 1. Numbers greater than 1 have logarithms greater than 0, and numbers smaller than 1 (but greater than 0) have logarithms less than 0. Thus, odds ratios larger than 1 have logarithms that are positive, whereas odds ratios smaller than 1 have negative logarithms.

The immediate advantage to talking about the logarithm of the odds ratio, instead of the odds ratio itself, is that if we reverse the order of the categories of one of the variables in a contingency table, the logarithm of the odds ratio changes sign (from positive to negative or vice versa), but the absolute value does not change. So let's go back to our example. We saw that the odds ratio corresponding to the contingency table in which we cross-classified high school seniors in 1992 according to their gender and reported use of marijuana was 1.370: The odds of having used marijuana were 1.37 times as high for boys as for girls. The logarithm of 1.370 is 0.137. If we change perspective with respect to gender, we can calculate that the odds ratio is 0.730: The odds of having used marijuana are 0.730 times as high for girls as for boys. The revised odds ratio, 0.730, is the reciprocal of the old odds ratio ($1/1.370 = 0.730$), but that is not obvious from inspection. On the other hand, the logarithm of the revised odds ratio is -0.137: the same absolute value as the logarithm of the old odds ratio, just a different sign.

A second advantage of using the logarithm of the odds ratio is that it makes it easier to compare the strengths of different relationships. The rule is that the larger the absolute value of the logarithm of the odds ratio, the stronger the relationship. To illustrate, let's go back to the relationship between region of residence and marijuana use. The various conditional odds that we saw before for the relationship of those two variables are shown in Table 11, along with the odds ratio. That is, the conditional odds of being a marijuana user are 0.829 times as high among those living in the South as among those living in the nonsouthern regions of the country, and the conditional odds of living in the South are 0.829 times as high among users as among nonusers. The logarithm of the odds ratio is -0.081.

The odds ratio is less than 1, telling us the nature of the relationship: The odds of being a user are lower for those living in the South than for those living elsewhere. But how do we compare the odds ratio for this

Table 11

Relationship Between Marijuana Use and Region of U.S.

Marijuana use	*n* Non-South	*n* South	Conditional odds for South
Never used	6,528	3,194	0.489
Ever used	3,307	1,342	0.406
Conditional odds that have used	0.507	0.420	
Odds ratio	0.829		
Log odds ratio	−0.081		

relationship with the odds for the relationship between gender and use? The value of the odds ratio for the relationship with gender is 1.370, the value for the relationship with region is 0.829, but which is closer to 1? To make that comparison, we would have to take the reciprocal of one of the odds ratios (for example, we have already noted that the reciprocal of the odds ratio for the relationship of gender and marijuana use is 1/1.370 = 0.730). If we report the logarithms of the two odds ratios, however, there is no need for any calculation to compare the strengths of the two relationships: The logarithm of the odds ratio for the relationship between gender and marijuana is 0.137 and that for the relationship between region and marijuana use is −0.081. The absolute value of the logarithm of the odds ratio for the relationship involving gender is larger than that for the relationship involving region, and we can conclude that gender is more strongly related to marijuana use than is region.

Is the Difference Real?

Pick a random student from the Monitoring the Future respondents in 1992, and the odds against that student being male are 6,767 to 7,604, or about 0.890. But suppose we were to pick a random student from all high school seniors in 1992: Does it follow that the odds would be against that student being male? That is, does the fact that there were fewer male than female respondents in the Monitoring the Future sample tell us anything about the population of all high school seniors? The brief answer is yes, the Monitoring the Future numbers do tell us something about the larger population, but not everything.

The high schools and the students chosen to participate in the Monitoring the Future study were selected to make them representative of all high school seniors by giving every high school senior the same chance of being selected. (Well, not quite: The design is a little more complex than that. For example, some high schools that were selected to participate did not do so, and some students were not in class the day the study came to their school. However, because this is not a chapter on study design or response rates, the simplified description is close enough to what really happened.) So yes, the sample of seniors does tell us something about the population of all seniors that the sample was selected to represent. But no, the sample does not tell us everything, simply because a sample is only a portion of the population.

If you were to flip a fair coin 10 times, you would expect it to turn up heads about 5 times, but you wouldn't be surprised if you got only 3 or 4 heads, or 6 or 7 heads. If you tossed the coin 100 times, you would expect it to land heads about 50 times, although it would be unusual if it were to do so exactly 50 times. It is very likely to land heads between 40 and 60 times, and if it landed heads 30 times or less on 100 tosses, you might well suspect that the coin was not really fair. The statistician can confirm your suspicion: The probability of getting 30 (or fewer) heads on 100 tosses of a fair coin is very small: about 1 in 25,000, or .00004. This probability is commonly referred to as the *p value* by statisticians and is often printed by statistical computer programs. Because the outcome of only 30 heads out of 100 tosses is so unlikely if the coin is fair (that is, because the *p* value is so low), you could feel quite certain that the coin was biased.

High school seniors and coins are two different things, of course, but the principle is very much the same whether we are talking about drawing a sample of high school seniors from the large population of all such people and counting how many of them are boys or flipping a coin many times and counting how often it lands heads up. And if you were to draw a sample of 14,371 students from a population that was half boys and half girls, you would expect your sample to contain an approximately equal number of boys and girls (i.e., about 7,185 of each). Consequently, the probability that no more than 6,767 of those in the sample would be boys is extremely small. Therefore, just as too small a proportion of heads when you toss a coin should lead you to the conclusion that the coin is biased, our sample data lead us to the conclusion that less than half of the population from which the Monitoring the Future sample was selected was male. (For the purposes of this chapter, we can stop with this conclusion rather than with trying to explain why it is that less than half the population was male. This need not keep us from speculating, however: Perhaps boys are more likely than girls to drop out of school before their senior year, or perhaps they are more likely to miss days of school and so to be absent from school on the day the study was conducted.)

In general, we can ask how likely it is that we would observe a particular odds in a sample of observations if a specific proportion holds in the population from which the sample is drawn. Exactly how to answer that question is a topic that we cannot get into in this chapter, but what is important is to understand that the question can be answered with a statistical test.

For the sake of illustration, imagine that the odds were even (i.e., 1 to 1) that someone who was an American high school senior in 1988 had ever smoked marijuana. We want to know whether the odds were the same in 1992 as they were in 1988. To answer this question, we first recall the data from the Monitoring the Future study, which found that in the 1992 sample of high school seniors, 4,649 reported that they had used marijuana at least once, compared with 9,722 who reported that they had never done so. The odds in 1992 were 4,649 to 9,722 or 0.478. Our question in this case can be restated as follows: Are the odds in the sample data for 1992 (0.478) sufficiently different from those in 1988 (1.000) for us to conclude that there was a change among high school seniors between 1988 and 1992? If we were to pose that question to the computer, we would get the following as an answer: Yes, the odds have almost certainly changed, because it would be extremely unlikely to observe the odds in the sample data from 1992 being as small as 0.478 if the odds in the population remained 1.000. (The computer program might print, for example, that the p value was less than .0001.)

Well, all right, we answered the question as it was posed, but let's face it, it was a strange question: Did the odds of a student being a marijuana user change from 1988 to 1992, given that the odds were even, at 1.000, in 1988? What is strange about the question is that it assumes that we know what the odds were in 1988 but had to estimate the odds from sample data in 1992. In fact, no census of all high school seniors was taken in 1988, so there is no way to know that the odds were really 1.000 in that year. The Monitoring the Future study did take place in 1988, however, just as it has in every year since 1975, and we can use the sample data to estimate the odds in 1988 just as we used the sample data from high school seniors in 1992 to estimate the odds in 1992. The data from both years are displayed in Table 12.

The sample estimate for 1988 is that the odds were 0.921 that a

Table 12

Relationship Between Marijuana Use and Year

Marijuana use	n		Conditional odds for 1992
	1988	1992	
Never used	7,825	9,722	1.242
Ever used	7,206	4,649	0.645
Conditional odds that have used	0.921	0.478	
Odds ratio	0.519		
Log odds ratio	−0.285		

randomly selected high school senior would have reported using marijuana. This certainly looks different from the estimated odds of 0.478 for high school seniors in 1992. The odds ratio is 0.478 divided by 0.921, or 0.519, and the logarithm of the odds ratio is −0.285. Could it be that the odds were really the same in these 2 years in the population of all high school seniors in 1992 and that we just happened to draw samples that had different odds? After all, if we were to toss each of two fair coins 100 times, one of them might come up heads 45 times and the other 56 times, just by chance. Could that be what has happened here? To answer this specific question, we could use a test with which you may well already be familiar: the Pearson chi-square test. But our purpose in this chapter is to consider a whole range of questions involving more than two variables and more than two categories on each variable, so we are going to use a general procedure, one that involves the specification and evaluation of a *log-linear model*.

Full consideration of log-linear models and the associated notation would take us well beyond the limits for this chapter, but the basic ideas are not difficult, and familiarity with them will allow you to understand analyses that evaluate such models and are published in journal articles and books. The intent of a log-linear model is to explain the number of observations in each cell of a cross-classification table. The model consists of a set of parameters corresponding to various types of effects, or factors, that are thought to be relevant to the cell frequencies.

To be concrete, let's get back to the cross-classification with respect to year and use of marijuana. We have implicitly specified a model for that classification, but now let's make that model explicit. The model includes one parameter that corresponds to the odds that a student is in the class of 1992 rather than that of 1988. Such a parameter is typically specified by using the first letter of the variable, so here we can use the letter Y, for year. The model includes another parameter that corresponds to the odds that a student has ever used marijuana; let's specify this effect by the letter U, for use. The model specification is often abbreviated by enclosing the notation for the parameters in brackets, so our model would be abbreviated as $\{Y\}\{U\}$.

We don't need to look at the specifics of the model, but you may find it helpful to realize that this model predicts the frequency of cases in each cell of the table as the product of three parameters: a parameter corresponding to the odds for year (the Y term), times a parameter corresponding to the odds for use (the U term), times a parameter that corresponds to the size of the sample. We are looking for evidence in the sample data about whether the odds differ from 1.000. But there is another way to express exactly the same model, and that is by taking the logarithms of the cell frequencies. The parameters of this modified model are just the logarithms of the parameters in the original model, but let me remind you of another fact about logarithms: The logarithm of the product of two or more numbers is just the sum of the logarithms of those numbers. Therefore the logarithm of a predicted cell frequency is the sum of three parameters: one corresponding to the logarithm of the odds for year, one corresponding to the logarithm of the odds for use, and one corresponding to the logarithm of the sample size. It turns out that it is easier for both statisticians and for computers to deal with addition than with multiplication of parameters, so standard practice is to estimate this modification of the model by taking logarithms. To a mathematician, such a model for the logarithm of cell frequencies is linear in the modified parameters, and so the name for this type of model is the *log-linear model*.

The question that we are trying to answer is whether the log-linear model that we have specified, $\{Y\}\{U\}$, does an adequate job of predicting the number of cases in each cell of the table. To test the hypothesis that such a model can indeed explain the cell frequencies, we need a test statistic (which we can calculate by hand or let a computer calculate for us): the likelihood ratio chi-square statistic, or G^2 as it is often called. (This is comparable to the Pearson chi-square statistic, or χ^2, that I mentioned earlier.) If we know the G^2 (or χ^2) value and the *degrees of freedom* (or *df*) associated with the test, we can obtain the p value. For this example, $G^2 = 746.55$, with degrees of freedom of 1, and the p value is less than .001. The low value of p tells us that the difference in the odds ratios for the 2 years is highly unlikely to be due to chance, in other words, that there almost certainly was a real change in marijuana use between 1988 and 1992. More formally, the test statistic tells us that the observed frequencies would have been highly unlikely if model {*Y*} {*U*} was correct, and we can safely conclude that the model is incorrect.

In this case, there is just one way to change the model to make it correct and that is by adding one more parameter: a parameter that corresponds to the odds ratio, or the *association* between year and marijuana use. Such an association effect is abbreviated by connecting the abbreviations for the two variables with an asterisk: {*Y***U*}, or simply {*YU*}. Altogether, then, the model includes parameters for the odds on the two individual variables, as well as the parameter for their association, and so can be abbreviated as {*Y*} {*U*} {*YU*}. Indeed, common practice is to leave implicit the parameters for the odds on the two individual variables. So-called *hierarchical models* automatically include lower level terms in models that have higher level terms that involve the same variables. In this case, because the model specifies the association between *Y* and *U*, a hierarchical model can be specified simply as {*YU*}, and it implicitly includes the terms, *Y* and *U*, that correspond to the distribution of cases across the categories of year and marijuana use.

A model for a cross-classification of just two variables that includes a parameter for the odds ratio (as well as parameters for the odds on the two individual variables and for the total number of cases) is called a *saturated* model. The frequencies predicted by a saturated model are necessarily identical to the observed frequencies, the test statistic is necessarily equal to zero ($G^2 = 0$), the number of parameters to be estimated (including the constant corresponding to sample size) is the same as the number of cells in the table, and there are no degrees of freedom ($df = 0$). In particular, this is true for the model {*YU*}.

We have seen that the model {*Y*} {*U*} does not adequately explain the sample frequencies in the cross-classification of year and marijuana use and that we need to specify the saturated model {*YU*} to obtain an adequate fit. What this all means is that for our cross-classification of year and marijuana use, the odds ratio is not equal to 1 and, thus, there is an association between the two variables. (Or, to say the same thing in terms of logarithms, the logarithm of the odds ratio is not equal to 0.)

There is one caution that should be kept in mind with respect to log-linear models and that is with respect to the frequencies that are predicted by a model. If some of the predicted frequencies are very small, the *p* value corresponding to the test statistic may be incorrect and so lead to the wrong conclusion about the model. For the present case, where we have cross-classified the students with respect to two dichotomous variables, all four of the predicted frequencies should be at least 5—and indeed they are in this example.

Is the Relationship Different Across Subgroups?

Thus far we have considered only the relationship between two characteristics, by examining the cross-classification of individuals with respect to their categorization on two variables. We did this three times, to look at the relationship between lifetime marijuana use by high school students and each of three different characteristics of those students: first, the gender of the student, then the region of the country in which he or she lived, and last the year in which she or he graduated from high school. Often we want to address questions that involve the relationships among three or more variables. For example, we may want to know whether the relationship between two variables is the same or different for individuals in different subgroups. To extend two of the examples we have already examined, we may want to know whether the trend toward fewer marijuana users between 1988 and 1992 is equally strong for residents of both the South and the non-South. In the following example, we use three variables: region, year, and marijuana use (see Table 13).

Let's look at the trend for residents of the South across these 2 years (we have added an additional row to this table, showing the logarithms of the conditional odds of having used marijuana, for a reason that will be explained later). The odds ratio for the relationship between marijuana use and year is 0.574 among students living in the South, that is, the odds of a student in the South being a marijuana user declined from 0.732 in 1988 to 0.420 in 1992, and the ratio of 0.420 to 0.732 is 0.574. This is the conditional odds ratio (i.e., conditional on living in the South), and its logarithm is -0.241; this is somewhat smaller in magnitude than the logarithm of the (unconditional) odds ratio of -0.285 that we observed earlier for all students: The trend toward decreasing use of marijuana from 1988 to 1992 for the South is not quite as pronounced as the corresponding trend computed for students from all regions combined.

For only students living outside of the South, we observe the following: The odds ratio for the relationship between marijuana use and year is 0.503 among students in the non-South, and the logarithm of this conditional log odds is -0.298, indicating that the trend toward decreas-

Table 13

Trends for Marijuana Use for 1988 and 1992, According to Region of Residence

Marijuana use	*n* 1988	*n* 1992	Conditional odds for 1992
Live in South			
Never used	2,443	3,194	1.307
Ever used	1,789	1,342	0.750
Conditional odds that have used	0.732	0.420	
Log conditional odds	−0.135	−0.377	
Conditional odds ratio	0.574		
Log conditional odds ratio	−0.241		
Live elsewhere			
Never used	5,382	6,528	1.213
Ever used	5,417	3,307	0.610
Conditional odds that have used	1.007	0.507	
Log conditional odds	0.003	−0.295	
Conditional odds ratio	0.503		
Log conditional odds ratio	−0.298		

ing odds of being a marijuana user is a little stronger outside of the South (−0.298) than it is among high school seniors living in the South (−0.241).

It is easier to understand what is going on by looking at Figure 1, which displays the logarithms of the conditional odds (the extra row of numbers shown in Table 13) for students in the South and non-South separately across the years 1988 and 1992. (Plotting the logarithms of conditional odds provides a clearer picture of the trends than would plotting the conditional odds themselves, for pretty much the same reason that statisticians prefer to specify models for the logarithms of cell frequencies, rather than for the simple cell frequencies.)

Figure 1 makes it clear that the logarithm of the conditional odds for ever using marijuana is higher for students outside of the South than for those in the South, in both years (the Non-South line is higher than the South line) and also that the log odds ratio was smaller in 1992 than it was in 1988 for residents of both regions (both lines slope downward). The two lines are not quite parallel, however: Although the trend is indeed downward for both regions, the lines converge somewhat between 1988

Figure 1

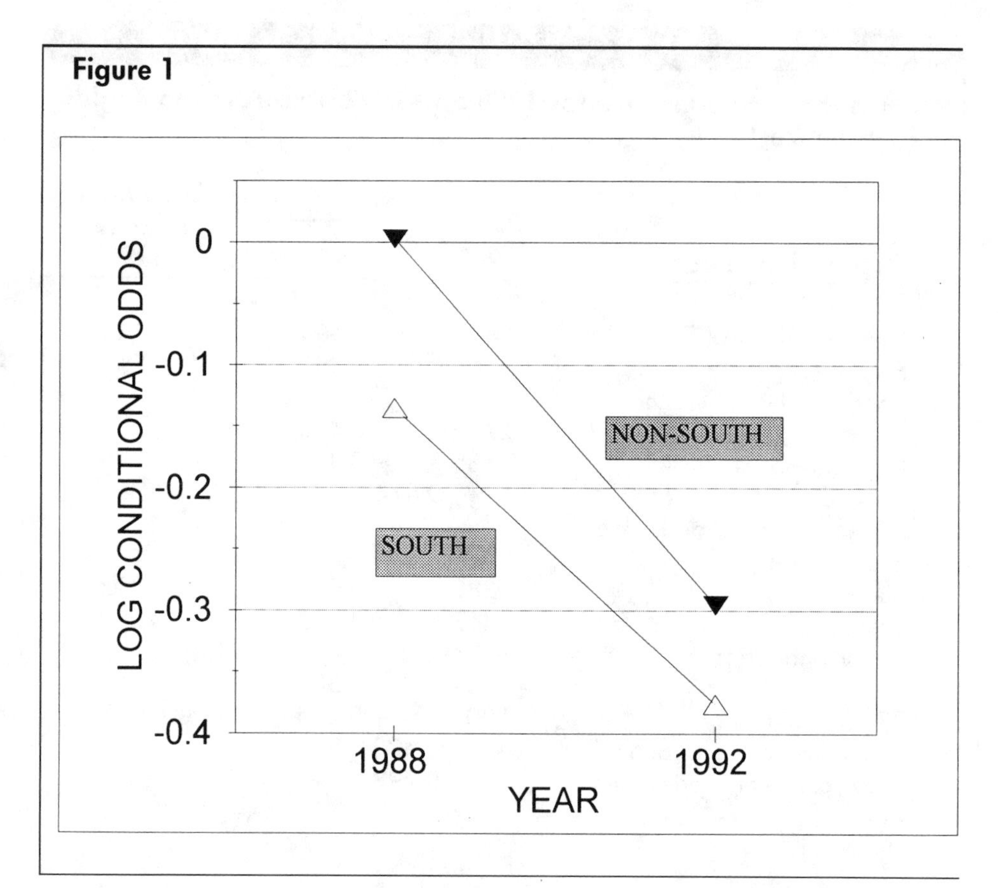

Changes in log odds of use from 1988 to 1992, by region.

and 1992. This convergence is the visual equivalent of the fact that the log odds ratio for the South (−0.241) is smaller in magnitude than that for the non-South (−0.298). The convergence indicates that although students in the South started out (in 1988) less likely than those in the non-South to be marijuana users, the odds did not fall as much for those in the South as they did for those in the non-South.

Again, however, we should ask whether the difference in the trends reflects anything more than sampling variability: Perhaps in the population of all high school seniors, the relative odds ratios are identical for those in the South and non-South, and the observed difference in the Monitoring the Future samples is only what might be expected by chance. To test the statistical significance of the difference in the trends, the procedure is to specify a log-linear model that *excludes* such a difference and find the probability that sample data would show a difference in trends as large as what we observe in this particular sample if that model is correct. We specify the model by using the conventions described earlier. We include parameters for the odds for each of the three variables: one for year {Y}, one for lifetime marijuana use {U}, and one for region of residence {R}. We also include parameters for the association between each pair of variables: one for the association between year and use {YU}, one for the association between year and region {YR}, and one for the association between region and use {RU}. Putting these terms together, the notation for the model is {Y} {R} {U} {YR} {YU} {RU}, or more simply {YR} {YU} {RU}, and the model has a total of seven parameters (including one for the sample size).

The test statistic, the likelihood ratio chi-square statistic, is $G^2 = 6.01$ with 1 degree of freedom, and the associated p value is .014: less than the .05 level that is often used as the cutoff for deciding whether to accept a model (the so-called *statistical significance*). This means that the model that we specified, {YR} {YU} {RU}, does not fit the data very well, the observed cell frequencies would be improbable if that model were correct. This in turn implies that we need to change the model to take into account the difference in the trends for the two regions, by adding a parameter for the *interaction* of the three variables, year, region, and use, {YRU}. Putting this additional parameter along with those we have already specified gives us the model, {Y} {R} {U} {YR} {YU} {RU} {YRU}, or simply {YRU}. This is a saturated model, so it is necessarily true that all of the predicted frequencies are identical to the observed frequencies and that $G^2 = 0.00$ with no degrees of freedom.

We could compare the trends for other groups, as well, to determine whether the odds against marijuana use fell more for some than for others, perhaps even finding some groups of students who bucked the overall trend by showing an *increase* in the odds of use between 1988 and 1992. Another example of such a comparison is instructive. The frequencies shown in Table 14 are again for high school seniors living in the South and for those in other regions of the country, but now for the years 1984 and 1988. For students living in the South, the odds against being a marijuana user dropped from 0.956 in 1984 down to 0.732 in 1988, so the log odds ratio for those living in the South was -0.116. Over the same period, the odds against a student living in other parts of the country being a marijuana user dropped from 1.357 down to 1.007, so the log odds ratio for those living in the non-South was -0.130. These numbers show that the odds fell in both the South and the non-South and to about the same extent in each part of the country. To test whether the trend across these 4 years differed between regions, we again specify

Table 14

Trends for Marijuana Use for 1984 and 1988, According to Region of Residence

Marijuana use	*n* 1984	*n* 1988	Conditional odds for 1988
Live in South			
Never used	2,536	2,443	0.963
Ever used	2,425	1,789	0.738
Conditional odds that have used	0.956	0.732	
Log conditional odds	−0.019	−0.135	
Odds ratio	0.766		
Log odds ratio	−0.116		
Live elsewhere			
Never used	4,011	5,382	1.342
Ever used	5,441	5,417	0.996
Conditional odds that have used	1.357	1.007	
Log conditional odds	0.132	0.003	
Odds ratio	0.742		
Log odds ratio	−0.130		

the model {*YR*} {*YU*} {*RU*}. The likelihood ratio chi-square for this model is $G^2 = 0.39$ with 1 degree of freedom, and the associated p value is .53, so in this case the model does provide an adequate description of the data, and we conclude that the data do not show a difference in trends between regions from 1984 to 1988. Figure 2 adds the pattern for the period from 1984 to 1988 to that from 1988 to 1992, which was shown in Figure 1, and inspection of the figure suggests that the two lines are indeed very close to parallel between 1984 and 1988 and then converge somewhat between 1988 and 1992. (And we can note in passing that this illustrates the reason that we have plotted the logarithms of the conditional odds: If we plotted the conditional odds themselves, the lines would appear to converge between 1984 and 1988.)

Can a Relationship Between Two Variables Be Explained by a Third Variable?

The odds against a student being a marijuana user increased between 1988 and 1992. We observed this change in the Monitoring the Future

Figure 2

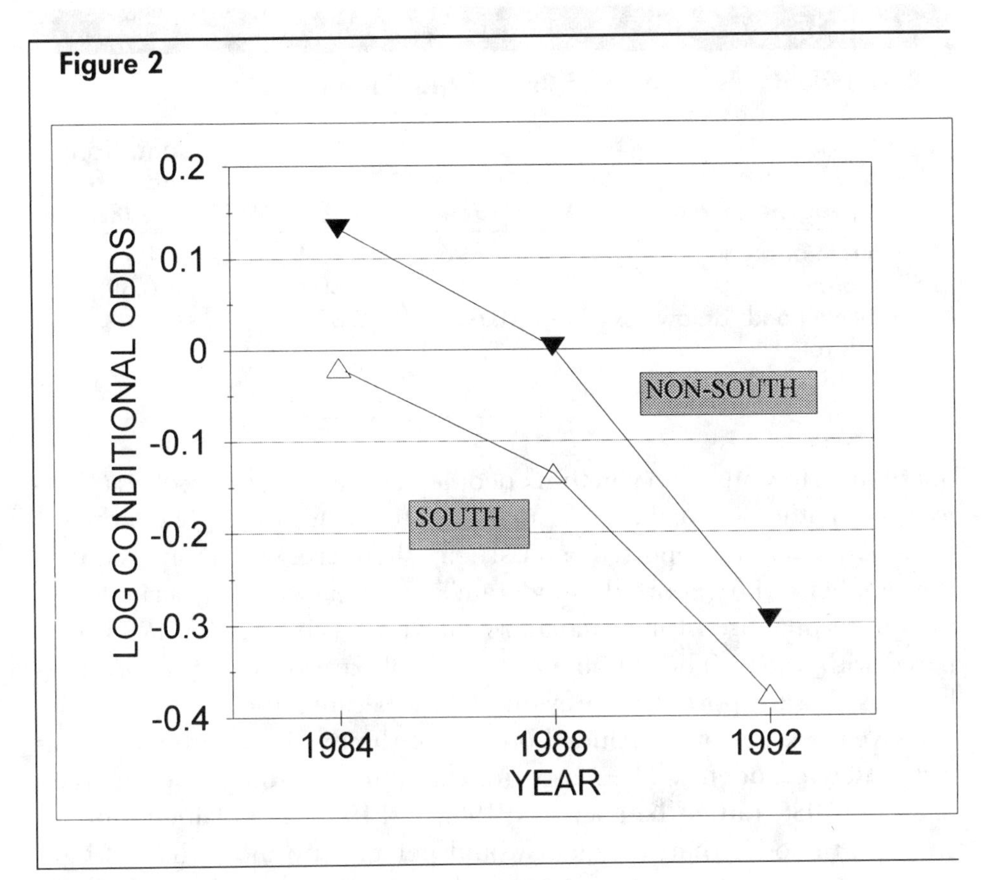

Changes in log odds of use from 1984 to 1992, by region.

sample data and concluded that it almost certainly reflected a real change in the population of all high school seniors. But why did it change? That is a more difficult, and also a more important, question to answer. Had it become more difficult to obtain marijuana? Had enforcement of laws against the purchase and use of marijuana become more successful? Or had students become more committed to college or other longer range goals with which drug use might be considered incompatible? The answers to such questions would be useful to policymakers, and probably to drug producers as well, in their efforts to either encourage or discourage the continuation of the decreasing trend into the 1990s.

One possible explanation for the decline in use of marijuana is that there has been an increase in the perceived risk associated with use. For such an explanation to be plausible, it first must be shown that there has indeed been such an increase in perceived risk. A random subsample of the Monitoring the Future participants each year is asked the following

Table 15

Respondents' Perception of Risk of Marijuana

Degree of risk	n 1984	n 1988	Conditional odds for 1988
Moderate or great	1,696	2,154	1.270
Slight or none	1,161	803	0.692
Conditional odds of low risk	0.685	0.373	
Odds ratio	0.545		
Log odds ratio	−0.264		

question: "How much do you think people risk harming themselves (physically or in other ways), if they try marijuana once or twice?" The numbers of respondents who reported "no risk" or "slight risk" are compared with the numbers who reported "moderate" or "great risk" associated with such occasional use of marijuana are shown in Table 15, for 1988 compared with 1984. (The numbers in this table are only about a fifth as large as those reported in previous tables because the question about perceived risk was only included on one of the five questionnaire forms distributed to students. The reason for examining the difference between 1984 and 1988, rather than between 1988 and 1992, is explained shortly.) In 1984, the odds that a student would perceive no more than a slight risk in occasional marijuana use were 0.685; these odds fell to only 0.373 in 1988. In other words, students were more likely to perceive a risk in use of marijuana in 1988 than they were 4 years earlier.

It is also true that those who perceive more risk in the use of marijuana are less likely to report having actually used marijuana. Combining the samples from 1984 and 1988, we obtain Table 16. Those who see little or no risk to occasional use have high odds of reporting actual use: 3.910, so they are almost four times as likely as not to be users. Those who see at least moderate risk, on the other hand, are unlikely to report use: Their odds are 0.678. The odds ratio is 5.771, so the odds are almost six times as high for those perceiving low risk as for those perceiving higher risk.

Finally, Table 17 shows the change in the odds between 1984 and 1988 for the two groups of students as defined by their perception of the risk of occasional use. For those who perceived moderate or high risk, the odds of reporting use were low in 1984 (0.698), and they were slightly lower in 1988 (0.662). The logarithm of the odds ratio is −0.023. The

Table 16

Cross-Classification of Perception of Risk of Marijuana Use by Actual Marijuana Use

Marijuana use	*n*: Moderate or great risk	*n*: Slight or no risk	Conditional odds of low risk
Never used	2,295	400	0.174
Ever used	1,555	1,564	1.006
Conditional odds that have used	0.678	3.910	
Odds ratio	5.771		
Log odds ratio	0.761		

Table 17

Change in Odds in Actual Use of Marijuana Between 1984 and 1988, As Defined by Perception of Risk

Use	*n*: 1984	*n*: 1988	Conditional odds for 1988
Moderate or high risk			
Never used	999	1,296	1.297
Ever used	697	858	1.231
Conditional odds that have used	0.698	0.662	
Log conditional odds	−0.156	−0.179	
Odds ratio	0.949		
Log odds ratio	−0.023		
Slight or no risk			
Never used	233	167	0.717
Ever used	928	636	0.685
Conditional odds that have used	3.983	3.808	
Log conditional odds	0.600	0.580	
Odds ratio	0.956		
Log odds ratio	−0.019		

Figure 3

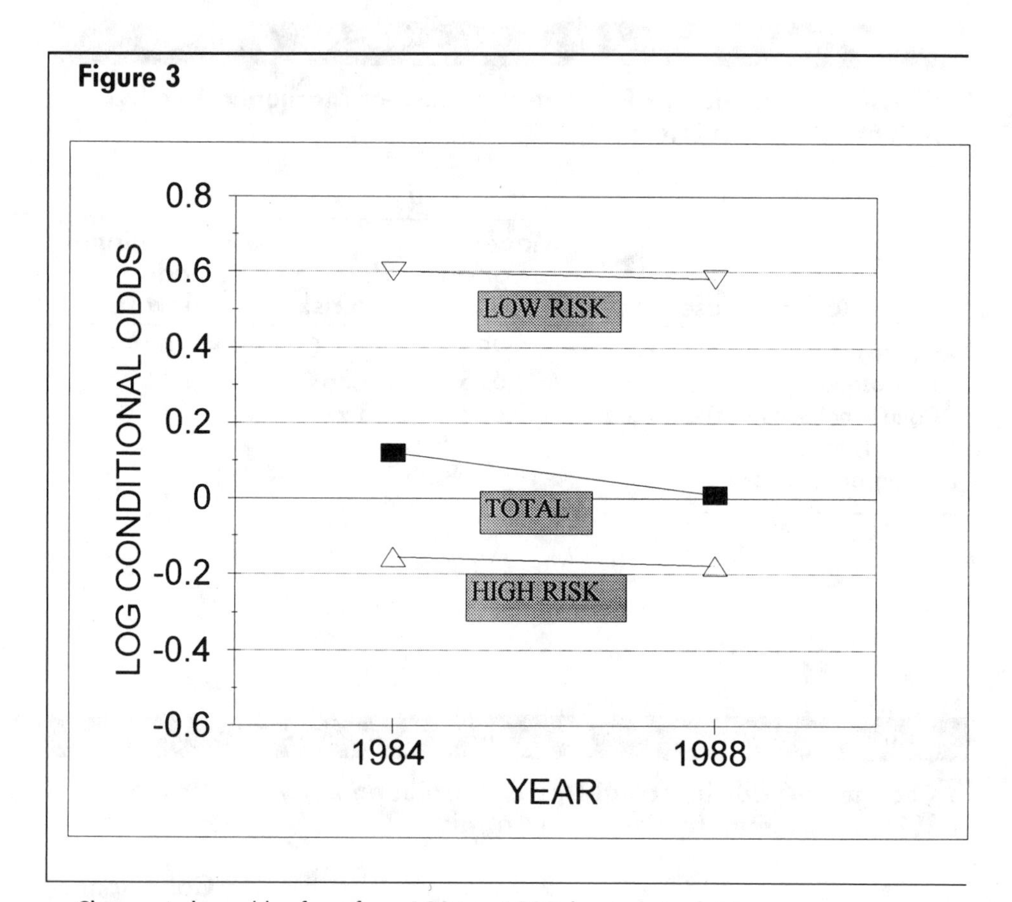

Changes in log odds of use from 1984 to 1988, by perceived risk.

same pattern is observed among those who perceived slight or no risk: Their odds of reporting use were much higher than those who perceived less risk, but they remained high in both years (3.983 in 1984 and 3.808 in 1988). The logarithm of the odds ratio for this group is -0.019. The pattern is shown more clearly in Figure 3: Among all students, there has been a decline in the odds of reporting marijuana use, but this trend all but disappears when we look separately at those who perceive low risk associated with use and also when we look at only those who perceive higher risk.

First let's test whether there was indeed an overall trend in marijuana use from 1984 to 1988, using data only from the subsample of respondents who answered the question about perceived risk but for the moment ignoring their answers to that question. That is, we specify a model that includes parameters for year and use but not for perception of risk: $\{Y\}\{U\}$. The likelihood ratio chi-square for this model is $G^2 = 23.61$ with

1 degree of freedom, and the associated p value is less than .01, so we conclude that these data do show an overall trend from 1984 to 1988.

To test whether the trend remains even within these groups defined by perceived risk, we go through a two-step procedure. First we test whether the trend is different for those who perceive little or no risk than for those who perceive considerable risk, and we do so in exactly the same way that we tested for whether there was a difference in trends for those living in different regions: by specifying a model that omits a parameter for a difference in trends: $\{YP\}\{YU\}\{PU\}$. (I have used the letter P to represent the variable perception of risk.) The likelihood ratio chi-square for this model is $G^2 = 0.00$ with 1 degree of freedom, and the associated p value is .95, so we conclude that the data do not show a difference in trends between groups from 1984 to 1988.

Second, we test whether the trend in marijuana use from 1984 to 1988 remains after we take perceived risk into account. We do so by modifying the model specified in the previous paragraph: We eliminate the term for the association between year and use (the $\{YU\}$ term), and the model becomes $\{YP\}\{PU\}$. The likelihood ratio chi-square for this model is $G^2 = 0.78$ with 2 degrees of freedom, and the associated p value is .68, so we conclude that the data do not show any trend from 1984 to 1988 after we take perceived risk into account.

My interpretation of this pattern is that the decline in marijuana use observed between 1984 and 1988 was almost if not entirely explained by the rise in the perceived risk of even occasional marijuana use. Those who perceive more risk are less likely to use marijuana, so when the mix of students shifts toward those who perceive more risk, we observe that there are fewer users as well.

Time out for a reality check: What is your reaction to this finding? Are you surprised to learn that the decline in marijuana use can apparently be explained by the rise in perceived risk associated with use? Or does it seem so obvious as to be trite? If the latter, you will be surprised to learn that although use declined even more between 1988 and 1992 than it did between 1984 and 1988, little or none of that decline could be explained by the shift toward higher perceptions of risk. The relevant sample frequencies, conditional odds, and odds ratios are shown in Table 18.

Recall from an earlier table that the odds ratio for the entire 1988 and 1992 samples was 0.519. If we estimate the odds ratio with only the data from the subsamples who were asked the question about perceived risk, we obtain a very similar value of 0.509. That is, the odds of reporting

Table 18

Change in Odds in Actual Use of Marijuana Between 1988 and 1992, As Defined by Perception of Risk

Use	*n* 1988	*n* 1992	Conditional odds for 1992
Moderate or great risk			
Never used	1,296	1,373	1.059
Ever used	858	393	0.458
Conditional odds that have used	0.662	0.286	
Log conditional odds	−0.179	−0.543	
Odds ratio	0.432		
Log odds ratio	−0.364		
Slight or no risk			
Never used	167	170	1.018
Ever used	636	409	0.643
Conditional odds that have used	3.808	2.406	
Log conditional odds	0.581	0.381	
Odds ratio	0.632		
Log odds ratio	−0.199		

marijuana use fell almost in half over that 4-year span. Table 18 shows that the odds ratio among those who perceived little or no risk was 0.632: not as steep as the decline for the full sample, but nonetheless a marked decline, in contrast to what we observed for comparably defined samples in 1984 and 1988. And the table shows that the odds ratio among those who perceived at least a moderate risk was 0.432: an even greater decline than is observed for the full sample. In other words, the decline in use between 1988 and 1992 remains clearly in evidence even after we take account of changes in perceived risk, which indicates that something else may account for the decline in use. This is not the place to pursue the issue, though it is an intriguing pattern. A much more thorough investigation of the factors explaining the trend in marijuana use from 1976 through 1986 has been reported in an article by Bachman, Johnston, O'Malley, and Humphrey (1988), and an extension of that analysis into the 1990s is needed to determine whether the 1990s are indeed different from the 1980s. The objective in this chapter is to illustrate the type of questions that can be addressed by cross-classification procedures, and what this example is intended to show is that cross-classification with respect to a third variable can suggest the explanation for the observed relationship between two variables.

Note, however, that the data examined in this chapter are observational, not experimental. You have learned how to compare odds to see if there is an association among variables. An association does not mean that the variables are causally related. One can offer several hypotheses to account for the association among variables. By including other variables in the analysis, as was illustrated with *perceptions of risk*, it is possible to support or rule out alternative explanations of the association between any pair of variables. By ruling out alternative hypotheses that could account for an association, the researcher can make a stronger case for a theoretical model that includes causal statements. Nonetheless, the ultimate test for causality resides in the manipulation and control of independent variables.

The next section shows how a more refined interpretation of cross-tabulated data may be possible by including a greater number of variables in the table.

Extension of the Analysis to Four or More Variables

The preceding sections of this chapter have described the basic steps involved in the analysis of cross-classifications of data. What remains is to suggest extensions of those basics in ways that expand the utility of cross-classification as a tool for data analysis. I illustrate two such extensions: first, taking more than three variables into account in the cross-classifications and, second, analysis of variables with more than two categories.

The extension of the analysis to the cross-classification with respect to four or more variables is straightforward in principle, though it may strain our imaginations to keep track of all the possible combinations of variables. Consider an example presented earlier: the cross-classification of the respondents from the Monitoring the Future study with respect to year of graduation (1988 vs. 1992), region of residence (South vs. non-South), and reported lifetime use of marijuana. Analysis of the frequencies for each of the eight cells defined by those three dichotomous variables indicated that although the odds of a student reporting marijuana use were lower for those in the South than for those living in other parts of the country, the difference became smaller in 1992 than it was in 1988.

Table 19

Change in Odds in Actual Use of Marijuana Between 1988 and 1992, As Defined by Region of Residence, for Students Who Perceive Little Risk

Use	*n* 1988	*n* 1992	Conditional odds for 1992
Live in South			
Never used	45	46	1.022
Ever used	137	100	0.730
Conditional odds that have used	3.044	2.174	
Odds ratio	0.714		
Log odds ratio	−0.146		
Live elsewhere			
Never used	122	124	1.016
Ever used	499	309	0.619
Conditional odds that have used	4.090	2.492	
Odds ratio	0.609		
Log odds ratio	−0.215		
Ratio of odds ratios	1.172		
Log of ratio of odds ratios	0.069		

Note. These students perceived no, or only slightly, risk in occasional marijuana use.

A second analysis indicated that the decline in marijuana users between 1984 and 1988 could be explained by the increase in the odds that a student would perceive a risk in the occasional use of marijuana but that this explanation could not account for the decline in use between 1988 and 1992. Now let's pose the following question: Is the convergence in the odds for the two regions between 1988 and 1992 explained by a convergence in perceived risk? To answer this question, we need to cross-classify the respondents according to all four variables: year, region, perceived risk, and reported marijuana use. The frequencies are shown first for those who perceive no, or only slight, risk in occasional marijuana use (see Table 19).

The odds that a student perceiving little or no risk and living in the South in 1988 would report ever using marijuana were 137 to 45, or 3.044. These odds fell to 100 to 46, or 2.174, in 1992; the odds ratio is 2.174 divided by 3.044, or 0.714. Over the same period, the odds for students perceiving little or no risk and not living in the South fell from 4.090 down to 2.492; the odds ratio is 0.609. That is, among those per-

Table 20

Change in Odds in Actual Use of Marijuana Between 1988 and 1992, As Defined by Region of Residence, for Students Who Perceive Greater Risk

Use	*n* 1988	*n* 1992	Conditional odds for 1992
Live in South			
Never used	404	452	1.119
Ever used	237	137	0.578
Conditional odds that have used	0.587	0.303	
Odds ratio	0.517		
Log odds ratio	−0.287		
Live elsewhere			
Never used	892	921	1.033
Ever used	621	256	0.412
Conditional odds that have used	0.696	0.277	
Odds ratio	0.399		
Log odds ratio	−0.399		
Ratio of odds ratios	1.294		
Log of ratio of odds ratios	0.112		

Note. These students perceive moderate or great risk in occasional marijuana use.

ceiving little risk, the odds fell less for those living in the South than for those outside the South: The ratio of the odds ratios is 0.714 divided by 0.609, or 1.172.

Next look at those who perceived a moderate or great risk in occasional marijuana use (Table 20). The odds that a student perceiving considerable risk and living in the South in 1988 would report ever using marijuana were 237 to 404, or 0.587. These odds fell to 137 to 452, or 0.303, in 1992; the odds ratio was 0.303 divided by 0.587, or 0.517. Over the same period, the odds for students perceiving considerable risk and not living in the South fell from 0.696 down to 0.277; the odds ratio was 0.399. That is, among those perceiving considerable risk, the odds fell less for those living in the South than for those outside the South: The ratio of the relative odds ratios was 0.517 divided by 0.399, or 1.294.

To answer our question about whether the convergence in the odds for students living in the two parts of the country with respect to marijuana use is explained by convergence in the perceived risk associated with such use, we again specify a series of two log-linear models. Because the cross-classification of interest is with respect to four variables, the models include a good number of parameters. First, there are four parameters to take account of the odds for each of the four variables: one for year {*Y*}, one for region {*R*}, one for perceived risk {*P*}, and one for use {*U*}. Second, there are six parameters to take account of associations between each pair of variables, for example between year and region: {*YR*}, {*YP*}, {*YU*}, {*RP*}, {*RU*}, and {*PU*}. Third, there are four parameters to take account of interactions involving each triplet of variables: {*YRP*}, {*YRU*}, {*YPU*}, and {*RPU*}. Altogether, then, the model includes 15 parameters (including one that corresponds to the size of the sample) and can be specified as follows: {*Y*} {*R*} {*P*} {*U*} {*YR*} {*YP*} {*YU*} {*RP*} {*RU*} {*PU*} {*YRP*} {*YRU*} {*YPU*} {*RPU*}. Again, because the model is hierarchical, it can be written more compactly by leaving out the association terms and those corresponding to the distributions of the individual variables because these are implicit in the interaction terms: {*YRP*} {*YRU*} {*YPU*} {*RPU*}. This model comes very close to predicting the observed cell frequencies: The likelihood ratio chi-square for this model is $G^2 = 0.09$, with 1 degree of freedom, and the associated *p*-value is .76. The conclusion is a little complex to say: The data do not show evidence that the difference in trends between regions from 1984 to 1988 is any different for those who perceive little or no risk than for those who perceive considerable risk. Another way to express this conclusion is as follows. We have observed that among those who perceive little or no risk in occasional marijuana use, the ratio of the odds ratio for those living in the South to that for those living outside of the South is 1.172 and that among those who perceive a moderate or great risk in occasional marijuana use, the ratio of the odds ratio for those living in the South to that for those living outside of the South is 1.294. The ratio of those ratios of odds ratios is 1.172 to 1.294, or 0.906. What we have found is that the data are not inconsistent with a value of 1.000 for that ratio of ratio of odds ratios. Or to say it a third way, the four-way interaction is not statistically significant.

We now come to the test of interest, for whether the convergence between regions that occurred from 1988 to 1992 can be explained by convergence in perceived risk. To test that hypothesis, we omit the parameter for that three-variable interaction, {*YRU*}, from the model, leaving us with the following model: {*YRP*} {*YPU*} {*RPU*}. This model also does a good job of predicting the observed cell frequencies: The likelihood ratio chi-square for this model is $G^2 = 3.07$, with two degrees of freedom, and the associated *p* value is .22.

Whatever else we learned from this example, one lesson is clear: We

Table 21

Summary of Log-Linear Models

No.	Model description	G^2	*df*	*p* value
1	*YRPU*	0.00	0	—
2	*YRP YRU YPU RPU*	0.09	1	.762
3	*YRP YPU RPU*	3.07	2	.215
	Model 3 vs. Model 2: *YRU* term	2.98	1	.084
4	*YPU RPU YR*	3.12	3	.373
	Model 4 vs. Model 3: *YRP* term	0.05	1	.823
5	*YPU YR RP RU*	3.90	4	.419
	Model 5 vs. Model 4: *RPU* term	0.78	1	.377
6	*YR YP YU RP RU PU*	10.70	5	.058
	Model 6 vs. Model 5: *YPU* term	6.80	1	.009
	Model 6 vs. Model 2: all three-variable interaction terms	10.61	4	.031
7	*YR YP YU RP RU*	814.97	6	.000
	Model 7 vs. Model 6: *PU* term	804.27	1	.000
8	*YR YP YU RP PU*	13.06	6	.042
	Model 8 vs. Model 6: *RU* term	2.36	1	.124
9	*YR YP YU RU PU*	29.77	6	.000
	Model 9 vs. Model 6: *RP* term	19.07	1	.000
10	*YR YP RP RU PU*	156.51	6	.000
	Model 10 vs. Model 6: *YU* term	145.81	1	.000
11	*YR YU RP RU PU*	19.70	6	.003
	Model 11 vs. Model 6: *YP* term	9.00	1	.003
12	*YP YU RP RU PU*	16.22	6	.013
	Model 12 vs. Model 6: *YR* term	5.52	1	.019

Note. *Y* = year, *R* = region, *P* = perception, *U* = usage.

are rapidly being overwhelmed by numbers—odds, conditional odds, odds ratios, the ratio of two odds ratios, and now the ratio of two ratios of odds ratios. If we were to introduce five, and even more, variables into our analysis (as is not uncommon in the analysis of cross-classifications), keeping track of all those numbers, let alone their interpretation, would be a daunting task. Fortunately, this is all much less confusing when we are careful to specify the appropriate log-linear model for testing a particular hypothesis.

A common practice is to summarize a cross-classification analysis in a table that compares several log-linear models. Table 21 is an example of such a table: It compares a total of 12 different log-linear models for the cross-classification we have been examining. Model 1 is the saturated model, containing parameters for the distributions of each of the four variables used in the cross-classification, parameters for the association between each pair of variables, parameters for the interaction of each subset of three of the four variables, and a parameter for the interaction of all four variables. The model description for this model is simply {*YRPU*}, because as a hierarchical model, the inclusion of the four-variable interaction implicitly includes all lower level terms as well.

We have already considered what are called Models 2 and 3 in Table 21. Model 2 omits the four-way interaction term, and as we have seen, this model fits the data very well ($G^2 = .09$, with 1 degree of freedom and with a p value of .76), indicating that the four-way interaction is not statistically significant. Model 3 omits the interaction among region, year, and marijuana use, and it also describes the data very well ($G^2 = 3.07$, with 2 degrees of freedom and with an associated p value of .22). The next line of Table 21 provides a test of the statistical significance of the three-way interaction term that is omitted from Model 2 to obtain Model 3: The test statistic, $G^2 = 2.98$, is simply the difference in the values of the test statistics for Models 3 and 2, respectively, and the associated degrees of freedom, $df = 1$, is simply the differences in the degrees of freedom for the same two models. The associated p value, .084, means that the three-way interaction term {*YRU*} is not significantly different from 0 in the sample data.

The remaining rows of Table 21 describe additional models that simplify Model 3 by omitting other parameters and testing for the statistical significance of those omissions. Models 4, 5, and 6 omit, in turn, the parameters for the other three-variable interactions: Model 4 omits the term for the interaction of region, year, and perceived risk {*YRP*} and continues to provide a good fit to the data; the comparison with Model 3 indicates that the omitted interaction is not statistically significant. Similarly, inspection of Model 5 and its comparison with Model 4 indicate that the interaction of region, marijuana use, and perceived risk {*RPU*} is not statistically significant.

Model 6 is more interesting. That model omits the final three-variable interaction, that involving year, marijuana use, and perceived risk {*YPU*}, and overall this model provides a marginally acceptable fit to the data: The p value associated with the test statistic is .058, and by one conventional rule of using a p value of .05 as the cutoff, we would not reject this model. However, that test applies to the overall comparison of Model 6 with the saturated Model 1 and so refers to the combination of all 4 three-variable interaction terms as well as the four-way interaction of all variables. If we test specifically for the statistical significance of the particular interaction, {*YPU*}, by comparing Model 6 with Model 5, we obtain the test statistic $G^2 = 6.80$ with 1 degree of freedom and with a *p* value of .009, so we would reject the null hypothesis of no three-way interaction {*YPU*} at conventional levels for *p* values of either .05 or .01. Finally, if we began our analysis by testing the statistical significance of the four-way interaction term and so found it to be nonsignificant (Model 2) and then tested the overall statistical significance of the 4 three-way interaction terms (Model 6 vs. Model 2), we would observe that the *p* value for this test is .031, and our conclusion would depend on the criterion *p* level that we had prespecified. If we had specified a *p* value of .01 or .001, we would conclude that Model 6 was acceptable: The data do not show strong evidence of any three-way interactions. If we had specified a *p* value of .05, we would conclude that there was at least one such interaction. The more detailed comparisons of Models 2, 3, 4, 5, and 6 shown in Table 21 suggest that the most important such interaction is that involving year, perceived risk, and marijuana use. In other words, whether we accept or reject Model 6 depends on the hypothesis that we are testing and the *p* value that we require for rejecting or failing to reject the null hypothesis. If we are following sound analytic practice rather than engaging in "data snooping," we will have specified our hypotheses and the criterion *p* value *before* looking at the data.

A final observation about the {*YPU*} interaction is with respect to the interpretation of that interaction, if we conclude that it is indeed anything more than sampling variability. Inspection of the conditional odds suggests that the trend toward lower odds of marijuana use between 1988 and 1992 was stronger among those who perceived great or moderate use than among those who perceived little or no risk.

If we do decide to accept Model 6 as providing an adequate fit to the sample data, the remaining models shown in Table 21 provide tests for the statistical significance of terms for the association between pairs of variables. More precisely, we can test what is called the *partial association* between each pair of variables, that is, the association between those two variables after taking into account their associations with the other variables included in the cross-classification. For example, Model 7 omits the term for the association between marijuana use and perceived risk (the {*PU*} association). Earlier in the chapter, we looked at the cross-classification of just those two variables and found that the odds ratio was 5.77: Those who perceive little or no risk are almost six times more likely to have used marijuana than are those who perceive a moderate or great risk. Now we are looking at the association of those variables after taking account of their associations with region and year, that is, the partial association, and we conclude that there still is a strong association between them because the Model 7 that omits that partial association provides much poorer fit to the data than does Model 6, which includes that association (G^2 = 804.27 with 1 degree of freedom, with a p value of less than .001).

Similarly, Models 10 and 11, lead us to conclude that there are partial associations between year and use of marijuana and between year and perceived risk, respectively. Model 10 omits the partial association between year and use of marijuana, and the fit of that model is much worse than that of Model 6, which includes their association (G^2 = 145.81 with 1 degree of freedom, with a p value of less than .001). Model 11 omits the partial association between year and perceived risk, and the fit of that model is somewhat worse than that of Model 6, which includes their association (G^2 = 9.00 with 1 degree of freedom and with a p value of less than .01).

Comparison of Model 9 with Model 6 indicates that there is a partial association between region of residence and perceived risk of marijuana use, and inspection of the conditional odds shows that students living in the South are more likely to think that there is considerable or great risk than are students living elsewhere.

Recall that earlier in the chapter (see Figure 1), we observed that students living in the South were less likely to use marijuana than students living elsewhere. Based on Model 8 in Table 1, we cannot reject the hypothesis that there is no partial association between region and marijuana use after taking year and perceived risk into account. This is an example of a simple association between two variables that is apparently explained by the association of those variables with other variables.

Finally, Model 12 shows that there is an association between region and year, an association that may indicate a demographic shift (a higher proportion of students lived in the South in 1992 than in 1988) that is not of any direct interest with respect to our understanding of marijuana use.

Extension of the Analysis to Variables With More Than Two Categories

All of the examples that we have considered to this point have used characteristics that divide the respondents into just two categories: boys and girls, whether or not the respondent has ever used marijuana, whether the student was a senior in 1988 or 1992, whether the perceived risk of marijuana use is low or high, and so on. In real life, of course, we are often interested in variables with more than two categories. Instead of looking at just two regions of the country, we might want to distinguish among four or more regions. Instead of looking at the distinction between those who have never used marijuana and those who have done so at least once, we might want to distinguish several categories according to amount or frequency of use. Instead of looking at just two senior classes, we might want to compare multiple classes.

As an example, let's go back to the trend in the reported lifetime use of marijuana and the extent to which this trend is explained by changes in the perceived risk of occasional use. We examined this trend over two different intervals, finding that changes in the odds of perceiving high risk accounted for essentially all of the decline from 1984 to 1988 but for virtually none of the decline from 1988 to 1992. Now let's take a more comprehensive and detailed look at the trends over time, starting in 1976 and carrying through at 4-year intervals to 1992. Two trends are plotted in Figure 4: The triangles show the logarithms of the odds that a student in each of the years reported ever having used marijuana; the squares show the logarithms of the odds that the students perceived moderate or great risk in occasional use. The odds of being a user increased between 1976 and 1980, but have been declining ever since, and the odds of perceiving considerable risk have been rising since 1976. The pattern is suggestive: Perhaps the decline in use during the 1980s is explained by the increase in perceived risk, although something else would have to explain the contrary rise in use between 1976 and 1980 despite increased perception of risk during the same time period.

The extent to which the trend in marijuana use is explained by the trend in perceived risk is addressed more directly in Figure 5, which shows the trend in reported use separately for each of three groups of students as defined by their perception of the risk. If we look only at the 1980s (specifically, from 1980 to 1988), this figure suggests that the decline in use over that period may be explained almost entirely by the rising perception of risk, because there is virtually no change observed for students who perceive slight or no, moderate, or great risk. On the other hand, the trend in risk does not appear to do much to explain either the rise in use between 1976 and 1980 or the decline in use between 1988 and 1992. The trend toward more users is actually stronger from 1976 to 1980 if we look within categories of perceived risk than if we combine

Figure 4

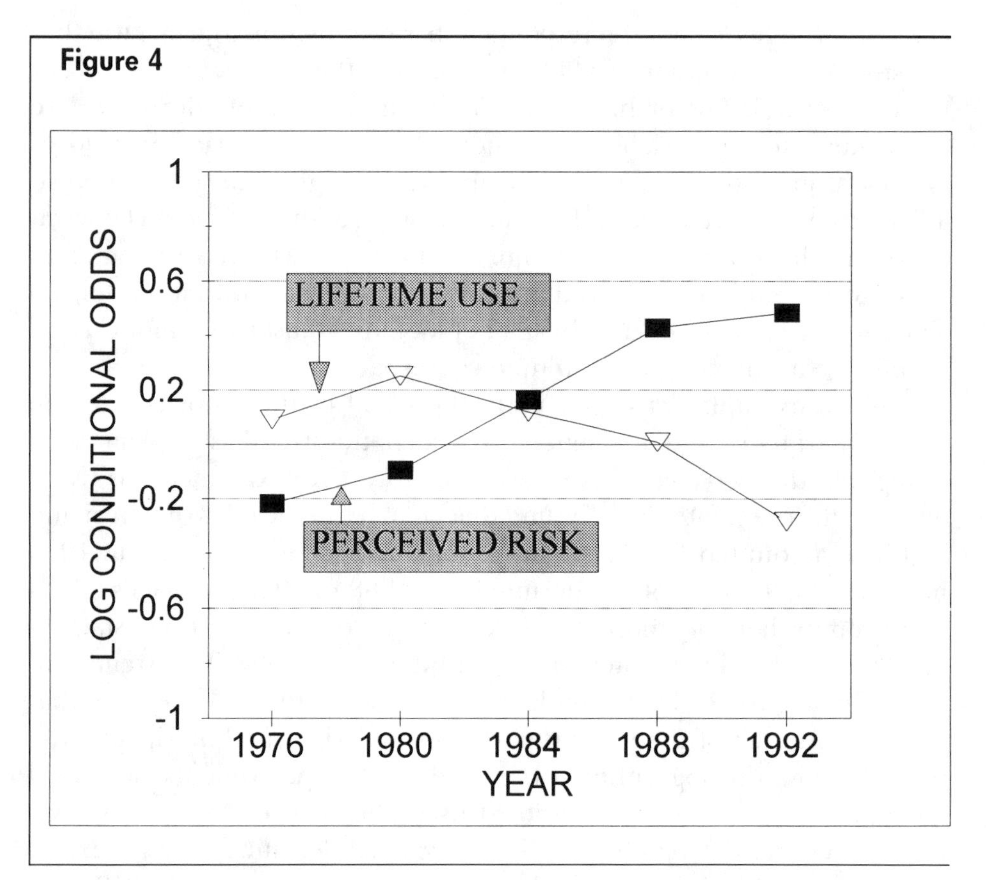

Trends in use and perceived risk of marijuana use.

across those categories, and the decline from 1988 to 1992 is about as strong within those categories as it is for the total group.

With multiple categories on these variables, there are more odds to consider: For example, we could calculate the odds of perceiving moderate risk relative to perceiving great risk, but we could also calculate the odds of perceiving moderate risk relative to perceiving slight or no risk or the odds of perceiving great risk relative to perceiving any risk (i.e., combining those who perceive slight or no risk and moderate risk). These different types of odds lead to the same conclusions about the associations and interactions of the variables, but the ambiguity makes the interpretation of particular odds and odds ratios less clear. Fortunately, it is much easier to deal with variables with more than two categories by specifying log-linear models.

If there are more than two categories for one or more of the variables that we include in a log-linear model, we confront an issue that does not

Figure 5

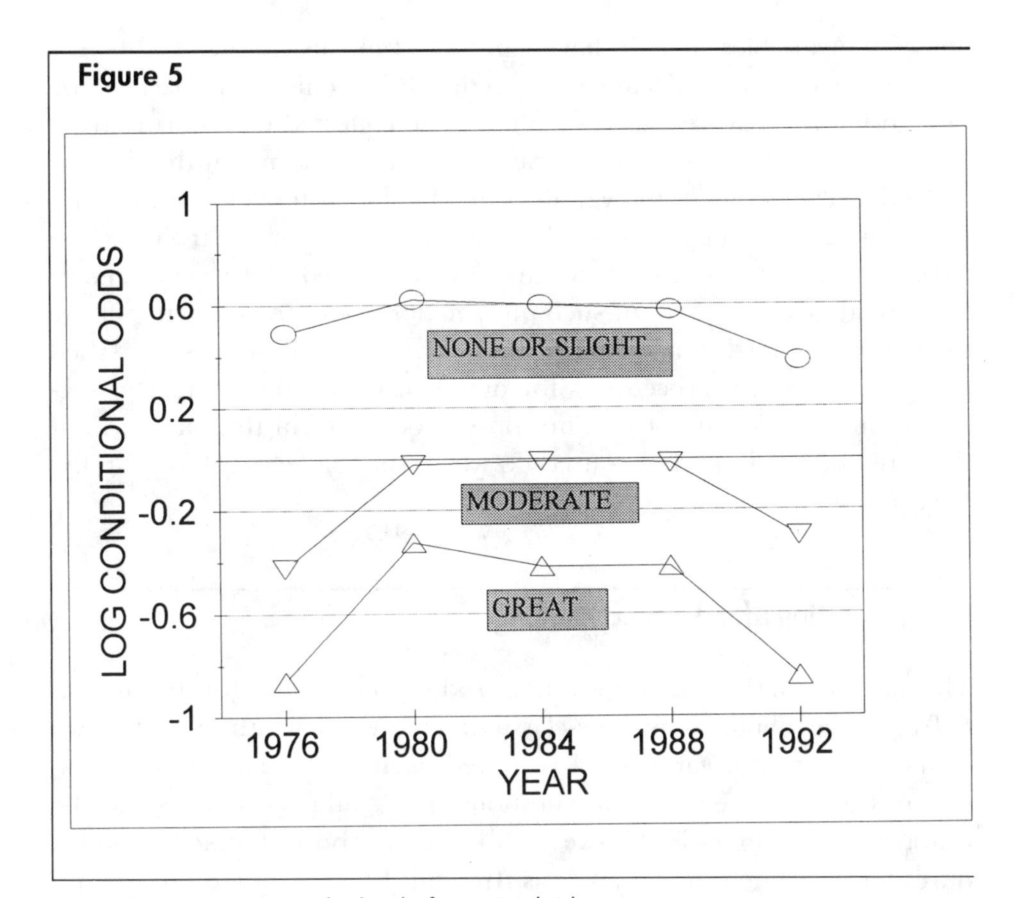

Trends in marijuana use by level of perceived risk.

arise when all of the variables are dichotomies: What is the relationship among the categories? For some variables, the categories are completely unordered; for example, we may want to consider the original four categories of region instead of collapsing the Northeast, North Central, and West into the single non-South category we have used in the models in this chapter. For other variables, there may be an ordering of the categories that should be taken into account when we analyze the data. For example, the question about the perceived risk of marijuana use asked the high school students to place themselves into one of four categories: no risk, slight risk, moderate risk, or great risk. Those who perceive a slight risk are in an important sense intermediate between those who perceive no risk and those who perceive a moderate risk, and those who perceive a great risk are even farther out on the same dimension. Sometimes we can say even more about the categories, by taking into account the relative distances between different pairs of categories. For example, one of the variables that we have considered in this chapter is the year in which data were collected from high school seniors. As part of the Monitoring the Future study, data has been collected each spring from 1975 through the present, and when we examine trends in the data by using the year as one of the variables in a log-linear model, it is clear not only that are the categories on the year variable ordered (from first to last), but also that the categories are equally spaced at 1-year intervals. We should take into account such information about the ordering of the categories and the distances among them when we analyze cross-classified data, and there are procedures for doing so within the context of log-linear models. We cannot go into those procedures in this chapter, but they are discussed in other sources to which the reader is referred in the next section.

Suggestions for Further Reading

The intention of this chapter was to introduce procedures for the analysis of frequencies obtained by cross-classifying individuals with respect to two or more categorical variables. I reviewed examples of such cross-classifications and of the types of questions that could be addressed. The chapter was not intended, however, to instruct you in the actual procedures for implementing an analysis. Instead, I have pointed out the impracticality of trying to answer many types of questions if we were to restrict ourselves to the calculation of odds, odds ratios, and so on, and how it is possible to test for the significance of various types of effects through the specification of log-linear models. A more thorough introduction to such models would require the use of notation to designate the specification of associations and interactions among the variables. An introduction to such notation and to log-linear models is available in a short book by Knoke and Burke (1980). More thorough expositions are available in books by Fienberg (1980) and Wickens (1989). For the more advanced reader, the most comprehensive recent treatment of the topic is a volume by Agresti (1990). An earlier book by the same author (Agresti, 1984) deals specifically with the analysis of categorical data when one or more of the variables has *ordered* categories.

There are other approaches to the analysis of cross-classified data. The weighted least squares approach was developed by Grizzle, Starmer, and Koch (1969), and a good introduction to this approach is available in a book by Forthofer and Lehnen (1981). Also, Reynolds (1977) wrote a book that describes and compares these two approaches.

Programs for estimating log-linear models are included in most major software packages designed for data analysis. These include SPSS (SPSS, 1988; the LOGLINEAR and HILOGLINEAR commands, available on both the mainframe version, SPSS-X and on the personal computer version, SPSS-PC; the HILOGLINEAR command permits the estimation of hierarchical models only, whereas the LOGLINEAR command permits nonhierarchical models to be estimated as well); BMDP (Dixon, 1983; the P4F command); SAS (SAS Institute, 1987; the CATMOD and FREQ commands); and SYSTAT (Wilkinson, 1990; the TABLES command). As with any such program, it is essential that the user understand the procedure before trying to use these programs or to interpret their output.

Glossary

ASSOCIATION Any lack of independence in the distribution of two variables, as reflected in an odds ratio different from 1.0 (or, equivalently, the log of the odds ratio different from 0.0).

CELL A cell of a cross-classification table is defined by specific categories on each of the variables. The number of cells is the product of the number of categories of all of the variables used in the cross-classification ($I \times J \times K \ldots$).

CONDITIONAL ODDS The odds of an event, conditional on the occurrence of another event. In the context of this chapter, the sample conditional odds for a cross-classification of two variables are defined as the number of cases in a particular cell divided by the number of cases in the remaining cells of the same row or column.

HIERARCHICAL LOG-LINEAR MODEL A model in which specification of an association or interaction among two or more variables implies the specification of lower level terms for all subsets of those variables.

INTERACTION Any term in a log-linear model that requires the simultaneous specification of at least three variables. If there is an interaction among three variables, the partial association of any two of those variables is different for cases in different categories of the third variable.

LIKELIHOOD RATIO CHI-SQUARE, OR THE G^2 STATISTIC A test statistic that is commonly used to assess the extent to which the frequencies predicted by a log-linear model correspond to the observed frequencies from sample data. If the model is correctly specified, the statistic follows the chi-square distribution with the appropriate degrees of freedom. The same type of statistic is used to test the significance of deleting a subset of parameters from a log-linear model.

LOG-LINEAR MODEL A model that is intended to predict the frequencies in each of the cells of a cross-classification table in terms of a set of parameters. These parameters are related to the distributions of cases across categories of individual variables (the odds), the associations between pairs of variables (the odds ratios); and higher order interactions among three or more variables.

MARGINAL ASSOCIATION The overall association between two variables, as obtained by simply cross-classifying the data on those two variables.

OBSERVED FREQUENCY The number of cases in a sample that fall into a particular cell in a cross-classification with respect to two or more variables.

ODDS The ratio of the probability of an event occurring to the probability of the event not occurring. In the context of this chapter, the sample odds are defined as the number of cases in a specific category of a variable to the number of cases in the remaining category(ies) of that variable. These are also sometimes called the *marginal* odds, to distinguish them from *conditional* odds (see above).

ODDS RATIO The ratio of two conditional odds. An odds ratio of 1.0 indicates that there is no association between two variables.

PARTIAL ASSOCIATION The association between two variables, conditional on one or more other variables.

PREDICTED FREQUENCY The number of cases that are predicted by a log-linear model to fall in a particular cell of a cross-classification.

SATURATED MODEL The model for a cross-classification of variables that includes parameters for the interaction of all of those variables, as well as for all possible subsets of them. A saturated model includes as many unknown parameters as there are cells in the cross-classification table. The frequencies predicted by a saturated model are necessarily identical to the observed frequencies, the test statistic is necessarily equal to zero ($G^2 = 0$), and there are no degrees of freedom ($df = 0$).

References

Agresti, A. (1984). *Analysis of ordinal categorical data*. New York: Wiley.

Agresti, A. (1990). *Categorical data analysis*. New York: Wiley.

Bachman, J. G., Johnston, L. D., & O'Malley, P. M. (1989). *Monitoring the future: questionnaire responses from the nation's high school seniors, 1988*. Ann Arbor, MI: Institute for Social Research.

Bachman, J. G., Johnston, L. D., O'Malley, P. M., & Humphrey, R. H. (1988). Explaining the recent decline in marijuana use: Differentiating the effects of perceived risks, disapproval, and general lifestyle factors. *Journal of Health and Social Behavior, 29*, 92–112.

Dixon, W. J. (Ed.). (1983). *BMDP statistical software*. Berkeley: University of California Press.

Fienberg, S. E. (1980). *The analysis of cross-classified categorical data* (2nd ed.). Cambridge, MA: MIT Press.

Forthofer, R. N., & Lehnen, R. G. (1981). *Public program analysis: A new categorical data approach*. Belmont, CA: Lifetime Learning Publications.

Grizzle, J. E., Starmer, C. F., & Koch, G. G. (1969). Analysis of categorical data by linear models. *Biometrics, 25*, 489–504.

Knoke, D., & Burke, P. J. (1980). *Log-linear models*. Beverly Hills, CA: Sage.

Reynolds, H. T. (1977). *The analysis of cross-classifications*. New York: Free Press.

SAS Institute. (1987). *SAS/STAT guide for personal computers* (Version 6). Cary, NC: Author.

SPSS. (1988). *SPSS-X user's guide* (3rd ed.). Chicago: SPSS.

Wickens, T. D. (1989). *Multiway contingency tables analysis for the social sciences*. Hillsdale, NJ: Erlbaum.

Wilkinson, L. (1990). *SYSTAT: The system for statistics*. Evanston, IL: SYSTAT.

Logistic Regression

Raymond E. Wright

Are clients with high scores on a personality test more likely to respond to psychotherapy than are clients with low scores? Do children have a better chance of surviving a severe illness than do adults? Do income, credit history, and education distinguish persons who are good credit risks from persons who are poor credit risks?

You might think that linear regression is an appropriate technique for determining whether a person is likely to respond to psychotherapy, to survive a disease, or to repay a loan. After all, each problem requires estimating the relationship among one or more independent (predictor) variables and a dependent variable. However, the research questions in the previous paragraph do not lend themselves to ordinary regression analyses in that the dependent variable is not *continuous*, as is assumed by linear regression, but rather is *dichotomous*. That is, the dependent variable has only two values—responder or nonresponder to therapy, survival case or death, good credit risk or poor credit risk—instead of several values. Logistic regression analysis is specifically designed for use in such situations.

Although logistic regression is used primarily with dichotomous dependent variables, the technique can be extended to situations involving outcome variables with three or more categories (*polytomous*, or *multinomial*, dependent variables). For example, polytomous logistic regression might be appropriate in a study of patient outcomes if patients are categorized as dead, alive with poor recovery, or alive with good recovery. For an introduction to polytomous logistic regression, see Hosmer and Lemeshow (1989).

Discriminant analysis (Klecka, 1980) can also be used to estimate the relationship among one or more predictors and a categorical dependent variable. However, discriminant analysis requires assumptions that are more restrictive than those for logistic regression.

In this chapter, I give an overview of the logistic regression model. First, I discuss the main similarities and differences between logistic regression and linear regression and the basic assumptions of logistic regression. Next, I use data from a hypothetical study to show how to interpret a logistic regression analysis. In particular, I review how to interpret model coefficients, test hypotheses, and interpret classification results. Then, I use data from actual research studies to show how to interpret logistic regression analyses that involve more than one predictor variable. Finally, I describe model-building procedures for studies that have many potential predictor variables.

The Logistic Regression Model

The logistic regression model overcomes the major disadvantages of the linear regression model for a dichotomous dependent variable (Aldrich & Nelson, 1984; Hosmer & Lemeshow, 1989). Like linear regression, the logistic model relates one or more predictor variables to a dependent variable, and the logistic model yields regression coefficients, predicted values, and residuals. Moreover, the predictors in a logistic model can be continuous or noncontinuous. If you are unfamiliar with basic concepts of linear regression, chapter 2 should be read before proceeding.

For example, consider an analysis of fictitious data for 54 students enrolled in a math course graded pass or fail. Before the students enrolled in the course, they received a score between 1 and 11 on a math pretest. A high pretest score meant that a student was proficient in the skills underlying the subject matter taught in the course, and a low score represented math deficiency. We will use logistic regression to determine the relationship between pretest scores and whether a student passed the course. Thus, a continuous measure is used as the predictor variable, and a dichotomous variable (pass/fail) is used as the predicted variable (also called the *dependent* or *criterion* variable). For the analysis, students who passed, that is, students in the target group, are arbitrarily coded 1 on the dependent variable; students who failed are coded 0.

In logistic regression, the relationship between the predictor and the predicted values is assumed to be nonlinear. Figure 1 shows the logistic

Figure 1

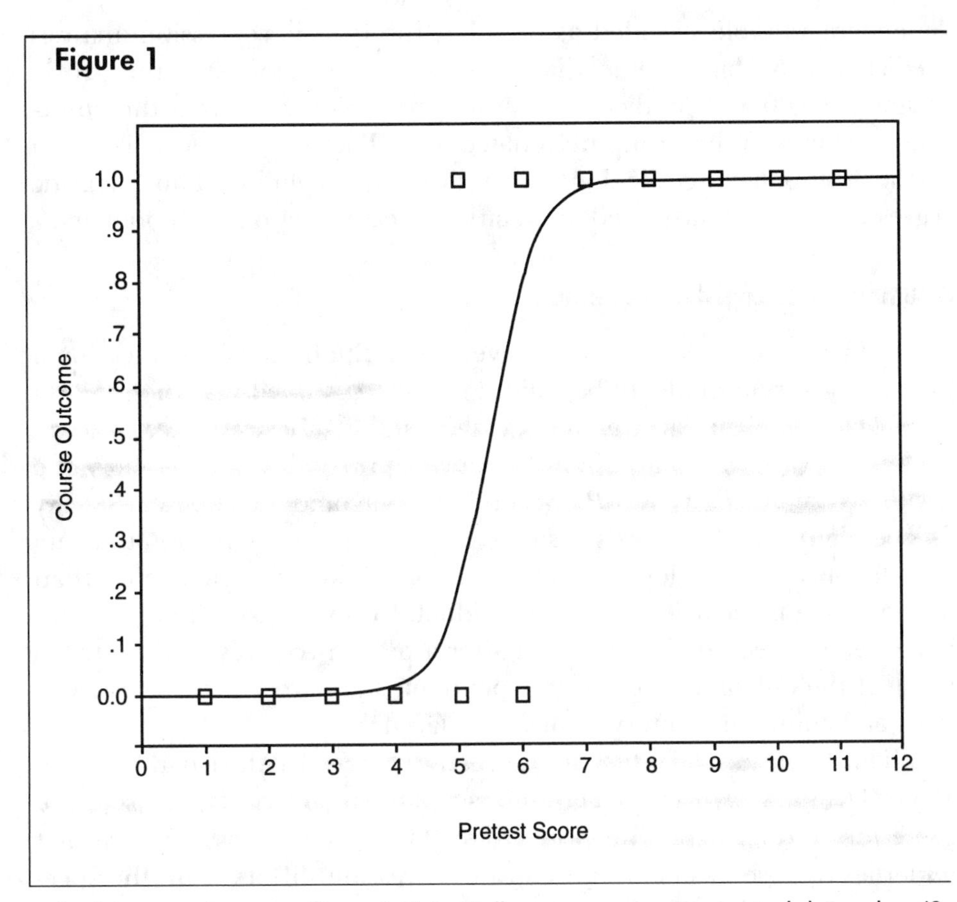

Logistic regression curve for math data. Hollow squares represent actual data values (0s and 1s). The curve represents probabilities predicted by the model. Each square represents more than one case.

curve for the math study. Note that the logistic curve is S shaped, or *sigmoidal*. Moreover, the curve never falls below 0 or reaches above 1, even for extreme values of the predictor. Thus, the predicted values obtained using the logistic model can always be interpreted as probabilities. Also, in Figure 1, note that the vertical distances between the actual and predicted values (i.e., the residuals) are generally small in the logistic regression plot, indicating that overall, the logistic model provides an excellent fit to the actual outcomes. However, the residuals are larger for students with pretest scores of 5 or 6 than they are for other students.

In linear regression, one attempts to predict a score on a continuous dependent measure. In logistic regression analysis for a dichotomous dependent variable, one attempts to predict the probability that an observation belongs to each of two groups. For example, if a dichotomous dependent variable is coded as 0 and 1, the logistic regression analysis predicts a probability value that an observation belongs to the group designated as 0 and predicts a separate probability value that the observation belongs to the group designated as 1. The observation is assigned to the group having the higher predicted probability. Thus, logistic regression is frequently used as a statistical classification methodology.

Assumptions of Logistic Regression

As in linear regression analysis, several conditions must be met for a logistic regression model to be valid. First, it is assumed that the random variable of interest, group membership status, for example, is a dichotomous variable taking the value 1 with probability P_1 and the value 0 with probability $P_0 = 1 - P_1$. Second, the outcomes must be statistically independent. In other words, a single case can be represented in the data set only once. The independence condition is violated when more than one outcome is recorded for an individual, for example, when presence of a disease is recorded before and after a person receives medical treatment. If the outcomes are not independent, standard errors, hypothesis tests, and confidence intervals may be inaccurate.

Third, the model must be correctly specified (Aldrich & Nelson, 1984). The *specificity assumption* requires that the model contain all relevant predictors and no irrelevant predictors. (The specificity assumption that underlies the use of linear and logistic regression differs from the specificity of the model, which is an index of classification performance.) If theoretically important predictors are omitted from the model, or if irrelevant predictors are included in the model, the analysis will give incorrect estimates of population coefficients for variables in the model. This problem is analogous to the problems of suppressor variables and mediating variables in linear regression analysis. In practice, however, the specificity assumption is rarely met.

Fourth, the categories under analysis must be *mutually exclusive* and *collectively exhaustive*. In other words, a case cannot be in more than one outcome category at a time, and every case must be a member of one of the categories under analysis. For example, *life* and *death* are mutually exclusive and exhaustive categories, because one cannot be alive and dead simultaneously and because everyone is either alive or dead. In contrast, the categories *baseball fan* and *football fan* are neither mutually exclusive (because you can be a fan of both baseball and football) nor exhaustive (because you can be a fan of another sport or a fan of no sport at all).

In the writing study, which is presented in the Multivariable Models section, a student is either a good writer or a poor writer (collectively exhaustive), yet he or she cannot be in both categories (mutually exclusive).

Finally, to test hypotheses involving the logistic regression coefficients, larger samples are required than for linear regression analysis. This is because standard errors for maximum likelihood coefficients are large-sample estimates. For small samples, hypothesis tests may be inaccurate. For most applications, a minimum of 50 cases per predictor variable is sufficient (Aldrich & Nelson, 1984).

Using the Logistic Regression Model

As in linear regression, a single-predictor logistic model yields a constant term (b_0) and a regression coefficient for the predictor variable (b_1). You can use these coefficients and the scores on the predictor variable to compute for any case the predicted probability of membership in the target group. Follow these steps.

First, add the constant to the product of the regression coefficient and the predictor to obtain the quantity $\hat{g}$:

$$\hat{g} = b_0 + b_1 (X).$$

Then insert $\hat{g}$ into the following equation, which in this example gives the predicted probability of passing the course:

$$e^{\hat{g}}/(1 + e^{\hat{g}}).$$

Note that in both the numerator and the denominator of the equation, e (the base of the natural logarithms) is raised to the power of $\hat{g}$.

When applied to the math course data, the logistic model yields a b_0 of -14.79 and a b_1 of 2.69. Thus, for cases with a pretest score of 5, $\hat{g}$ is $-14.79 + 2.69$ (5), or -1.34, and the predicted probability of passing is

$$e^{-1.34}/(1 + e^{-1.34}) = .21.$$

For a student with a math score of 7, $\hat{g}$ equals $-14.79 + 2.69$ (7), or 4.04, and the predicted probability of passing is

$$e^{4.04}/(1 + e^{4.04}) = .98.$$

These predicted probabilities are located on the logistic regression curve shown in Figure 1. To compute the value of the curve for other values of the pretest, follow the same steps but change the value of the predictor.

You can use the predicted probabilities from the logistic model to compute predicted group memberships for each case. Generally, if the predicted probability for a case of .50, the case is classified as a member of the target group, and if the predicted probability is less than .50, the case is classified as a member of the other group. Thus, using the predicted values computed in the preceding paragraphs, the model predicts that a case with a score of 5 will fail, and a case with a score of 7 will pass. You can also compute the 50% cutoff value: the smallest value of the predictor needed for a predicted probability of .50 or greater. Simply reverse the sign of the constant term and divide the result by the predictor coefficient (you can use this method to compute a cutoff value only when the model contains one predictor). For example, for the math study, the 50% cutoff is (14.79/2.69), or 5.50. Thus, the model predicts that students with pretest scores of 5.50 or greater will pass the course and students with lower scores will fail.

As in linear regression, residuals are computed as the actual value of the dependent variable minus the predicted value. For example, for a case with a pretest score of 5 who passed the course, the residual is 1 − .21, or .79. The residual for a case with the same pretest score but who failed the course is 0 − .21, or −.21. In logistic regression of a dichotomous dependent variable, only two residual values are possible for any level of the predictor. This does not mean that there are two residuals for a single instance of prediction. In this example, if a student with a pretest score of 5 passes, there is one residual. If a student with the same pretest score fails, then there is a different value for the residual.

Interpreting the Coefficient for the Predictor Variable (b_1)

In linear regression, the model coefficients have a straightforward interpretation. The coefficient for the predictor variable estimates the change in the dependent variable for any one-unit increase in the independent variable. The constant term estimates the value of the dependent variable for a case that has a predictor value of 0.

To interpret the logistic regression coefficients, you need to understand the concept of *odds*. For a dichotomous variable, the odds of membership in the target group are equal to the probability of membership in the target group divided by the probability of membership in the other

ODDS: target grp. membership = prob of target grp membership / prob of membership in other grp

group, both of which were illustrated in the previous section. For example, if the probability of membership in the target group is .50, the odds are 1 (.50/.50); if the probability is .80, the odds are 4 (.80/.20); and if the probability is .25, the odds are .33 (.25/.75). Notice that the odds are 1 when both outcomes are equally likely, greater than 1 when the target event is more likely than the other event, and less than 1 when the target event is less likely than the other event. Note that the concept of *odds* is different than *probability*. A probability value can range from 0 to 1; an odds value can range from 0 to infinity. Odds tell you how much more likely it is that an observation is a member of the target group rather than a member of the other group.

Another important concept in logistic regression analysis is *odds ratio* (*OR*). The odds ratio estimates the change in the odds of membership in the target group for a one-unit increase in the predictor. An odds ratio is computed by using the regression coefficient of the predictor variable as the exponent of e. In the math study, $b_1 = 2.69$. Thus the odds ratio equals $e^{2.69}$, or 14.73. Therefore, the odds of passing are 14.73 times greater for a student who had a pretest score of 5 than for a student whose pretest score was 4. The odds ratio is the same for any other one-unit increase in pretest score. Thus, the odds of passing are 14.73 times greater for a student who had a pretest score of 6 than for a student whose pretest score was 5. To compute the odds ratio for a predictor increase of a different size, multiply the regression coefficient by the size of the increase before you raise e to the power of the coefficient. For example, for a two-unit increase in pretest score, the odds ratio is $e^{(2 \times 2.69)} = e^{(5.38)}$, or 217.02.

The raw coefficient of the predictor variable (b_1) represents the change in the natural logarithm of the odds ratio, which is harder to interpret than an odds ratio. However, the raw coefficient does have a useful function: A positive predictor coefficient means that the predicted odds increase as the predictor values increase; a negative coefficient indicates that the predicted odds decrease as the predictor increases; and a coefficient of zero means that the predicted odds are the same for any value of the predictor. That is, the odds ratio is 1.

Why not interpret the logistic regression coefficient in terms of probabilities rather than odds? That is, why don't we determine how much the *probability* increases for a particular increase in the predictor? The reason is that the change in the probability depends on the level of the predictor, so to estimate how much the probability increases for an increase in a predictor, you need to know not only the value of the logistic regression coefficient but also the level of the predictor. For example, in Figure 1, note that the curve is much steeper in the middle range of pretest scores than for large or small values. This means the estimated change in probability is much larger for pretest increases in the middle range than for other changes: For a pretest increase from 5 to 6, the predicted probability increases from about .20 to about .80; for a change from 0 to 1 on pretest score, the change in the predicted probability is negligible. In contrast, the estimated change in odds is the same for all levels of the predictor variable.

The Logistic Regression Curve

The logistic regression curve obtained for the math data is only one of many possible logistic regression curves. Although the logistic regression curve is generally S shaped, each combination of coefficients for the predictor and the constant term produces a different logistic regression curve.

The coefficient for the predictor controls the steepness and direction of the curve. For example, Figure 2 shows hypothetical logistic regression curves derived from two samples. The curves use the same value for the constant term (0) but a different value for the predictor coefficient. You can see that the curve that has the larger predictor coefficient (3) is steeper than the curve whose coefficient is 1, especially near the 50% point. The curve having the larger coefficient represents a stronger position association between the predictor and the odds of membership in the target group. The odds ratio is smaller for the curve with the smaller coefficient.

Because each predictor coefficient is positive, each curve increases from left to right. A logistic curve having a negative predictor coefficient is also S-shaped, but the curve decreases from left to right. This means that the predicted probabilities are smallest for relatively large values of the predictor. When the coefficient is 0, the curve loses its S shape and becomes horizontal, as shown in Figure 3. This means that the predicted probability is the same for any value of the predictor.

The constant term controls the location of the logistic curve along the X axis. Curves having the same predictor coefficient but different constant terms have the same shape, but as the constant increases, the curves shift left on the X axis. Thus, in the model with the larger constant

Figure 2

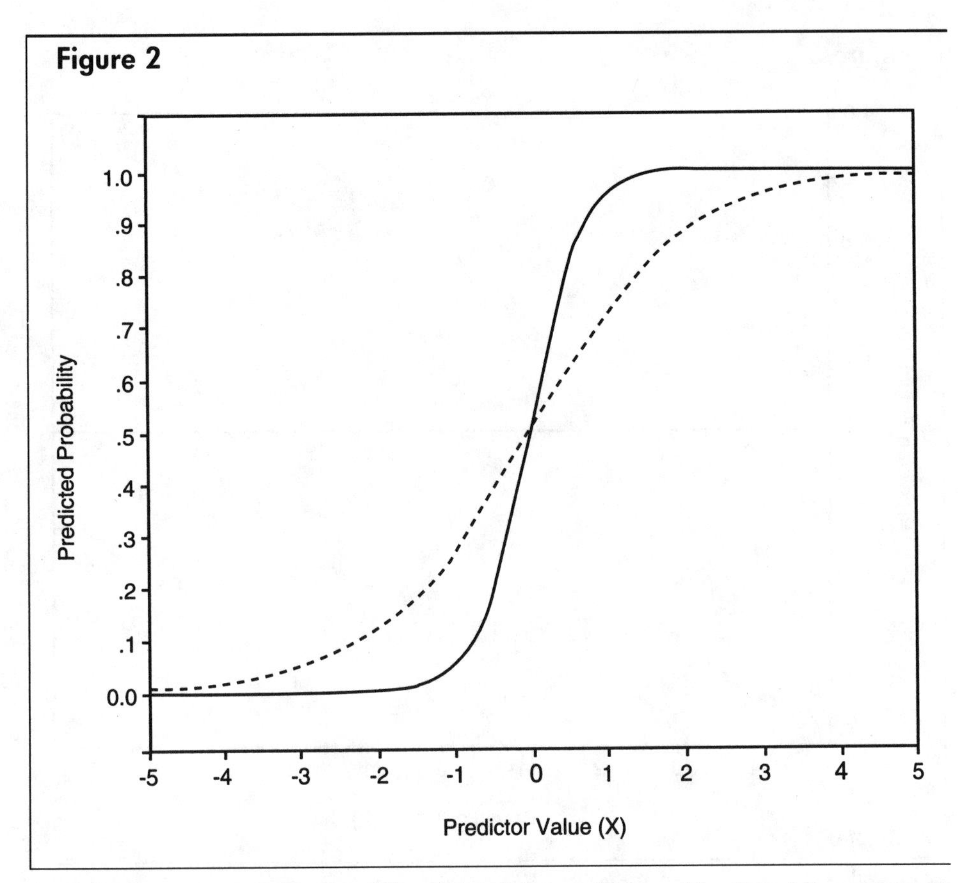

Logistic regression curves for different predictor coefficients. The dotted line represents predicted probabilities for various *X* values when the predictor coefficient (*b*) is 1; the solid line shows predicted probabilities when the predictor coefficient is 3. For both curves, the constant term is 0.

term, a smaller value of the predictor is needed to predict the same probability.

Estimation of Coefficients

In linear regression analysis, the model coefficients minimize the sum of the squared differences between the actual values and predicted values of the dependent variable. That is, the coefficients are chosen according to the least squares criterion. In logistic regression, the *maximum likelihood* (*ml*) criterion is generally used for selecting parameter estimates. The coefficients maximize the probability (*likelihood*) of obtaining the actual group memberships for cases in the sample. Thus, the logistic regression coefficients are known as *maximum likelihood parameter estimates*.

Figure 3

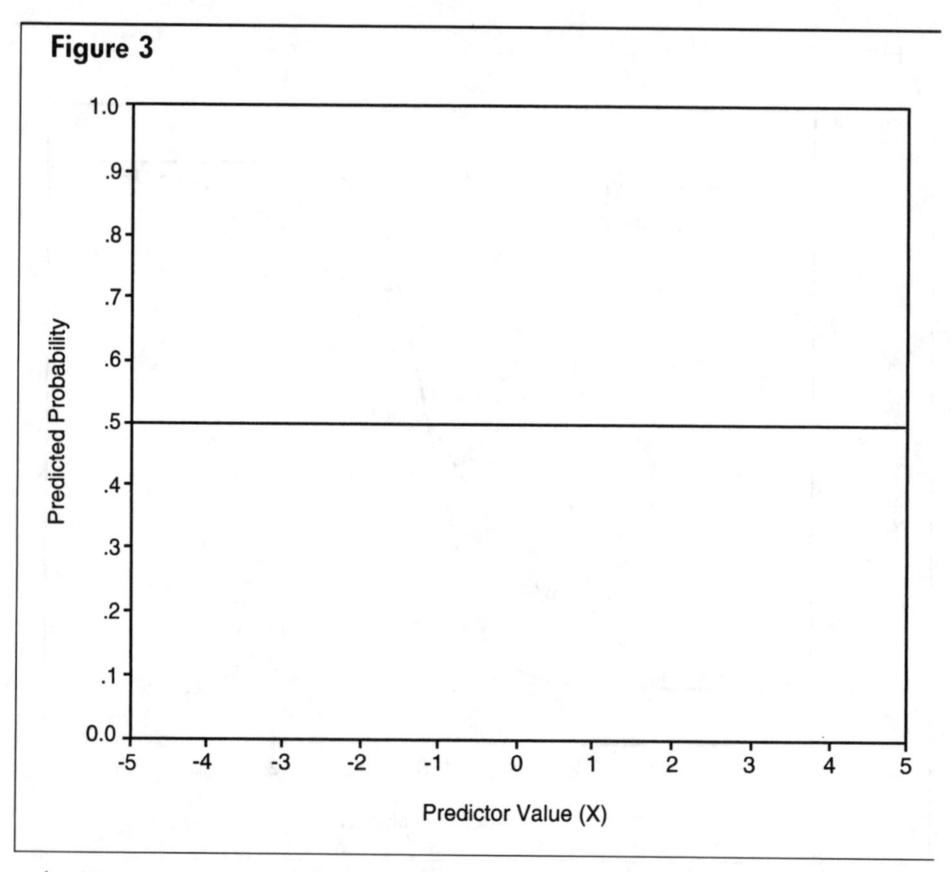

Logistic regression curve when predictor coefficient (*b*) is 0. The predicted probability is .50 for all *X* values. For this curve, the constant term is 0.

Instead of reporting the actual sample likelihood, researchers usually report one of two measures computed from the sample likelihood: the *log likelihood* (*LL*) or the *deviance* ($-2LL$). The log likelihood is generally negative. The deviance, equal to negative two times the log likelihood, is generally positive. The log likelihood and the deviance are 0 when the sample likelihood is 1. Coefficients that maximize the likelihood also maximize *LL* and minimize deviance (McCullagh & Nelder, 1989).

Hypothesis Testing

In the math study, the nonzero pretest coefficient means that a relationship exists between pretest score and the odds of passing the course for cases in the sample. However, we don't know whether this association generalizes to other groups of students. To determine whether the regression coefficient is different from zero in the population of math students from which the present sample was drawn, hypothesis tests are performed. As in linear regression, hypothesis testing in logistic regression involves reasoning by contradiction. We first assume that the predictor coefficient is 0 in the population (this assumption is our *null hypothesis*). Hypothesis tests tell us whether there is sufficient evidence in the sample data to reject the null hypothesis and, therefore, to accept the *alternative hypothesis* that the predictor coefficient differs from zero.

One way of testing whether the population coefficient for a predictor variable is zero is to use a *likelihood ratio statistic* (G). Like an F statistic in linear regression, a large likelihood ratio statistic means that the population coefficient probably differs from zero. The criterion for "large" is the probability (p) associated with the likelihood ratio statistic. If the observed likelihood ratio statistic occurs in 5% or fewer of random samples from a population in which the coefficient is zero, the null hypothesis is rejected. (The cutoff probability for the hypothesis test, .05, is known as *alpha*.)

For models containing a single predictor, the probability for the likelihood ratio statistic is obtained from a chi-square distribution with 1 degree of freedom. The likelihood ratio statistic for the math sample is 56.17, with $p < .00005$, which indicates that the population regression coefficient probably differs from 0. Moreover, because the sample regression coefficient is positive, it can be concluded that students with high pretest scores are more likely to pass the math course than are students whose pretest scores are lower.

Another popular test used to determine whether the population coefficient differs from zero is the z test. To compute z, the predictor coefficient is divided by its estimated *standard error* (SE), a measure of the expected variability in the coefficient from sample to sample. For example, the z score for pretest is 2.69/0.91, or 2.96. The probability of obtaining a z score of 2.96 or greater is less than .0003, so the hypothesis that the population coefficient is 0 would be rejected. (In general, the probability is less than .05 for z scores with absolute values of 1.96 or greater.) Thus, as with the likelihood ratio test, the result of the z test indicates that a positive association exists between pretest score and the odds of passing. The distinction between G and z is largely arbitrary: Both assume large samples, are frequently reported, and typically give parallel findings when used on the same set of data (cf. Hauck & Donner, 1977).

Interval Estimation for Coefficients and Odds Ratios

The regression coefficient for pretest score is only a point estimate of the population coefficient. A 95% confidence interval uses sample data to estimate the likely range of values for the population coefficient. The approximate lower limit for the interval is computed as the value of the coefficient minus 2 times its standard error. The approximate upper limit is computed as the value of the coefficient plus 2 times the standard error. (For a more precise estimate of the lower and upper bounds, use 1.96 times the standard error.) For example, for pretest score, the lower limit is 2.69 − 2 (0.91), or 0.87, and the upper bound is 2.69 + 2 (0.91), or 4.51. Thus, the population coefficient is estimated to be between 0.87 and 4.51. The 95% confidence interval has the same interpretation as in linear regression: If one computed 95% confidence intervals for an infinite number of random samples from a population, 95% of the intervals would contain the population value on average. In addition, when the 95% confidence interval includes 0, the coefficient is *nonsignificant* at the 5% alpha level.

You can also compute a 95% confidence interval for the odds ratio if you raise e to the power of the upper and lower confidence bounds for the regression coefficient. In the present example, the lower bound for the odds ratio is $e^{0.87}$, or 2.39, and the upper bound is $e^{4.51}$, or 90.92. Note that the confidence interval is skewed: The lower limit is much closer to the sample odds ratio (14.73) than is the upper limit. Skewed confidence intervals for odds ratios are common because the odds ratio cannot be less than 0, but it has no upper limit. Also, note that the interval is very wide. For samples from the same population, confidence intervals generally become narrower as the sample size increases. Finally, in the present example, for any one-unit increase in the pretest score, the increase in odds of membership in the target group is estimated to be between 2.39 and 90.92 for the population, with 95% certainty. Other confidence intervals could be computed for a two-unit increase in pretest scores, a three-unit increase, and so on. Confidence intervals can be used for hypothesis testing here, as was the case for regression coefficients. However, if the 95% confidence interval contains the number 1, the odds ratio is nonsignificant at the 5% level of significance. This would mean that a change from one score to another does not reliably increase the odds of membership in the target group.

Table 1

Classification Table for Math Study

	Predicted group	
Actual group	**Failed**	**Passed**
Failed	24	2
Passed	2	26

Classification Analysis

A classification table summarizes the fit between the actual and predicted group memberships. The classification table for the math study is shown in Table 1. The number of cases in the *diagonal cells* (the failed–failed and passed–passed cells) is large compared with the number of cases in the other (*off-diagonal*) cells. This pattern indicates a good match between the observed outcomes and those predicted by the model. The overall percentage of cases correctly classified by the model, that is, the percentage accuracy in classification, or *PAC*, is 92.59%: the number of accurately classified cases (50) divided by the total number of cases classified (54).

Four other measures of classification accuracy are often computed. Suppose the task is to classify sick (the target group) and healthy (the other group) individuals. Moreover, assume that 80% of the sample is sick. If the model accurately classifies half of the target group, sensitivity is 50%. *Sensitivity* is the percentage of the target group accurately classified, also known as the correct identification of true positives. *Positive predictive value* is the percentage of individuals that the model classifies as belonging to the target group that are actually in the target group. *Specificity* refers to the percentage of the other group that is correctly classified, also know as the *correct identification of true negatives*. Thus, in this example, specificity is the percentage of the 20% healthy individuals correctly classified. *Negative predictive value* is the percentage of individuals that the model classifies in the other group that are actually in the other group. In practice, the positive and negative predictive value are of greatest concern because these indexes tell how useful the model is for making actual decisions. That is, these measures tell what percentage of the time the model's predictions are likely to be true.

In general, overall classification results should be interpreted with caution. One reason is that the correct logistic regression model doesn't always yield good classification for both groups (Hosmer & Lemeshow, 1989). For this reason, the average of the positive and negative predictive values, the *mean predictive values across classes*, is often of greatest interest to researchers whose goal is a model that predicts well for both groups. Also, best mean classification accuracy is typically obtained when the proportions of cases in each group are approximately equal. For example, 80% classification accuracy can be obtained without even formulating a logistic regression model if 80% of the cases are in the target group (you simply assign all cases to the target group—at the cost of misclassifying all 20% of the cases in the other group). Moreover, classification accuracy is usually lower in a new sample of cases than it was for the original sample. Ideally, a study whose goal is classification accuracy should fit the model on one group of cases and then apply the model to another group (called a *holdout sample* or *cross-validation sample*), to get an idea of the generalization of classification accuracy across samples.

Presentation of Results

Results from a study that uses logistic regression are typically presented in a format similar to that of linear regression studies. Generally, an overall goodness-of-fit measure is given: the log likelihood (*LL*), the deviance ($-2LL$), or the likelihood ratio (*LR*) for the model. Next, model coefficients are presented with test statistics (z or likelihood ratios), probabilities, and confidence intervals. (Sometimes odds ratios (e^b) are reported because they are easier to interpret than model coefficients.) Finally, a classification table and summary measures of classification accuracy (e.g., *PAC*, sensitivity) are given. However, presentation of results varies widely across studies; some studies may report only a few of the statistics described above.

Multivariable Models

Our analysis of the math course data found that students proficient in math were more likely to pass the course than were less proficient students. However, we didn't consider other variables that might be related to the odds of passing the course. For example, the chances of passing the course might increase for students who study a lot. Unless our analysis took into account the interrelationships between time spent studying, pretest score, and whether a student passed, we could not conclude that the reason successful students passed was that they knew more about math than the unsuccessful students. It might be that students with high pretest scores passed because they also happened to be the hardest workers.

However, the fact that 50 of 54 subjects (*PAC* = 92.59%) were correctly classified on the basis of a single predictor is quite impressive. Nevertheless, it may be possible to improve this classification performance—to correctly classify some or all of the remaining 7.4% of the sample who were misclassified—by using one or more additional predictor variables. The decision regarding which variables to select as predictors is one faced by anyone using a multivariable regression technique. The best approach for selecting predictors is based on knowledge of established relations between predictor and criterion variables reported in the literature. In addition, predictor variables are often selected because the researcher is guided by a theory that suggests relevant predictors of a dependent variable.

As in linear regression analysis, you can formulate a multivariable model to determine the relationship between a set of predictors (as a whole) and the odds of being in the target group, as well as the relationship between each predictor and the odds, while statistically adjusting for the other predictors in the model. The logistic regression model can contain continuous or categorical predictors, or both.

To illustrate logistic regression analysis with several predictors, I use data from a study of children's writing ability (Wright, 1992). In the study, 63 elementary school students wrote expository essays. On the basis of their essay performance, students were categorized as good writers or poor writers. Predictors of writing performance included a student's grade level (fourth, fifth, or seventh), sex, and score on a paragraph-reviewing task, a measure of the ability to identify errors in brief texts. The Reviewing scale has a range of 0 to 24 (24 was the highest possible score).

I test whether a student's ability to identify errors in texts is associated with good writing, when controlling for grade level and sex. I take into account grade level and sex for the analysis because it is possible that any relationship between error identification ability and writing skill will be spurious—perhaps students who identify the most errors, and who are the best writers, simply are the oldest students or are members of the same sex. This analysis will help determine whether a relationship exists between error detection and writing skill that is independent of grade level and sex. (However, grade level and sex might be of primary interest if our analysis had a different aim; for example, if we wanted to know whether children's writing skill and error detection ability improve with age or differ for males and females.)

We can include the two categorical predictor variables (sex and grade level) in the analysis if these variables are first converted into dummy-coded variables. Because sex is a dichotomous predictor, we treat one category (girls) as a baseline, or reference category, and the other category (boys) as our comparison group. By convention, the reference group is arbitrarily coded 0, and the comparison group is coded 1. To dummy-code grade level, which has three levels, I define two new variables, each comparing one grade level to fourth graders. (The number of dummy variables to be created equals the number of categories of the original variable minus 1.) Fourth grade, our reference category, is arbitrarily coded 0 for each new variable. One of the variables (Level 1) compares fifth graders (coded 1) with fourth graders. The other variable (Level 2) compares seventh graders (coded 1) to fourth graders. Note that this method of coding represents a contrast between grades. If it turns out that one or both of the levels are found to be significant predictors, it means that writing performance is related to the grade levels of the students. (Many statistical software programs can dummy code categorical variables automatically for the user.)

Here it is important to recall the prior discussion concerning the choice of G or z to evaluate the likelihood of the obtained results. For multivariable problems, only g can be used to test the overall model (i.e., the system of coefficients for all predictor variables), which is analogous to using F to evaluate an overall model in linear regression. However, either z or G can be used to evaluate individual coefficients in a multivariable model. Recall that when using the z test, one compares a model coefficient to its standard error; for an individual coefficient, G is analogous to partial F in multiple regression analysis.

Although we are interested primarily in the coefficient of the Reviewing scale, we can use an overall likelihood ratio test (G) to determine whether any predictor coefficient differs from zero. The degrees of freedom for the overall likelihood ratio test are 4, the number of predictor variables in the model. For the writing study, the likelihood ratio statistic (27.01) has a probability of less than .00005, which means that at least one of the population coefficients differs from zero (the overall test doesn't tell us which of the coefficients differ from zero).

Coefficients and z scores for the four-variable model are shown in Table 2. The z score for Reviewing is positive and has a probability of less than .05, so students with high Reviewing scores tend to be better writers than students with low Reviewing scores, when taking into account

Table 2

Results for Writing Study

Variable	*b*	*SE*	*z*	*p*	e^b	CI (odds)
Sex	0.46	0.65	−0.71	.479	0.63	0.17–2.32
Level 1	0.88	0.83	1.06	.289	2.42	0.46–12.68
Level 2	0.16	0.78	0.21	.837	1.17	0.25–5.58
Review	0.80	0.23	3.48	.001	2.22	1.40–3.53
Constant	−16.71					

Note. CI = confidence interval.

a student's grade level and sex. That is, there is a positive relationship between error identification and writing skill that does not depend on whether a student is in fourth grade, fifth grade, or seventh grade or on whether the student is male or female. Raising *e* to the power of the Reviewing coefficient gives the odds ratio for Reviewing when statistically adjusting for the other predictors in the model. Thus, the model estimates that when controlling for sex and grade level, the odds of passing increase by a factor of 2.22 for each one-unit increase in Reviewing score.

On the basis of the *z* tests in Table 2, there is insufficient evidence to reject the hypothesis that the population coefficient is 0 for sex and grade level. Thus, these results cannot be generalized to other samples, or to the population. Nevertheless, one may still examine these coefficients to uncover any hints of sex or grade level trends *in this particular sample.* For example, both grade level coefficients are positive, which means that when statistically controlling for the other predictors, fifth graders and seventh graders tended to be better writers than were fourth graders. (Our dummy-coding scheme doesn't compare fifth graders and seventh graders. However, this comparison can be made by using a different coding scheme.) The negative coefficient for sex means that girls (the reference category, dummy coded as 0) were better writers than were boys (coded as 1), when taking into account the other predictors in the model. If the coding scheme had been reversed, that is, if boys were coded 0 and girls were coded 1, the coefficient for sex would have been positive.

As for a single-predictor model, two steps are followed to compute a predicted probability for a multivariable model. First, compute $\hat{g}$ by multiplying each predictor by its coefficient, then these products and the constant term. From Table 2, $\hat{g}$ equals

$$-16.71 + 0.80\,(\text{Review}) + 0.46\,(\text{sex}) + 0.88\,(\text{Level 1}) + 0.16\,(\text{Level 2}).$$

Table 3

Classification Table for Writing Study

Actual group	Predicted group: Poor	Predicted group: Good
Poor	22	8
Good	6	27

(The basic method for computing $\hat{g}$ for the multivariable model is the same as that for the one-variable model, except that now there are more variables and coefficients to sum.)

Then $\hat{g}$ is inserted into the same formula as before to obtain the predicted probability:

$$e^{\hat{g}}/(1 + e^{\hat{g}}).$$

For example, let's compute the predicted probability for a male seventh grader who scored 24 on the Reviewing scale. Recall that boys are coded 1 for sex and that seventh graders are coded 1 on Level 2. Thus, for this student, $\hat{g}$ equals

$$-16.71 + 0.80(24) - 0.46(1) + 0.16(1) = 2.19$$

Thus, the predicted probability is

$$e^{2.19}/(1 + e^{2.19}), \text{ or } .90.$$

Because the predicted probability of passing is greater than .50, this student is classified as a good writer.

The classification table for the model is shown in Table 3. *PAC* for the sample is 77.78% [(22 + 27)/(22 + 6 + 8 + 27)]. The prior probabilities, 52% [(6 + 27)/(22 + 6 + 8 + 27)] for good writers and 48% [(22 + 8)/(22 + 6 + 8 + 7)] for poor writers, are roughly equal. Sensitivity is 81.82% [(27)/(6 + 27)], and specificity is 73.33% [(22)/(22 + 8)]. The positive predictive value is 77.1% [(27)/(27 + 8)], and the negative predictive value is 78.6% [(22)/(22 + 6)].

Research Examples

Logistic regression was used in two research studies to determine the relationship between a set of predictors and the probability of membership in a target group.

Table 4

Results for Severe Head Injury Study

Variable	*b*	*SE*	*z*	*p*	e^b	CI (odds)
Oculo.	0.08	0.62	0.13	.897	1.08	0.31–3.74
Size	0.86	0.70	1.23	.219	2.37	0.58–9.58
Response	2.18	0.65	3.35	.001	8.81	2.41–32.46
Constant	−1.50					

Note. CI = confidence interval. Oculo. = oculocephalic response, Size = pupil size, Response = pupil response to light.

Severe Head Injury

What is the prognosis for someone who has suffered a severe head injury? Stablein, Miller, Choi, and Becker (1980) used logistic regression to estimate the probability of death for 115 patients with severe head injury (patients whose outcome was permanent vegetative state were included among the deaths). Predictor variables included three dichotomous measures thought to indicate whether brain damage had occurred: pupil size, pupil response to light, and oculocephalic response. For each predictor, normal response (coded 0) was the reference category, and abnormal response (coded 1) was the comparison category. Stablein et al. tested whether persons with abnormal response on these measures were more likely to die than were patients with normal response on these variables.

The overall likelihood ratio statistic (G) for the three-variable model is 38.70, with 3 degrees of freedom. The probability associated with this statistic is less than .00005, indicating that at least one of the population coefficients differs from zero. Coefficients and related statistics are shown in Table 4. The coefficient for pupil light response differs from zero when taking into account the effects of other variables in the model. Thus, when controlling for the other predictors, patients with abnormal pupil light response are more likely to die after severe head injury than are patients with normal pupils. The odds ratio for pupil light response estimates that when pupil size and oculocephalic response are statistically adjusted for, the odds of death are 8.81 times greater for patients with abnormal pupil light response than for patients with normal pupil light response. Neither of the other model coefficients differs from zero.

Confidence intervals for odds ratios are also shown in Table 4. Recall that to compute a 95% confidence interval for an odds ratio, you first compute a 95% confidence interval for the logistic regression coefficient.

For example, for pupil light response, the lower confidence limit is 2.18 − 2(.65), or 0.88, and the upper limit is 2.18 + 2 (.65), or 3.48. Raising e to the power of the upper and lower limits gives a 95% confidence interval for the odds ratio: 2.41 to 32.46. Thus, with 95% confidence, one can infer that in the population, the odds of dying are between 2.41 and 32.46 times greater for a patient with abnormal pupil light response than for a normal patient.

Let's compute predicted probabilities of death for various combinations of predictor values. For a patient with normal responses on all predictors,

$$\hat{g} = -1.50 + 0.08(0) + 0.86(0) + 2.18(0) = -1.50.$$

Thus, the predicted probability of death for such a patient is

$$e^{-1.50}/(1 + e^{-1.50}) = .18.$$

Because the predicted probability is less than .50 and the target group is deaths, we classify this patient as a survivor. (In the sample, the actual percentage of survivors for patients with this combination of responses was 13/68, or .19.) For a patient with abnormal measurements on each predictor,

$$\hat{g} = -1.50 + 0.08(1) + 0.86(1) + 2.18(1) = 1.62.$$

The estimated probability of death is

$$e^{1.62}/(1 + e^{1.62}) = .83.$$

This patient is expected to die. (The actual percentage of deaths for this combination of responses was 17/18, or .94.)

In this example, we used the overall model to obtain predicted probabilities, even though not all predictors were statistically significant. This is a common practice, however, and finds justification, in a sense, by referring to the significant G obtained when evaluating the overall model. Some researchers, however, insist that only models for which all individual coefficients are significant should be used to obtain predicted values.

The classification table for the study by Stablein et al. (1980) is shown in Table 5. Overall, the model correctly classified 80% [(64 + 28)/(64 + 28 + 15 + 8)] of the cases. However, the prior probabilities, 37% for deaths [(15 + 28)/(15 + 28 + 64 + 8)] and 63% for survivors [(64 + 8)/(64 + 8 + 15 + 28)], were unequal. Sensitivity, 65.12% [(28)/(28 + 15)] is low compared with specificity, which is 88.89% [(64)/(64 + 8)]. The positive predictive value, computed for death, is 77.78% [(28)/(28 +

Table 5

Classification Table for Severe Head Injury Study

	Predicted group	
Actual group	**Survivor**	**Death**
Survivor	64	8
Death	15	28

8)]; the negative predictive value, computed for survival, is 81.01% [(64)/(64 + 15)].

The results from the Stablein et al. (1980) study indicate that abnormal pupil light response is positively associated with the likelihood of death after severe head injury. However, Stablein et al. followed patients for only 3 months after injury. If patients were followed for a longer period of time, the researchers might have obtained different model coefficients and predicted probabilities. Moreover, the eye data were recorded within an hour or so of admission to the hospital. The authors noted that more accurate probabilities might be obtained by including patient data obtained at a longer time interval after injury, such as follow-up assessments by a physician.

Hypnosis and Smoking Cessation

Smokers often try a variety of methods in their attempts to quit smoking. An increasingly popular method is self-hypnosis therapy, in which patients are taught to enter a state of hypnosis to change their perceptions about smoking. In Spiegel, Frischholz, Fleiss, and Spiegel's (1993) study, 223 patients received self-hypnosis treatment as part of a smoking cessation program. Overall, 23% of the participants quit smoking for 2 years or more. The authors used logistic regression to identify characteristics of individuals who were most likely to quit smoking after self-hypnosis.

Spiegel et al. (1993) reported a model containing three predictor variables. Two of the variables, induction score and self-rated hypnotizability, were continuous. High scores on these variables were thought to indicate that a patient was highly hypnotizable (induction scores ranged from 0 to 10; self-ratings ranged from 1 to 11). The third measure, social support, was a dichotomous variable, coded 1 if a patient lived with a spouse or partner and coded 0 if a patient lived alone. The authors predicted that highly hypnotizable patients would be more likely to quit

Table 6

Results for Hypnosis Study

Variable	*b*	*SE*	*z*	*p*	e^b	CI (odds)
Induction	1.40	0.33	4.24	.0001	4.05	2.14–7.69
Support	1.44	0.46	3.13	.003	4.22	1.71–10.36
Self-ratings	0.26	0.08	3.25	.003	1.29	1.10–1.52
Constant	−6.96					

Note. CI = confidence interval. From "Predictors of Smoking Abstinence Following a Single-Session Restructuring Intervention With Self-Hypnosis" by D. Spiegel, E. J. Frischholz, J. L. Fleiss, & H. Spiegel, 1993, *American Journal of Psychiatry, 150,* p. 1094. Copyright 1993 by the American Psychiatric Association. Adapted with permission.

smoking after self-hypnosis therapy than would patients who were less hypnotizable. The authors also hypothesized that social support would help a patient maintain his or her motivation to quit, so patients who lived with a spouse or partner would be more likely to quit smoking than would patients who lived alone.

Model coefficients and related statistics are shown in Table 6, which is adapted from Spiegel et al. (1993). (No *G* statistic or classification table was reported.) Examination of the *z* scores and their probabilities suggests that the population coefficient is probably greater than 0 for each predictor. Thus, when controlling for the other predictors in the model, each predictor is positively related to the odds of quitting for at least 2 years. The odds ratio for induction score estimated that the odds of quitting for 2 years would increase by a factor of 4.05 for each one-unit increase in induction score, when taking into account social support and self-rated hypnotizability. Similarly, the other predictors were adjusted for, the estimated odds of quitting were 4.22 times greater for subjects who live with a spouse or partner than for subjects who did not, and the odds of quitting increased by a factor of 1.29 for each one-unit increase in self-rated hypnotizability.

Using all of the model coefficients, $\hat{g}$ is computed as

$$-6.96 + 1.40\,(\text{induction}) + 1.44\,(\text{support}) + 0.26\,(\text{self}).$$

For a patient with the lowest scores on all predictors (that is, for a patient who had an induction score of 0, lived alone, and had a self-rated hypnotizability score of 1),

$$\hat{g} = -6.96 + 1.40(0) + 1.44(0) + 0.26(1) = -6.70.$$

Thus, the estimated probability of successfully quitting for such a patient is

$$e^{-6.70}/(1 + e^{-6.70}), \text{ which is effectively zero (.001).}$$

For a patient who had an induction score of 5, lived with a partner, and had a self-rated hypnotizability score of 2,

$$\hat{g} = -6.96 + 1.40(5) + 1.44(1) + 0.26(2) = 2.00.$$

Thus, the probability of quitting for 2 years or more is

$$e^{2.00}/(1 + e^{2.00}) = .88.$$

Spiegel et al. (1993) concluded that patient characteristics moderated the effect of the intervention: Highly hypnotizable individuals with an intact social support network were most likely to quit smoking after self-hypnosis therapy. However, as in the study by Stablein et al. (1980), the logistic model might differ if the study were conducted over a different time period. In fact, Spiegel et al. found that when using smoking status 1 week after the hypnosis treatment as the dependent variable, the hypnotizability measures and a motivational measure predicted smoking cessation, but social support did not. (As you might expect, the success rate, 52%, was also higher 1 week after treatment.)

Although the 2-year success rate (23%) in the study by Spiegel et al. (1993) is low, it is more than twice the 1-year abstinence rate (11%) found in a study of people who quit smoking on their own. Moreover, the authors used a strict criterion for smoking cessation: total abstinence. Patients who smoked even one cigarette were classified as recidivists. If the study used a more lenient criterion of success (e.g., a reduction in the number of cigarettes smoked per day), the overall success rate would have been higher. In addition, the patients received only a single-session hypnosis treatment. Although this helped reduce the cost of treatment, an intervention that provided several follow-up visits, and that incorporated other techniques for quitting smoking, might have a higher success rate.

Iterative Multivariable Logistic Regression

In applied research, one often has available too many variables to legitimately include in a single logistic model, on the basis of the sample size (recall that there should be at least 50 times as many subjects as predictor variables). *Iterative logistic regression* is a popular technique that is used to try to find a good subset of variables, that is, a subset that includes only statistically significant predictors and that results in good negative and positive predictive values. For example, if one has 200 cases, there should be no more than 4 (200/50) variables in the model. Thus, the researcher would use iterative logistic regression techniques to search for a good combination of 4 variables.

The most popular forms of iterative-model building strategies include *forward*, *backward*, and *stepwise logistic regression*. These are analogous to model-building procedures used in multiple regression.

In forward selection, variables are tested, one at a time, for entry into the model. The first variable added is the variable whose likelihood ratio is smallest among the statistically significant predictors. Other variables are added to the model if their likelihood ratios are also significant when taking into account the variables already in the model. Model building stops when all variables have been entered, or when the likelihood ratio is nonsignificant for all variables that have not been entered.

In backward logistic regression, the model starts with all predictors, whether or not they are statistically significant. Variables are tested, one at a time, for removal from the model. The first variable that is removed is the variable whose likelihood ratio statistic has the largest probability that is greater than alpha. The procedure continues to remove variables from the model until the model contains only variables that are statistically significant.

Stepwise logistic regression is a combination of forward and backward model building. Each variable is tested for entry into the model. Whenever a predictor is entered into the model, other variables in the model are tested for removal. The model-building process continues until no more variables can be entered or removed.

Note that these iterative procedures involve many tests of statistical hypotheses (i.e., tests of individual coefficients) and, therefore, dramatically increase the Type I error rate (e.g., see Multivariate Analysis of Variance chapter) for the overall study. In such analysis, cross-validation samples are highly recommended.

Summary

Logistic regression is a statistical procedure used to estimate the relationship between one or more predictor variables and the likelihood that an individual is a member of a particular group. The procedure also gives the probability associated with each prediction. Discriminant analysis (Klecka, 1980) can also be used to predict group membership, but it requires assumptions about the data that are more restrictive than those for logistic regression. For example, it requires that the predictors have a multivariate normal distribution for each category of the grouping variable and that each category have the same variances and covariances for the predictors. This implies that, in particular, discriminant analysis should not be used with categorical predictors. Nonetheless, logistic regression and discriminant analysis usually give similar results, except when the prior probability of membership in the target group is near 0 or 1 (Press & Wilson, 1978).

Logistic regression has many of the same strengths as linear regression. Both techniques determine whether scores on a dependent variable tend to increase or decrease for increasing values of a predictor. Moreover, hypothesis test and confidence intervals for model coefficients can be computed for both procedures, and iterative-model-building programs are available for both procedures, which attempt to find the best subset of a large set of predictor variables. However, only logistic regression is appropriate for analysis of a categorical dependent variable, because unlike linear regression, which can give predicted values that are negative or greater than 1, the predicted values from a logistic regression analysis can always be interpreted as probabilities of membership in the target group.

Unfortunately, logistic regression requires many of the same assumptions as linear regression analysis, including independence of observations, and a completely specified model—conditions that are often difficult to meet in practice. Furthermore, for hypothesis tests to be accurate, logistic regression requires large samples. Before drawing conclusions from a study that uses logistic regression, one should consider whether these assumptions have been met.

Suggestions for Further Reading

This chapter gives only an overview of logistic regression. Fortunately, a growing body of literature is available to help you to extend your knowledge of logistic regression. For introductory treatments that show how logistic regression is used in a variety of disciplines, see Fleiss, Williams, and Dubro (1986), who applied stepwise logistic regression to psychiatric data; Shott (1991), who discussed logistic regression in the context of veterinary medicine and who compared logistic regression and discriminant analysis; and Walsh (1987), who applied logistic regression to sociological data. For a comparison of logistic regression and linear regression, see Dwyer (1983) or Neter and Wasserman (1974). These introductory treatments require only a basic understanding of linear regression.

For a comprehensive treatment of logistic regression, see Aldrich and Nelson (1984) and Hosmer and Lemeshow (1989). These books include an introduction to polytomous logistic regression. Mathematical treatments of logistic regression, which require an understanding of matrix algebra, are given by McCullagh and Nelder (1989), who described logistic regression in the context of the general linear model; Press and Wilson (1978), who compared logistic regression and discriminant analysis; and Agresti (1990), who gave a general treatment of categorical data analysis.

Glossary

COEFFICIENT FOR THE PREDICTOR (b) Parameter estimate for a predictor variable. Controls the steepness and direction of the logistic regression curve.

CONSTANT TERM Parameter estimate that determines the location of the logistic regression curve on the X axis.

DEVIANCE ($-2LL$) -2 times the natural logarithm of the sample likelihood. Generally a positive number, the deviance decreases as the likelihood increases (when the likelihood is 1, the deviance is 0).

LIKELIHOOD (l) The probability of obtaining the actual group memberships given the values of the model coefficients. The likelihood ranges from 0 to 1.

LIKELIHOOD RATIO STATISTIC (G, G^2, χ^2) Also known as the *likelihood ratio chi-square statistic.* Used to test whether one or more model coefficients are different from zero. When the likelihood ratio statistic is large compared with its degrees of freedom, you reject the hypothesis that the population coefficients are 0. A likelihood ratio test is equivalent to an F test in linear regression.

LOG LIKELIHOOD (LL or L) The natural logarithm of the sample likelihood. Generally a negative number, the log likelihood increases as the likelihood increases (the log likelihood is 0 when the likelihood is 1).

MAXIMUM LIKELIHOOD METHOD Criterion generally used for selecting logistic regression coefficients. The coefficients maximize the probability of obtaining the actual group memberships for cases in the sample.

MAXIMUM LIKELIHOOD PARAMETER ESTIMATE (*mle*, *b*) Model coefficient chosen according to the maximum likelihood method. These coefficients have the highest likelihood of any coefficients.

MEAN PREDICTIVE VALUE ACROSS CLASSES The average of the positive and negative predictive values. Often the most important measure to researchers whose goal is a model that predicts well for both groups.

NEGATIVE PREDICTIVE VALUE The proportion of cases predicted by the model to be in the other group that really were members of the other group.

ODDS The probability of membership in the target group divided by the probability of membership in the other group. The odds are always 0 or greater, and are 1 when both groups are equally likely.

ODDS RATIO (e^b, OR, ψ) The quantity e raised to the power of a model coefficient. Estimates the increase in odds of membership in the target group for a one-unit increase in the predictor while controlling for other predictors in the model.

PAC Percentage of cases correctly classified by the logistic model. Measures the goodness of fit between the actual group memberships and those predicted by the model.

POSITIVE PREDICTIVE VALUE The proportion of cases predicted by the model to be in the target group that really were members of the target group.

PREDICTED PROBABILITY The model's estimate of the probability that an individual is a member of the target group.

RESIDUAL Difference between the actual outcome (0 or 1) and the predicted probability. A small residual (in absolute value) means that the model fits well for a case.

SENSITIVITY Percentage of the cases in the target group correctly classified by the model, also known as the *correct identification of true positives*. A measure of classification accuracy.

SPECIFICITY The percentage of cases in the "other" group correctly classified by the model, also known as the *correct identification of true negatives*. A measure of classification accuracy.

SPECIFICITY ASSUMPTION Requirement that the logistic regression model contain all relevant predictors and no irrelevant predictors. If the model is incorrectly specified, parameter estimates will be inaccurate.

STANDARD ERROR (*SE*) Estimates the variability from sample to sample in a model coefficient. Used in computing z scores and confidence intervals.

z TEST Tests whether a model coefficient differs from zero in the population. Large z values (in absolute value) mean that the population coefficient probably differs from 0. To compute z, divide a parameter estimate by its standard error.

References

Agresti, A. (1990). *Categorical data analysis.* Wiley: New York.

Aldrich, J. H., & Nelson, F. D. (1984). *Linear probability, logit, and probit models.* Beverly Hills, CA: Sage.

Dwyer, J. H. (1983). *Statistical models for the social and behavioral sciences.* New York: Oxford Press.

Fleiss, J. L., Williams, J. B. W., & Dubro, A. F. (1986). Logistic regression of psychiatric data. *Journal of Psychiatric Research, 20,* 145–209.

Hauck, W. W., & Donner, A. (1977). Wald's test as applied to hypotheses in logit analysis. *Journal of the American Statistical Association, 72,* 851–853.

Hosmer, D. W., & Lemeshow, S. (1989). *Applied logistic regression.* New York: Wiley.

Klecka, W. R. (1980). *Discriminant analysis.* Beverly Hills, CA: Sage.

McCullagh, P., & Nelder, J. A. (1989). *Generalized linear models.* New York: Chapman and Hall.

Neter, J., & Wasserman, W. (1974). *Applied linear statistical models.* Homewood, IL: Irwin.

Press, S. J., & Wilson, S. (1978). Choosing between logistic regression and discriminant analysis. *Journal of the American Statistical Association, 73,* 699–705.

Shott, S. (1991). Logistic regression and discriminant analysis. *Journal of the American Veterinary Medical Association, 198,* 1902–1905.

Spiegel, D., Frischholz, E. J., Fleiss, J. L., & Spiegel, H. (1993). Predictors of smoking abstinence following a single-session restructuring intervention with self-hynosis. *American Journal of Psychiatry, 150,* 1090-1097.

Stablein, D. M., Miller, J. D., Choi, S. C., & Becker, D. P. (1980). Statistical methods for determining prognosis in severe head injury. *Neurosurgery, 6,* 243–246.

Walsh, A. (1987). Teaching understanding and interpretation of logit regression. *Teaching Sociology, 15,* 178–183.

Wright, R. (1992). *Semantic evaluation criteria and the expository writing skills of grade school children.* Unpublished doctoral dissertation, University of Illinois at Chicago.

8 Multivariate Analysis of Variance

Kevin P. Weinfurt

Maybe you have had this dream before: You are earnestly reading a research article you need to digest in order to work on a paper, grant application, or thesis. You successfully wade through the jargon found in the Introduction and Method sections and reach the climax of the article, the Results section, where you are confronted with something like this[1]:

> A 4 (Group) × 2 (Time) repeated measures multivariate analysis of variance revealed a significant multivariate main effect for Time [Wilks's Λ = .406, F (4, 113) = 41.28, $p < .001$, η^2 = .59], but no significant effect for Group [Wilks's Λ = .865, F (12, 229) = 1.41, $p > .05$] or the Group × Time interaction [Wilks's Λ = .856, F (12, 299) = 1.51, $p > .05$].

Although it may seem intimidating, such text need not be the stuff of nightmares. In short, multivariate analysis of variance (MANOVA) is used to assess the statistical significance of the effect of one or more independent variables on a set of two or more dependent variables.

I begin this chapter with an example of a research situation in which a MANOVA is used, followed by a discussion of some basic statistical concepts and the general purpose of a MANOVA. The assumptions underlying a MANOVA, as well as the consequences of violating those assumptions, are discussed next. Following this is a brief, nontechnical

1. This tongue-in-cheek paragraph was based on the Results section of an actual article published in a legitimate psychology journal, with very few modifications made. Unfortunately, the difference between exaggeration and reality is small in this case.

explanation of how a MANOVA is performed and how it relates to a traditional analysis of variance (ANOVA). Next, the ways in which MANOVA results are presented in journal articles and various analytical methods that often follow multivariate significance are reviewed. The latter sections of the chapter are devoted to multivariate analysis of covariance (MANCOVA), repeated measures MANOVA, and power analysis. I also provide a list of recommended readings and a glossary of terms and symbols used throughout the chapter.

A Hypothetical MANOVA Design

Imagine that a clinical psychologist interested in panic disorders (see Antony, Brown, & Barlow, 1992, for a recent review) designed and published the following study. One hundred subjects suffering from panic disorder were solicited to participate in a program to improve their condition. The author of the study wanted to determine the effectiveness of two different forms of therapy: relaxation training and cognitive–behavioral therapy. Subjects were randomly assigned to one of four groups: relaxation training only (relaxation), cognitive–behavioral therapy only (cog–behav), relaxation and cognitive–behavioral therapies combined (combined), and a control group with no therapy (control). Thus, the study is a 2 (relaxation) × 2 (cog–behav) factorial design (Campbell & Stanley, 1963). On the basis of a structured interview, the clinician assigned each subject a rating from 1 to 10 to index the severity of each subject's disorder at the outset of the study, with higher scores reflecting greater severity of the condition. Subjects participated in the program for 8 weeks, after which they were tested using three subscales of a panic disorder questionnaire. The subscales tapped different theoretical components of panic disorder: cognitive, emotional, and physiological. Each scale was scored on a 1–20 scale and then converted to standard scores, with higher scores indicating greater intensity of panic disorder. Table 1 displays each group's mean and standard deviation for the three panic subscales. This example is referred to throughout the chapter to demonstrate various aspects of a MANOVA.

Some Preliminary Statistical Concepts

Knowledge of a few key statistical concepts is required to understand MANOVA and issued related to its interpretation. They are Type I error, the Bonferroni inequality, effect size, and statistical power.

Table 1

Mean Hypothetical Standard Scores on the Cognitive, Emotional, and Physiological Subscales for the Four Experimental Groups

	Cognitive		Emotional		Physiological	
Group	*M*	*SD*	*M*	*SD*	*M*	*SD*
Control	1.75	1.10	1.63	1.11	1.44	0.90
Relaxation	−0.19	0.66	−0.07	1.07	0.56	1.20
Cog–behav	0.03	0.92	0.33	0.90	0.20	1.02
Combined	0.45	0.96	−0.18	0.89	0.08	0.97

Note. There were 25 participants in each group. Cog–behav = cognitive–behavioral.

Type I Error

The Type I error rate, expressed as alpha (α), is the probability of rejecting the null hypothesis when it is true. In other words, it is the probability of detecting a significant effect when there is no real effect in nature (Kleinbaum, Kupper, & Muller, 1988). Statisticians speak of two types of alphas: actual and nominal. The actual alpha level is the actual (true) probability of making a Type I error. The nominal alpha level is the Type I error rate that the researcher desires (e.g., .05). The nominal alpha level is based on assumptions, and if these assumptions are tenable, the nominal alpha is equal to the actual alpha. However, if the assumptions are violated to some degree, the actual alpha may be different from the nominal alpha. Hence, it is important to evaluate a MANOVA's assumptions to better determine what the actual alpha might be in a given situation.

One can also speak of alphas at three different levels of analysis. Most researchers are familiar with the alpha for a single statistical test (e.g., a *t* test), indicating the probability of falsely rejecting the null hypothesis for that particular test of significance. There is also a *familywise alpha* (or comparisonwise alpha; Ryan, 1959) in situations in which multiple levels of an independent variable are being compared with respect to a dependent variable, as in a one-way ANOVA. The familywise alpha is the probability of falsely rejecting the null hypothesis for at least one of the statistical comparisons being made. The comparisons are considered a "family" of tests. There are many multiple-comparison procedures available for this situation, including the Tukey, Sheffé, the least significant difference, Newman-Keuls, and Duncan's multiple range (Keselman, Keselman, & Games, 1991; Kleinbaum et al., 1988; Seaman, Levin, & Serlin, 1991).

Finally, there is the *experimentwise alpha*, which is the probability of falsely rejecting the null hypothesis for at least one statistical test when several tests are used in the same study. Hence, in a study with two ANOVAs, the experimentwise alpha would be the probability of obtaining erroneously significant results for any test performed in the study, including the ANOVA omnibus tests, as well as any analysis of main and interaction effects done for each ANOVA (see the Glossary for definitions of main and interaction effects).

The Bonferroni Inequality

The Bonferroni inequality (Stevens, 1986) is an extremely important concept for understanding the familywise and experimentwise alphas, because it defines the maximum value of alpha for a given set of statistical tests. Basically, the Bonferroni inequality states that the overall alpha for a set of tests will be less than or equal to the sum of the alpha levels associated with each individual test. Therefore, if six *t* tests are performed using .05 as the criterion for rejecting the null hypothesis for each, then the overall alpha level is approximately 6(0.5) = .30. (Actually, the overall alpha is .265; see Maxwell & Delaney, 1990.) Hence, there is a 30% chance of committing at least one Type I error across the six *t* tests.

When performing MANOVAs, there are often many statistical tests involved. If the null hypothesis is correct in every case, then the more tests one performs, the greater the possibility of committing a Type I error. In other words, the experimentwise alpha increases. When reading a research article, be aware of how many tests are being performed, the alpha level used for each test, and what the maximum experimentwise alpha could be using the Bonferroni inequality. It is surprising how many studies are published that base their conclusions on "significant" statistical results, when in actuality the experimentwise alpha level is so high that erroneous results are almost guaranteed.

Effect Size

In addition to determining whether a finding could have occurred by chance, it is also useful to know the magnitude of a finding. For example, consider a simple pretest–posttest study with some intervening treatment between testing periods. A paired *t* test could be used to ascertain whether the mean pretest score differed from the mean posttest score at a statistically significant level. One might also ask how large the difference between the two means is. In other words, what is the magnitude of the intervening treatment's effect? This is precisely the issue addressed by the notion of *effect size*.

Typically, an effect size is expressed as a number between 0 and 1, with higher values reflecting a larger effect. As discussed later, one measure of effect size in MANOVA is eta-square, which is roughly equivalent to the R^2 used in multiple regression (see chapter 2 in this book; also see Pedhazur, 1982). Cohen's (1977) classification of effect sizes has become somewhat of a standard in social research. The most often quoted standard refers to effect sizes measured as mean differences (as in the pretest–posttest example given earlier): Effect sizes around 0.20 are small, those around 0.50 are medium, and those larger than 0.80 are large. Although this classification scheme is often used to judge the magnitude of eta-squares, the more proper standard should be the one Cohen suggested for effect sizes measured via R^2 or other such indices: 0.01 is small, 0.09 is medium, and 0.25 or greater is large. The majority of social research produces small to medium effect sizes.

Statistical Power

Power is the probability of detecting a significant effect when the effect truly does exist in nature. A simple metaphor for understanding power is that of a flashlight. If a statistical test is a flashlight shining in the dark, then power is the brightness of the beam. One is able to see more of what is really there with a more powerful beam. Power is a function of sample size, effect size, and the nominal alpha level set by the researcher (Cohen, 1977). The larger the sample, the more power there is. The bigger the effect as it exists in nature, the greater the power to detect it. Power is expressed as a number ranging from zero to one, with zero indicating no power and one indicating perfect power. Power is an important notion when a MANOVA is concerned, because it is helpful to know how able a MANOVA is to reject the null hypothesis for a given sample size and effect size. In a research situation, if there is low power and no significant effects are obtained, then there is reason to believe that effects might truly exist in nature, but they are undetectable because of the research design or number of subjects. I say more about the power of a MANOVA in a later section.

MANOVA Basics

The Purpose of a MANOVA

Perhaps the best way to begin a discussion about a MANOVA is to consider its more familiar relative, the univariate ANOVA. In the ANOVA there is a continuous dependent variable (e.g., IQ) and one or more categorical independent variables (e.g., social class, gender). The purpose of the ANOVA is to determine whether the means of the dependent variable for each level of an independent variable are significantly different from each other. Therefore, if the independent variable social class has three levels (lower, middle, and upper), an ANOVA could tell whether one social class had a higher mean IQ than another class. This relationship between the independent and dependent variables is sometimes referred to as the "influence" or the "effect" of the independent variable on the dependent variable, even though the direction of the causal chain may not necessarily flow from independent to dependent variable. If a second independent variable was added to the model—gender, for instance—a 3 (social class) × 2 (gender) ANOVA could be performed to determine the influence of social class on IQ (as before), the influence of gender on IQ (e.g., Is there a difference between males and females with regard to IQ?), and the joint influence of social class and gender (e.g., Does the difference between the mean male and mean female IQ depend on the level of social class?). This last variable describes a statistical interaction. An interaction addresses whether the influence of one independent variable is altered by the level of another independent variable.

The MANOVA allows one to examine the effects of the independent variables in much the same way as the univariate ANOVA, including main effects, interaction effects, contrast analyses, covariates, and repeated measure effects. The meanings of these concepts are dealt with shortly. For now, note that the ANOVA is confined to situations in which there is only one dependent variable, hence the term *univariate analysis of variance. Multivariate analysis of variance* is a technique used for situations in which there is more than one dependent variable. In addition to having more than one dependent variable, the MANOVA design also requires that the dependent measures be correlated. If the variables being used are statistically correlated with one another, then there is an empirical relation between them, and they could be analyzed using a MANOVA.

Table 2

Correlations Between the Dependent Measures for the Hypothetical Panic Disorder Study

Subscale	Cognitive	Emotional	Physiological
Cognitive	—		
Emotional	.36	—	
Physiological	.25	.21	—

Note. $N = 100$. All $ps < .05$.

Ideally, the dependent variables should be theoretically correlated as well as empirically correlated. In the panic disorder example, there are three dependent variables (i.e., the three subscales of the panic disorder questionnaire). They are theoretically related in that they all assess panic disorder. Also, the variables are intercorrelated, as Table 2 illustrates, suggesting that this variable set may be a candidate for a MANOVA.

Why Multivariate Analyses?

At this point, a question may arise as to why a researcher would want to analyze two or more dependent variables at once. Why not use separate univariate tests for each dependent variable and avoid confusing people? Researchers in the social and biological sciences usually have one of two reasons for using a multivariate approach: controlling Type I error and providing a multivariate analysis of effects by taking into account the correlations between dependent measures.

Controlling Against Type I Error

As Huberty and Morris's (1989) survey demonstrates, MANOVAs are most often done with the intent of keeping the Type I error rate at the nominal alpha level. One school of thought, supported by Hummel and Sligo's (1971) research, maintains that a MANOVA should be conducted when there are multiple dependent variables, and if the multivariate test is significant, then univariate ANOVAs are conducted for each of the dependent measures. The theory is that by performing an overall omnibus test of significance first—the MANOVA—one is guarding against the chance of committing a Type I error that might occur as a result of unwarranted multiple ANOVAs. In recent years, however, this approach has come under attack (Huberty & Morris, 1989; Wilkinson, 1975), and these criticisms are reviewed later when discussing follow-up procedures.

Providing a Multivariate Analysis of Effects

If there are multiple dependent measures and they are intercorrelated, then the intercorrelations can be taken into account to provide a much richer multivariate analysis of the data. This is true for two reasons. First, intercorrelations between outcome measures suggest that the measures may be partially redundant. That is, there may a degree of conceptual overlapping. As an example, suppose that a psychologist performed a study on affect intensity (Larsen & Diener, 1987) that used the Affect Intensity Measure (AIM; Larsen, 1984) and galvanic skin response (GSR) as two of several dependent measures. Because scores on the AIM are positively correlated with GSR to emotional stimuli, it is reasonable to suppose that the two measures are tapping the same concept. Hence, if separate ANOVAs for each measure were performed and each produced significant results, could it really be said that significance was obtained for two completely separate dependent variables? Although there certainly were two separate measures, it is questionable whether the two tapped conceptually separate constructs. A MANOVA avoids this question by taking the correlations between the dependent measures into consideration. As long as the effects being tested are multivariate effects (i.e., taking all dependent variables at once), there will be no redundant information in the results of the MANOVA.

A second reason why a multivariate approach can offer a richer analysis of the data is that a MANOVA can detect when groups differ on a system of variables (Huberty & Morris, 1989). Taken individually, the dependent variables may not show significant group differences, but taken as a whole—as a system defining one or more theoretical constructs—differences caused by the independent variables may be revealed. This is accomplished by finding a *linear composite* of the dependent measures that maximizes the separation between the groups defined by the independent variable, resulting in the most statistically significant value of the MANOVA test statistic. A linear composite refers to some combination of the dependent measures (e.g., .46AIM + .24GSR − .10gender, etc.). This demonstrates how univariate tests using the AIM and GSR might not yield significance, but a linear combination of the dependent variable would (see Stevens, 1986). By examining such dependent variable systems, a researcher is sometimes better able to discern the effects of independent variables.

To summarize, MANOVA is a procedure used to test the significance of the effects of one or more categorical independent variables on two or more continuous dependent variables. In the panic disorder study, the researcher wants to test the effect of relaxation training and cognitive–behavioral therapy (two categorical independent variables) on the cognitive, emotional, and physiological subscales of the panic questionnaire (three continuous dependent variables). Having defined the MANOVA and some of the reasons it is used, I review the assumptions inherent in the MANOVA.

Assumptions of the MANOVA

All parametric statistical procedures are inferential procedures (i.e., they make inferences about populations). Mathematics and logic dictate that inferences be based on assumptions, and so like any other parametric statistical technique, the MANOVA has assumptions with which scientists must be concerned. These assumptions are not esoteric mathematic theory but conditions of the data that must be assessed (and hopefully satisfied) before trying to interpret the results of a MANOVA. The three necessary conditions are (a) multivariate normality, (b) homogeneity of the covariance matrices, and (c) independence of observations. For each of the three assumptions, I present the ANOVA analog to that assumption, a definition of the assumption, how it is tested, and what effect violating the assumption will have on the Type I error rate and the statistical power of the MANOVA.

Multivariate Normality

In the case of univariate ANOVA, the statistical tests are based on the assumption that the observations on the dependent variable be normally distributed for each group defined by the independent variables. Multivariate normality is a much harder criterion to satisfy than univariate normality. This is clearly seen when one considers some properties of data that are multivariate normal (Stevens, 1986). First, all of the individual dependent variables must be distributed normally. Second, any linear combination of the dependent variables must also be normally distributed. Finally, all subsets of the variables must have a multivariate normal distribution. Unfortunately, most researchers fail to report how well their data approximate a multivariate normal distribution, and so one will seldom find anything in an article using MANOVA that mentions this assumption. Still, there are procedures for assessing multivariate normality, as discussed by Stevens (1986, chapter 6). Because space limitations prevent a detailed discussion regarding these techniques, the interested reader should consult Stevens's thorough and systematic treatment of this topic.

What if the distribution of the dependent measures is not multivariate normal? In terms of Type I error rate, the MANOVA appears to be fairly robust. That is, violation of the multivariate normality assumption has a small effect on the actual alpha level with which the researcher is working (Stevens, 1986). With regard to statistical power, Olson (1974) found that extremely platykurtic distributions (i.e., those that are flat compared with the more peaked normal distribution) can reduce the power of the analysis. In practice, MANOVAs tend to be performed on data regardless of whether the data violate this assumption, because the general consensus is that the MANOVA is a robust procedure. How correct this belief is can be determined only by more studies on the MANOVA's ability to withstand violations of the multivariate normality condition.

Homogeneity of the Covariance Matrices

Univariate ANOVA requires that the variance of the dependent variable be the same for all groups defined by the independent variables. In the multivariate context, the variances for all of the dependent variables must be equal across the experimental groups defined by the independent variable. Additionally, MANOVA demands that the covariance—the variance shared between two variables—for all unique pairs of dependent measures be equal for all experimental groups. Table 3 shows the components of a covariance matrix for an experimental group in a study with three dependent measures (A, B, and C). Observe that each variable's variance is on the diagonal, and the covariance for each pair of variables make up the rest of the matrix. Each of these numbers is referred to as an element of the matrix; therefore, this matrix has nine elements.

Note also that this is a covariance matrix for only one experimental group. In order to say the covariance matrices are homogeneous across experimental groups, the covariance matrices for each group are exam-

Table 3

Components of a 3 × 3 Covariance Matrix (Three Dependent Variables) for One Experimental Group

Variable	A	B	C
A	Variance A	Covariance AB	Covariance AC
B	Covariance BA	Variance B	Covariance BC
C	Covariance CA	Covariance CB	Variance C

ined and checked to see that each matrix element (e.g., Covariance AB, Variance A, Variance B, etc.) is equal for all of the groups. To put this into more concrete terms, consider the hypothetical panic disorder data. There are three dependent measures, and so the covariance matrix for each group will be a 3 × 3 table of variances and covariances. Table 4 shows the covariance matrices for the four experimental groups that are formed when the two independent variables (relax and cog–behav) are crossed.

In examining Table 4, one might ask whether the variance for the

Table 4

Within-Groups Covariance Matrices for the Hypothetical Panic Disorder Study

Group	Cognitive	Emotional	Physiological
Control			
Cognitive	1.22	0.10	0.24
Emotional	0.10	1.25	−0.07
Physiological	0.24	−0.07	0.82
Relax			
Cognitive	0.44	0.05	−0.02
Emotional	0.05	1.13	0.24
Physiological	−0.02	0.24	1.44
Cog–behav			
Cognitive	0.85	0.02	−0.34
Emotional	0.02	0.80	−0.29
Physiological	−0.34	−0.29	1.05
Combined			
Cognitive	0.92	−0.01	0.16
Emotional	−0.01	0.78	−0.09
Physiological	0.16	−0.09	0.94

Note. Cog–behav = cognitive–behavioral.

cognitive subscale is significantly different across the four groups. (Observe that these variances are 1.22, 0.44, 0.85, and 0.92.) Moreover, it would be desirable to know whether the matrices of variances and covariances are equal for all four groups. Two indices used to test the homogeneity of covariance matrices are Box's *M* (Norusis, 1988) and Bartlett's chi-square (Green, 1978). Both tests are highly sensitive to violations of the multivariate normality assumption, and it is therefore recommended that the tenability of that assumption be thoroughly investigated. For illustration, assume that the panic disorder data conform reasonably well to the normality requirement, and so the results of the chi-square are presented here so that the reader might recognize such a test. The null hypothesis is that the groups have equal covariance matrices. Therefore, if the test yields statistical significance, the groups are not homogeneous with respect to covariance matrices. For the panic disorder study, Bartlett's test produces the following: χ^2 (18, N = 100) = 18.83, $p > .05$. Hence, it can be concluded that the panic data satisfy the homogeneity of covariance matrices assumption.

Now consider what happens when this assumption is not tenable. Summarizing the studies done on the subject, Stevens (1986) concluded that violation of the homogeneity of covariance matrices assumption when the number of subjects in each experimental group is approximately equal will lead to a slight reduction in statistical power. For cases in which the number of subjects in each group is markedly unequal, the Type I error rate will be either inflated or deflated depending on which matrices are the most different (see Stevens, 1986, p. 227, for more detail).

Independence of Observations

The last and most important assumption is exactly the same for a MANOVA as it is for a univariate ANOVA. Both procedures assume that observations are independent of one another. This means that a subject's scores on the dependent measures are not influenced by the other subjects in his or her experimental group. Hence, if a study involved having subjects interact in an experimental condition, it is possible that the subjects are affecting each other's scores. As an example, consider what might happen if schoolchildren were asked to answer survey questions aloud in a small group setting. In almost every group of children, there will be those who simply follow the lead of others; therefore, one would expect that these more passive children's responses to the survey questions would be influenced by a more dominant child's answers.

Such dependence among observations can be assessed by an *intraclass correlation* (Fleiss, 1986; Guilford, 1965). Stevens (1986) noted that even a small intraclass correlation—indicating a small degree of dependence among observations—can inflate the actual alpha to seven times the nominal alpha level the experimenter observes. Hence, although it is possible to speak of a MANOVA's relative robustness with regard to the first two assumptions, there is little room for violation of this last assumption. The critical journal reader can determine whether the tenability of the assumption is suspect by looking at the research methods. Were the experimental conditions administered on an individual basis or to a group? Could other subjects have affected a person's responses at the time the dependent variables were measured? In the panic disorder study, there is little chance of dependence among observations, because the therapies were individual therapies. If the relax and cog–behav programs were delivered in a group therapy context, then it would be wise to investigate the assumed independence of observations with an intraclass correlation. If dependence was found, then the group means would be the logical units of analysis, not the individual scores. Techniques for analyzing dependent data are being developed. (For data of dyads, e.g., see Mendoza and Graziano, 1982.)

The MANOVA Procedure

Having reviewed the conditions necessary for a MANOVA, I now present a conceptual discussion of how a MANOVA is performed, beginning with the null hypothesis that is tested in MANOVA and how it relates to the null hypothesis tested in univariate ANOVA. Then I show how matrix algebra is used in a MANOVA to carry out the same basic procedure that takes place in a univariate ANOVA. The various multivariate test statistics are then introduced, as well as the measure of explained variance, eta-square.

Null Hypothesis Testing

To begin, consider the simplest version of a univariate ANOVA, the t test for independent samples (Kleinbaum et al., 1988). The t test is used when two groups are compared on a single continuous dependent variable. As an example, imagine that 40 males and 40 females took the revised Wechsler Adult Intelligence Scale (WAIS–R; Kaplan & Saccuzzo,

1989) and that a verbal IQ score was obtained for each subject. A *t* test could be performed to test the null hypothesis that the mean IQ for the male population is equal to the mean IQ for the female population. If the *t* value exceeds the appropriate critical value, then the means for the two populations are not equal. Note that two single means (average male IQ and average female IQ) are compared.

What if a researcher were interested in comparing males and females on the six verbal subscales of the WAIS–R and wanted to do so in a multivariate analysis to account for the correlations between the six scales? Then one is no longer comparing two means. Rather, two sets of six means are being compared. The formal name for such a set is a *vector*. In this example, the male and female groups each have a vector of means consisting of the mean scores for each of the six WAIS–R subscales. Thus, the null hypothesis being tested is that the vector of means for the male population is equal to the vector of means for the female population. MANOVA will produce a test statistic (in this case the multivariate version of a *t* test, Hotelling's T^2) that will be compared with a critical value to obtain a significance level. If that probability is below the predetermined criterion for significance (e.g., $p < .05$), then it is concluded that the male and female populations have unequal vectors of means. Therefore, the important difference to remember between a univariate null hypothesis and a multivariate null hypothesis is that the univariate hypothesis considers single means, whereas the multivariate hypothesis considers vectors of means.

Just as a univariate ANOVA can test hypotheses concerning the effects of multiple independent variables, so too can a MANOVA. The null hypothesis being tested in a MANOVA is the same as it would be for an ANOVA, except that individual means are replaced with vectors of means. In the 2 (relax) × 2 (cog–behav) panic disorder study, there are two main effects and one interaction effect that can be tested. The null hypothesis for the first main effect is that the vector of the three subscale means (cognitive, emotional, and physiological) for the people in relaxation training is equal to the vector of means for those who are not in relaxation training. The null hypothesis for the cog–behav main effect is that the vector of subscale means for the people in the cognitive–behavioral therapy group will be same as the vector of means for those not in the cognitive–behavioral group. Finally, the null hypothesis for the interaction effect is that the four groups defined by crossing relax and cog–behav will have equal vectors.

Calculating MANOVA Test Statistics

To explain how a MANOVA test statistic is computed, it would be necessary to cover complex matrix operations and countless equations. Such an exposition is beyond the scope of this chapter, and I therefore limit this section to explaining how, at a conceptual level, the derivation of a MANOVA test statistic is the same as the derivation of an ANOVA test statistic. A natural way to start is to review how a univariate ANOVA is performed.

An ANOVA is an examination of means and mean differences, and the logic of the analysis is fairly straightforward. To begin, it is clear that not everyone has the exact same score on a dependent measure. If everyone did have the same score, say a 6.8 out of 10, then the overall (or *grand*) mean would be 6.8 with variance of zero. When not everyone scores exactly the same, then it is possible to express each person's score as a *deviation* from the grand mean. Statistical variance is a function of the sum of the squares of these deviations from a mean for an entire group. This is referred to as *sum of squares* (*SS*).

The ANOVA seeks to determine how much total variation can be explained by the variables in an experiment by dividing the total variance into two parts: (a) the variance attributable to the variables in the study and (b) the variance attributable to variables not included in the study, or "error." These variance components are called the *between sums of squares* (SS_{between}) and the *within sums of squares* (SS_{within}), respectively. Recall that sums of squares refer to the sum of the squared deviations about a mean, which is the primary component that is used to compute variance. One can readily see that the elements of an ANOVA are expressed in terms of differences, differences between an individual's score and the grand mean (SS_{total}), between an individual's score and his or her group's mean (SS_{within}), and between a group mean and the grand mean (SS_{between}). Hence, ANOVA tests whether the amount of variance explained by the independent variable (SS_{between}) is a significant proportion relative to the variance that has not been explained (SS_{within}).

In MANOVA, the same approach is taken, but the sum of squares is replaced with sum of the square and cross-product (SSCP) matrices. An SSCP matrix is similar to the matrix discussed when reviewing the homogeneity of covariance matrices assumption. An SSCP matrix consists of the sums of the squares (representing variances) for every dependent variable along the diagonal and cross-products (representing covariances) taking up the off-diagonal elements (recall that covariance is the amount

Table 5

Relationships Between Some Primary MANOVA Components

Effect size	B	W	Λ	*p*	η^2
No effect	Smaller	Larger	Larger	Larger	Smaller
Effect present	Larger	Smaller	Smaller	Smaller	Larger

Note. B = between-groups sum of squares cross-product (SSCP) matrix; W = within-groups SSCP matrix; *p* value = the probability of obtaining a particular value of Wilks's lambda; η^2 = the proportion of variance explained. MANOVA = multivariate analysis of variance.

of common variance shared by two variables). Just as a univariate ANOVA posits a total sum of squares and partitions it into between and within sums of squares, the MANOVA posits similar components, but in matrix form. Thus, there is a total (T) SSCP matrix that is divided into a within-groups (W) SSCP matrix and a between-groups (B) SSCP matrix. Matrix algebra allows one to derive a single number that expresses the amount of generalized variance (the variability present in a set of variables) for a particular matrix, call the *determinant*. Hence, one is able to compare the generalized variance of one matrix with another.

What kind of relationship should exist between T, B, and W if there is a significant effect? Because B addresses the amount of variance and covariance the effects can explain, and W is the variance and covariance remaining, then one would expect B to have a larger generalized variance than W. By the same token, one would expect W to be smaller when an effect is present. It is from the latter point that the oldest and most popular multivariate test statistic, Wilks's lambda (Λ), is derived (Tatsuoka, 1971). Basically, Wilks's lambda is a ratio of W to T. Hence, when lambda is small, the variance not explained by the independent variables is small. In order to determine how statistically significant lambda is, it is transformed into either an *F* or a chi-square statistic (Pedhazur, 1982).

If lambda is the proportion of variance not explained, then it follows that $1 - \Lambda$ is the proportion of variance explained by the independent variable's effect. This index is called eta-square (η^2; Huberty & Smith, 1982) and is analogous to other measures of explained variance such as R^2 in multiple regression (Pedhazur, 1982). Unlike some formulations of R^2, eta-squares for several variables are not additive (i.e., they will not add up to one), because different linear composites of the dependent variables are used to determine the effect of each independent variable. Table 5 demonstrates the relationship between the size of an independent

Table 6

2 × 2 MANOVA Results From the Hypothetical Panic Disorder Data

Effect	Λ	F^a	*df*	*p*	η^2
Relaxation	0.65	17.21	3, 94	.0001	.35
Cog–behav	0.71	12.82	3, 94	.0001	.29
Interaction	0.65	16.91	3, 94	.0001	.35

Note. Cog–behav = cognitive–behavioral.
[a]Rao's *F* transformation of Wilks's lambda.

variable's effect, the generalized variance of W and B, Wilks's lambda, the probability associated with the *F* or chi-square transformation of lambda, and eta-square.

There are other test statistics besides Wilks's lambda, all of which are some function of the T, W, and B matrices. Hotelling's T^2 is used when there are only two groups being compared on a set of dependent measures, and it can be transformed to an *F* statistic just as Wilks's lambda can in order to determine its statistical significance. Other test statistics for cases in which there are more than two levels of the independent variable or more than one independent variable include the Hotelling–Lawley trace, Roy's largest root, and the Phillai–Bartlett trace (Tatsuoka, 1971). Regarding which of these latter three statistics should be used, Stevens (1986) reviewed the literature and concluded that "as a general rule, it won't make that much of a difference which of the statistics is used" (p. 187).

As a way of further summarizing, Table 6 shows the preliminary results of the 2 (relaxation) × 2 (cog–behav) MANOVA performed in the panic disorder example. These are preliminary results, because no follow-up procedures were used to interpret the significant group differences observed. Such procedures are discussed in the next section. For now, the reader can observe three tests of significance, each addressing one of the three multivariate effects in the 2 × 2 MANOVA. The first significance test evaluates whether the vector of means for those subjects in relaxation training is equal to the vector of means for those not in relaxation training. The second significance test determines whether the group receiving cognitive–behavior therapy is equal to the group not receiving that therapy with respect to the vector of dependent variable means. Finally, the last test examines the interaction between relaxation and cog-behav to determine whether the four groups formed (relaxation, cog–behav, combined, and control) differ on the vector of dependent means.

For each effect shown in Table 6, there are two indices one should use to interpret the significance of an effect (Huberty & Smith, 1982). The first is the *p* value, which indicates the probability of obtaining a given effect if there was indeed no real effect in nature. In the panic disorder study, each multivariate effect was statistically significant at the .0001 level. The other criterion that should be used for assessing the significance of a multivariate effect is eta-squared, shown in the last column of Table 6. As I mentioned earlier, the eta-squares for each variable cannot be added together, even though the ones in Table 6 almost add to one. The proportions of explained variance shown in the table—.35, .29, and .35—would be extremely large according to Cohen's (1977) social science standards. Thus, it can be concluded that the three multivariate effects from the 2 × 2 panic disorder MANOVA are all significant because (a) the small *p* values suggest that the chance of the results being attributable to Type I error is small and (b) the eta-squares indicate that the effects are explaining nontrivial portions of the variance in the dependent measures.

Follow-Up Analyses for Significant Multivariate Effects

The most difficult part of performing and interpreting a MANOVA is determining what to do if a significant multivariate effect has been obtained. In the panic disorder example, that fact that there was a multivariate main effect for relaxation is interesting, but what does it mean in terms of that treatment's effect on the subjects' panic disorders? There are several procedures available for following up multivariate significance. Five of these procedures are reviewed here: multiple univariate ANOVAs, stepdown analysis, discriminant analysis, dependent variable contribution, and multivariate contrasts.

Multiple Univariate ANOVAs

By far the most popular way of proceeding from a significant effect in MANOVA is to perform univariate ANOVAs for each of the dependent variables (Bray & Maxwell, 1982). The reasoning behind this approach is that the preliminary MANOVA will control for Type I error. If the MANOVA yields significance, then it is considered acceptable to carry out multiple ANOVAs without undue inflation of the experimentwise alpha. Hummel and Sligo (1971) performed a simulation study to compare various methods for analyzing data in a MANOVA context and found that a MANOVA followed by univariate ANOVAs kept the experimentwise error rate the lowest. This often cited study serves as the major justification for using this technique.

This popular approach, however, has been criticized for three major reasons. The first is that Hummel and Sligo's (1971) simulation used data that do not resemble data found in real research situations. "Their use of equicorrelation matrices, in which all off-diagonal elements are equal, made their results applicable to almost no real data" (Wilkinson, 1975, p. 409). Hence, the findings from their study are highly suspect.

More important is the fact that a preliminary MANOVA protects the experimentwise alpha level only when the null hypothesis is true. Bray and Maxwell (1982) described the serious problem that occurs when the multivariate null hypothesis is rejected:

> In many cases the null hypothesis of no group differences is in fact false for one or more but not all variates. Hence, the multivariate test will produce significant results with a high probability if power is sufficient. However, univariate tests will then be performed for those variates for which there is no true differences as well as for those variates for which there is a difference. Because the individual alpha levels are not adjusted despite performing multiple significance tests, the overall multivariate test does not provide 'protection' for each of the p univariate tests. Consequently, in such cases the experimentwise error rate for the set of p univariate F ratios may be inflated above the nominal alpha level, even if the initial MANOVA test was significant. (p. 343)

A third reason for not following a significant MANOVA with univariate ANOVAs is that the separate ANOVAs ignore the correlations between the dependent variables, and as Bray and Maxwell (1982) noted, this means that valuable information concerning redundancies and conceptual relationships is left out. On the same note, Huberty and Morris (1989) distinguished between univariate questions and multivariate questions. Briefly defined, univariate research questions pertain to individual outcome variables, whereas multivariate questions are concerned with experimental effects for multiple dependent variables taken as a set. Huberty and Morris showed how separate univariate ANOVAs are inadequate for addressing multivariate questions.

Of course, some researchers may reply that they were never interested in the multivariate aspect of their data (i.e., in looking at the dependent variables as a set) but were only concerned with minimizing Type

I error. As pointed out earlier, the MANOVA–multiple ANOVA technique does not protect the experimentwise alpha when there is a significant multivariate effect present. If protection against Type I error is the concern, there are more appropriate techniques available, such as the various versions of the Bonferroni correction (see de Cani, 1984; Holland & Copenhaver, 1987, 1988; Holm, 1979; Maxwell & Delaney, 1990).

Stepdown Analysis

Recall that a MANOVA can be used when there is a theoretical relationship between dependent variables. If there is a logical a priori causal ordering of these variables, then a stepdown analysis may be appropriate (Bray & Maxwell, 1982; Stevens, 1986). Consider a study with three dependent variables, A, B, and C, wherein it is theorized that A causes B and B causes C. For such an analysis, a stepdown F is calculated for each dependent variable. For the variable that comes first in the causal ordering (A), the stepdown F is identical to the univariate ANOVA F for that variable. The stepdown F for the second variable (B) is calculated by performing an analysis of covariance (ANCOVA), whereby the influence of A is factored out, so that the unique contribution of B to group separation can be identified. (A more detailed discussion of ANCOVA is presented later when multivariate analysis of covariance [MANCOVA] is considered.) The stepdown F for the last variable, C, is derived from an ANCOVA that removes the effects of A and B, leaving only the variance uniquely associated with C. In this way, each dependent variable's contribution to the multivariate effect can be ascertained. Keep in mind when reading the results of a stepdown analysis that the ordering of the variables is extremely important for interpreting the results correctly.

As an example, consider the three dependent variables for the panic disorder study. Imagine that the experimenter has a theory which posits a biological cause of panic disorder, which produces a physiological state. This state initiates an emotional reaction, which is then conceptualized cognitively. Hence, the logical ordering of the dependent variables would be physiological, emotional, and cognitive. The stepdown F for each variable would give the relative contribution of each variable under the assumption of the causal ordering that was specified.

Discriminant Analysis

Another way of examining significant multivariate effects is to perform a discriminant analysis (DA) for each of the significant effects (Tatsuoka,

1971). A DA is a procedure that maximizes the separation between groups on some categorical variable by finding the optimal linear combination of several continuous variables. Recall that in a MANOVA, the same basic task is accomplished. In a MANOVA, the linear combination of the dependent variables that best separates the levels comprising the independent variables is determined. The DA allows one to discern subsets of the dependent variables that might constitute some underlying dimensions or constructs on which the experimental groups differ. A full discussion of DA is not possible here because of space limitations, and so the reader is referred to chapter 6 of this book, as well as other excellent treatments of the topic in relation to MANOVA (e.g., Huberty, 1984; Huberty & Smith, 1982; Pedhazur, 1982; Stevens, 1986).

Dependent Variable Contribution

Two lesser known procedures attempt to determine how much each dependent variable contributes to a multivariate effect. These techniques differ from stepdown analysis in that they look at the decrease in the multivariate effect as a function of removing a particular variable. Stepdown analysis tests one dependent variable at a time, controlling for the variables that precede it in the a priori causal ordering.

The first of these two techniques that evaluate the contribution of each dependent variable is from Wilkinson (1975). He suggested performing successive MANOVAs in which one dependent variable is left out in each analysis. The change in the multivariate F is examined to determine which variables are contributing the most to the multivariate effect and which are not. In other words, the procedure attempts to reveal which variable a researcher cannot do without. Of course, such a decision is not purely statistical. It might be the case that a particular variable does not contribute as strongly to the multivariate effect as others do, but the variable in question is essential to the model from a theoretical standpoint. A researcher might decide that a variable's theoretical necessity may supersede its statistical necessity.

The other approach for determining variable contribution is from Huberty and his associates (Huberty & Morris, 1989; Huberty & Smith, 1982). For a given set of dependent variables, Huberty recommended using an F-to-remove statistic for each dependent variable: "An F-to-remove tests the significance of the decrease in group separation if variable i is removed from the entire set of variables" (Huberty & Smith, 1982, p. 421). The F-to-remove values for the dependent variables are then rank ordered to determine the relative importance of each. There are strong similarities between Wilkinson's (1975) and Huberty's methods, both in mathematical form and practical function.

Although neither of these two methods are yet frequently seen in the journals, I discuss them here because they are viable approaches. Hence they might appear in future articles with increasing frequency.

Multivariate Contrasts

Whereas the previous procedures focused on the dependent variables (e.g., which contributes to the overall effect, what dimension underlies the variates, etc.), contrast analyses focus on the groups defined by the independent variables. Multivariate contrasts compare groups over a set of dependent variables simultaneously (Huberty & Morris, 1989). In other words, multivariate contrasts compare vectors of means. As Stevens (1986) observed, a contrast analysis is essentially a comparison between two groups. For simple (or pairwise) contrasts, the two groups being compared are two levels of some independent variable (e.g., male vs. female for the variable of gender). For complex contrasts, the two groups being compared can be combinations of levels of several dependent variables. Consider a study with four different treatments: T1 through T4. A simple contrast might be to compare T1 and T3. A complex analysis might compare T4 with a group formed by combining T1, T2, and T3. The latter contrast is complex because a group has been formed from the preexisting groups T1, T2, and T3.

The multivariate test statistic for comparing two groups in a multivariate contrast is Hotelling's T^2, which is transformed to an F to obtain a probability level. There are two basic ways of following up a significant multivariate comparison. The first is to use discriminant function analysis or obtain F-to-remove statistics as in the follow-up procedures for a significant MANOVA effect (Huberty & Smith, 1982). This method has already been discussed earlier. Note that this first approach retains the "multivariate" aspect of the data because it considers all of the dependent variables at once.

A second contrast procedure examines each dependent variable separately. Hence, a significant T^2 is followed by one of several methods for univariate pairwise comparisons such as multiple t tests, Tukey confidence intervals, Roy–Bose simultaneous confidence intervals, and modified Bonferroni procedures (Bray & Maxwell, 1982; Stevens, 1986). Each of these techniques have specific pros and cons regarding the control of

Type I error and power that are beyond the limits of this chapter (the interested reader can consult Keselman et al., 1991; Kleinbaum et al., 1988; Seaman et al., 1991; Stevens, 1986).

MANCOVA, Repeated Measures MANOVA, and Power Analysis

MANCOVA

In an ANOVA design, it is often desirable to eliminate as much error variance as possible so that a larger portion of the remaining variance is attributable to treatment effects or group differences (Kleinbaum et al., 1988). One way of reducing error variance is to measure a *covariate*—a continuous variable known to affect the dependent measures—whose effect on the dependent variable can then be factored out of the total variance. Hence, the levels of the independent variables are being compared using means on the dependent variable that have been adjusted using the covariate. In a multivariate analysis, covariates can be used in the same manner, changing a MANOVA to a MANCOVA—a multivariate analysis of covariance. Instead of comparing vectors of means, the MANCOVA compares vectors of adjusted means.

Consider the panic disorder study. The researcher assigned each subject a severity rating before the treatment period. It would be expected that the more severe cases of panic disorder would have generally higher scores on the cognitive, emotional, and physiological subscales, and so this severity rating might be an effective covariate. A MANCOVA could be performed to determine whether the vectors of adjusted means differed across the experimental groups.

Before conducting the analysis, however, the covariate needs to be carefully examined. A covariate should only be used if (a) there is a statistically significant linear relationship between the covariate and the dependent measures and (b) an additional assumption—the homogeneity of the regressions—is satisfied. The first condition can be tested with a simple correlation between the covariate and each dependent measure. For the panic disorder study, the severity rating correlated significantly with each of the dependent measures at the .05 level. The second condition, the homogeneity of regression hyperplanes assumption, requires the experimental groups to have equal regression slopes for the covariate. Another way of saying this is that the relationship between the covariate and the dependent measures must be equal for all of the groups. If this

Table 7

2 × 2 MANCOVA Results From the Hypothetical Panic Disorder Data

Effect	Λ	*F*	*df*	*p*	η^2
Relaxation	0.78	8.96	3, 93	.0001	.22
Cog–behav	0.87	4.74	3, 93	.005	.13
Interaction	0.74	10.65	3, 93	.0001	.26

Note. MANCOVA = multivariate analysis of variance; Cog–behav = cognitive–behavioral.

assumption was not satisfied, there would be a Group × Covariate interaction that would be overlooked. In the panic disorder MANCOVA, there was no interaction for the Severity × Relaxation or Severity × Cog-Behav conditions, and so the assumption is tenable.

Tests of the multivariate effects of relaxation, cog–behav, and the interaction between the two are performed. These tests, along the eta-squares, are shown in Table 7. Reading the results of a MANCOVA is exactly the same as reading the results of a MANOVA, except that in the former, variance attributable to individual differences on some covariate is partitioned out prior to examining the effects. Note that the *F* values for each effect are smaller than the corresponding *F* values for the 2 × 2 MANOVA presented earlier. Introducing a covariate can influence the *F* statistic in two ways. To the extent that the covariate accounts for variance in the dependent measures that is not explained by the independent variables, the *F* statistic will be larger. This is because the within-groups SSCP matrix W (reflecting unexplained variance) will be smaller, thereby making Wilks's lambda smaller. To the extent that the covariate accounts for variance in the dependent measures that is shared by an independent variable, the *F* statistic will decrease. This happens because the mean vectors of the dependent variables for the groups defined by the independent variable are brought closer together after adjusting for the covariate. In this example, the latter influence was evidently more predominant.

That the results of the MANCOVA are the same as they are for the regular MANOVA will not always be the case. In many instances, when the effect of a covariate on the set of dependent measures is removed, the effect of the remaining independent variables may be negligible. Such a finding may suggest one of the following: (a) The independent variable has an indirect effect on the dependent measures by working through the covariate; (b) the covariate has an indirect effect on the dependent measures via the independent variable, such that the independent variable does not explain variation in the dependent variables absent the covariate's effect on the independent variable; and (c) the covariate and independent measures are measures of the same variable, and so the presence of one will render the other's effect redundant.

Repeated Measures MANOVA

One of the most powerful and efficient research designs is the repeated measures design (Shaughnessy & Zechmeister, 1990). The term *repeated measures design* refers to a situation in which subjects are measured on more than one occasion. This design is powerful because error variance is reduced substantially and is efficient because fewer subjects are needed than in nonrepeated measures experimental designs. This can happen when subjects are assessed before and after a treatment (pretest–posttest) or when each subject participates in more than one experimental condition. To facilitate discussion, imagine that the author of the panic disorder study measured each subject on the cognitive, emotional, and physiological subscales before the intervention, as well as after. This would transform the study into a repeated measures design.

Researchers use various terms when describing repeated measures designs, and so it is helpful to review some of them here. Independent variables such as time (T1, T2, T3, etc.) or trial (Trial1, Trial2, etc.) are called *within* or *within-subjects* variables. For a within-subjects variable, each subject is measured on the dependent variables for every level of the within-subjects variable. Hence, a within-subjects variable in one study might be time, consisting of four points during the study when subjects were assessed on the dependent variables. A *between* or *between-subjects* variable is a grouping variable such as age or gender. A study can involve both within- and between-subjects independent variables. Hence, the modified panic disorder study constitutes a 2 (time) × 2 (relaxation) × 2 (cog–behav) repeated measure design, with repeated measures on the first variable (time). The other two variables—relaxation and cog–behav—are between-subjects variables.

In the standard MANOVA, vectors of means across levels of the independent variables are compared. For the repeated measures MANOVA, vectors of *mean differences* are compared across levels of the independent variables. These mean differences refer to differences in the value of the dependent measures between levels of the within-subjects variable. Thus, if GSR was measured four different times (T1 through T4) for each subject, the three difference scores for each subject would be T1–T2, T2–T3, and T3–T4. The mean difference variables would comprise the dependent measure vector, not the original scores.

A within-subjects effect such as time or trial can be analyzed in a manner analogous to a between-subjects effect. Returning to the GSR example, suppose that a researcher wanted to compare GSR with stressful stimuli over four time periods for male subjects and female subjects. This would constitute a 4 (time) × 2 (gender) repeated measures MANOVA. The MANOVA would produce three significance tests corresponding to three effects that can be tested: (a) the main effect for time (i.e., Do the subjects' GSRs differ over time?); (b) the main effect for gender (i.e., Do males and females differ with respect to GSR, ignoring the effect of time?); and (c) the interaction effect (i.e., Does the GSR of males change over time at a different rate than females?).

When conducting a repeated measures MANOVA, an additional assumption must be met. This *sphericity assumption* (Huynh & Mandeville, 1979) concerns the difference variables that are created from the original dependent variables. In the GSR and gender example, there were three such transformed variables, with each reflecting the difference between one pair of the original four dependent measures. The sphericity condition requires that the covariance matrix for the transformed variables be a diagonal matrix in which the values along the diagonal are equal and all the off-diagonal elements must be zero. In other words, the sphericity assumption requires that the variances of the transformed variables will be equal and the transformed variables are not intercorrelated. Failure to meet the sphericity condition will result in an inflated Type I error rate (Stevens, 1986).

One of the more sophisticated within-subjects analyses is the so-called *doubly multivariate repeated measures MANOVA* (Norusis, 1988, p. 283; SAS Institute, 1990, p. 988). A typical repeated measures MANOVA has only one dependent measure, such as GSR, which is measured multiple times. In the doubly multivariate repeated measure design, two or more dependent variables are measured at two or more points in time. Hence, the modified version of the panic disorder study constitutes such a design. There are three dependent measures (the three subscales) assessed twice for each subject (pretest and posttest). Recall that in the standard repeated measures MANOVA, a vector of mean differences for each group is compared. In the doubly multivariate design, a matrix of difference vari-

Table 8

2 × 2 × 2 Doubly Multivariate Repeated Measures MANOVA Results From the Hypothetical Panic Disorder Data

Multivariate effect	Λ	*F*	*df*	*p*	η^2
Time	0.35	58.90	3, 94	.0001	.65
Relaxation	0.77	9.30	3, 94	.0001	.23
Cog–behav	0.81	7.13	3, 94	.0005	.19
Relaxation × Cog–Behav	0.74	10.99	3, 94	.0001	.26
Time × Relaxation	0.75	10.34	3, 94	.0001	.25
Time × Cog–Behav	0.80	7.94	3, 94	.0001	.20
Time × Relaxation × Cog–Behav	0.83	6.49	3, 94	.001	.17

Note. MANOVA = multivariate analysis of variance, Cog–Behav = cognitive–behavioral.

ables is used, because there is more than one dependent measure being analyzed.

Table 8 shows the results of the doubly multivariate repeated measures MANOVA for the panic disorder study. Six different multivariate effects can be tested. The first is the main effect for time, which addresses the question of whether the dependent measures, taken as a set, change over time. The next three tests for the relaxation, cog–behav, and relaxation × cog–behav effects are somewhat superfluous, because they test for group differences collapsing across the two testing periods (although the reader may observe that each of these effects were significant). Of primary interest is the question of which groups did better on the posttest compared with the pretest. Hence, the Time × Relaxation, Time × Cog–Behav, and Time × Relaxation × Cog–Behav interactions should be examined. As Table 8 shows, all three interactions are highly significant, indicating that the change in pretest–posttest scores for the three subscales is different for the four experimental groups. Remember that the eta-squares cannot be combined.

Power Analysis in MANOVA

Earlier discussion concerned the notion of statistical power. In this section, the issue of power is addressed once again, focusing specifically on how power can be assessed for a MANOVA. It is desirable to know the power of a MANOVA for two reasons. First, one may wish to know the power a priori to determine the sample size needed to detect a given effect. This would be the case when one is designing a study or reviewing a proposal for a study. Second, one may be interested in a post hoc estimation of power to ascertain how well a MANOVA could detect an effect in a given study that was already done.

Recall that power is a function of sample size, the nominal alpha level specified, and effect size. Sample size and the nominal alpha level can easily be obtained from either a univariate or multivariate analysis, but little work has been done on measuring effect sizes for a MANOVA. The best work in this area has come from Stevens (1980, 1986). The effect size for a two-group univariate analysis is typically expressed as d and is a function of the difference between the two group means. Stevens showed that for the two-group multivariate case, the effect size can be measured as D^2, a function of the difference between the two vectors of means (see Stevens, 1986, p. 140). Determining the effect size when there are more than two experimental groups is more complicated. Those interested in determining the power of a MANOVA are strongly advised to consult Stevens (1980, 1986) for explicit information on how it is done, including easy-to-understand tables showing the power of a MANOVA for various numbers of dependent variables, groups, and subjects.

Conclusion

Recall the Results section paragraph that appeared on the first page of this chapter. What might have seemed intimidating and confusing earlier should now be more clear. The 4 (group) × 2 (time) MANOVA describes a repeated measures MANOVA with one between-subjects variable (group) and one within-subjects variable (time). The group variable has four levels corresponding to four groups, and the time variable has two levels corresponding to two testing periods. The values of Wilks's lambda, their F transformations, and their respective p values indicate that the only statistically significant independent variable was time. This denotes that the vector of dependent variable means for the first time period was significantly different from the vector of means for the second time period. For instance, one might say that scores on the dependent measures were higher at Time 2 than at Time 1. The eta-square listed for the time effect (.59) suggests that the effect explained over half of the variance in the dependent measures. The absence of a significant group effect means that the mean vectors for each of the four groups were equivalent. The absence of a significant Group × Time interaction indicates that the effect of one variable did not depend on the level of the other variable.

Suggestions for Further Reading

For introductory-level treatments of research designs and univariate ANOVA, see Campbell and Stanley (1963), Grimm (1993), Guilford (1965), Kerlinger (1986), Kleinbaum, Kupper, and Muller (1988), Pedhazur (1982), Shaughnessy and Zechmeister (1990), and Stevens (1990).

More advanced discussions of power analysis can be found in Cohen (1977) and Stevens (1980, 1986). Advanced discussions of Type I error rates and multiple comparison procedures can be found in de Cani (1984), Holland and Copenhaver (1987, 1988), Holm (1979), Keselman, Keselman, and Games (1991), Maxwell and Delaney (1990), and Seaman, Levin, and Serlin (1991).

Detailed treatments of matrix algebra and calculation of MANOVA test statistics can be found in Green (1978), Pedhazur (1982), Stevens (1986), and Tatsuoka (1971). For comprehensive review of MANOVA follow-up techniques, see Bray and Maxwell (1982), Maxwell and Delaney (1990), and Stevens (1986). For detailed discussions of particular follow-up procedures, see Huberty and Morris (1989), Huberty and Smith (1982), Tatsuoka (1971), and Wilkinson (1975).

Glossary

When appropriate, the symbol associated with a particular term is included in parentheses after the term.

ALPHA (α) The probability of rejecting the null hypothesis when it is true.

BETWEEN GROUPS The variation accounted for by the independent variables.

BETWEEN SUBJECTS A variable on which each subject can be found on only one level of the variable, such as age or gender.

CHI-SQUARE (χ^2) An inferential test statistic that multivariate statistics can be transformed to in order to derive a probability level.

COVARIANCE The variation in one variable that is shared by another variable.

Covariate A continuous variable that has a significant linear relationship with some dependent variable. Covariates are used to reduce error variance in an ANCOVA and MANCOVA.

Effect Size The magnitude of an independent variable's effect, usually expressed as a proportion of explained variance in the dependent variables.

Eta-Square (η^2) An index of the proportion of explained variance; can vary from 0 to 1.

F Statistic A test statistic to which multivariate indices can be transformed, to derive a probability level.

Hotelling's T^2 A multivariate test statistic used when there is one independent variable with only two levels.

Interaction Effect An interaction occurs when the effect of an independent variable on some dependent variable depends on the level of another independent variable.

Main Effect The effect of a single independent variable on one or more dependent variables.

Null Hypothesis (H_0) The hypothesis that states that there is no difference in the mean values of one or more dependent variables across levels of one or more independent variables.

Omnibus Test A test of the null hypothesis that none of the independent variables has an effect on any of the dependent variables.

Power The probability of detecting a significant effect when the effect truly exists in nature.

Probability (p) The probability of falsely rejecting the null hypothesis.

Repeated Measures An experimental design and corresponding analysis in which each subject is measured on the dependent variable for more than one level of an independent variable (e.g., Time 1 and Time 2).

SSCP A matrix with sums of squares for each dependent variable on the diagonal and sums of cross-products for every pair of variables filling the rest of the elements.

Sum of Cross-Products For a set of observations, it is the sum of the products of each subjects' squared deviations from the mean on one variable and the squared deviation from the mean of another variable. It is an index of covariance.

SUM OF SQUARES The sum of the squared deviations about a mean for a set of observations. It is an index of variance used in analysis of variance and covariance procedures.

TYPE I ERROR Rejecting the null hypothesis when it is really true.

TYPE II ERROR Accepting the null hypothesis when it is really false.

VARIANCE The average of the squared deviations from a mean for a set of observations. It reflects the dispersion of values around a mean.

WILKS'S LAMBDA (Λ) A multivariate test statistic that expresses the proportion of unexplained variance in the dependent measures.

WITHIN GROUP The variation in the dependent measure that is not explained by the independent variables.

WITHIN SUBJECT An independent variable used in such a way that dependent variable values are obtained for every level of the within-subject variable.

References

Antony, M. M., Brown, T. A., & Barlow, D. H. (1992). Current perspectives on panic and panic disorder. *Current Directions in Psychologial Science, 1*, 79–82.

Bray, J. H., & Maxwell, S. E. (1982). Analyzing and interpreting significant MANOVAs. *Review of Educational Research, 52*, 340–367.

Campbell, D. T., & Stanley, J. C. (1963). *Experimental and quasi-experimental designs for research*. Boston: Houghton Mifflin.

Cohen, J. (1977). *Statistical power analysis for the behavioral sciences*. San Diego, CA: Academic Press.

de Cani, J. S. (1984). Balancing Type I risk and loss of power in ordered Bonferroni procedures. *Journal of Educational Psychology, 76*, 1035–1037.

Fleiss, J. L. (1986). *The design and analysis of clinical experiments*. New York: Wiley.

Green, P. E. (1978). *Analyzing multivariate data*. Hinsdale, IL: Dryden Press.

Grimm, L. (1993). *Statistical applications for the behavioral sciences*. New York: Wiley.

Guilford, J. P. (1965). *Fundamental statistics in psychology and education* (4th ed.). New York: McGraw-Hill.

Holland, B. S., & Copenhaver, M. D. (1987). An improved sequentially rejective Bonferroni procedure. *Biometrics, 43*, 417–423.

Holland, B. S., & Copenhaver, M. D. (1988). Improved Bonferroni-type multiple testing procedures. *Psychological Bulletin, 104*, 145–149.

Holm, S. (1979). A simple sequentially rejective multiple test procedure. *Scandanavian Journal of Statistics, 6*, 65–70.

Huberty, C. J. (1984). Issues in the use and interpretation of discriminant analysis. *Psychological Bulletin, 95*, 156–171.

Huberty, C. J., & Morris, J. D. (1989). Multivariate analysis versus multiple univariate analyses. *Psychological Bulletin, 105*, 302–308.

Huberty, C. J., & Smith, J. D. (1982). The study of effects in MANOVA. *Multivariate Behavioral Research, 17*, 417–482.

Hummel, T. J., & Sligo, J. R. (1971). Empirical comparison of univariate and multivariate analysis of variance procedures. *Psychological Bulletin, 76*, 49–57.
Huynh, H., & Mandeville, G. K. (1979). Validity conditions in repeated measures designs. *Psychological Bulletin, 86*, 964–973.
Kaplan, R. M., & Saccuzzo, D. P. (1989). *Psychological testing: Principles, applications, and applications* (2nd ed.). Pacific Grove, CA: Brooks/Cole.
Kerlinger, F. N. (1986). *Foundations of behavioral research* (3rd ed.). New York: Holt, Rinehart & Winston.
Keselman, H. J., Keselman, J. C., & Games, P. A. (1991). Maximum familywise Type I error rate: The least significant difference, Newman-Keuls, and other multiple comparison procedures. *Psychological Bulletin, 110*, 155–161.
Kleinbaum, D. G., Kupper, L. L., & Muller, K. E. (1988). *Applied regression analysis and other multivariable methods* (2nd ed.). Boston: PWS-KENT.
Larsen, R. J. (1984). Theory and measurement of affect intensity as an individual difference characteristic. *Dissertation Abstracts International, 85*, 2297B. (University Microfilms No. 84-22112)
Larsen, R. J., & Diener, E. (1987). Affect intensity as an individual differences characteristic: A review. *Journal of Research in Personality, 21*, 1–39.
Maxwell, S. E., & Delaney, M. D. (1990). *Designing experiments and analyzing data: A model comparison perspective*. Belmont, CA: Wadsworth.
Mendoza, J. L., & Graziano, W. G. (1982). The statistical analysis of dyadic social behavior: A multivariate approach. *Psychological Bulletin, 92*, 532–540.
Norusis, M. J. (1988). *SPSS advanced statistics guide* (2nd ed.). Chicago: SPSS.
Olson, C. L. (1974). Comparative robustness of six tests in multivariate analysis of variance. *Journal of the American Statistical Association, 69*, 894–908.
Pedhazur, E. J. (1982). *Multiple regression in behavioral research* (2nd ed.). Fort Worth, TX: Holt, Rinehart & Winston.
Ryan, T. A. (1959). Multiple comparisons in psychological research. *Psychological Bulletin, 56*, 26–47.
SAS Institute. (1990). *SAS/STAT user's guide* (4th ed.). Cary, NC: SAS Institute.
Seaman, M. A., Levin, J. R., & Serlin, R. (1991). New developments in pairwise multiple comparisons: Some powerful and practicable procedures. *Psychological Bulletin, 110*, 577–586.
Shaughnessy, J. J., & Zechmeister, E. B. (1990). *Research methods in psychology* (2nd ed.). New York: McGraw-Hill.
Stevens, J. P. (1980). Power of the multivariate analysis of variance tests. *Psychological Bulletin, 88*, 728–737.
Stevens, J. (1986). *Applied multivariate statistics for the social sciences*. Hillsdale, NJ: Erlbaum.
Stevens, J. (1990). *Intermediate statistics: A modern approach*. Hillsdale, NJ: Erlbaum.
Tatsuoka, M. M. (1971). *Multivariate analysis: Techniques for educational and psychological research*. New York: Wiley.
Wilkinson, L. (1975). Response variable hypothesis in the multivariate analysis of variance. *Psychological Bulletin, 82*, 408–412.

Discriminant Analysis

A. Pedro Duarte Silva and Antonie Stam

Descriptive discriminant analysis (DA) is a statistical technique that allows one to identify *variables* (also known as *attributes*) that best discriminate members of two or more groups from one another. *Predictive* or *prescriptive* DA allows one to predict the group membership status of *subjects* (also known as *observations*, *cases*, or *entities*) of which the group status is unknown. For example, imagine that a researcher had a sample of patients (subjects) who had undergone a heart transplant operation (this sample of subjects is known as the *training* or *development* sample). For each patient, information was available regarding blood pressure, age, number of white blood cells, and percentage of normal weight. In addition, it was known which patients had survived for at least a year after the operation and which patients had died before 1 year (thus, the groups were alive or dead). Descriptive DA would be used to determine how well the variables allowed one to discriminate between the two groups. In this instance, the investigator could use a combination of the blood pressure, age, number of white blood cells, and percentage of normal weight variables to discriminate among those patients who had died and those who had survived. Some variables might be found to be very important, whereas others might turn out to be irrelevant. Now suppose that a new patient was considered for a transplant operation. In predictive DA, the classification rule derived using the training sample would be used to combine information concerning this new patient's blood pressure, age, and so forth, to predict whether the patient would survive the operation for at least 1 year.

In this chapter, we introduce a class of techniques for discriminating among distinct groups and techniques to predict group membership of cases for which the true membership is unknown. We first present three examples in which discrimination and classification methods can be very useful in understanding and analyzing the decision situation or problem. One of these examples will be used later to numerically illustrate the concepts.

Example 1 (Porebski, 1966)

A psychologist advises entering students at a technical college about the most suitable program of study. Students can choose between four majors: engineering, building, art, or commerce. The psychologist uses the perceived similarity between entering students and those students already enrolled in programs at the college as the principal criterion for his recommendations. Before entering the college, each student is required to take three tests: one in arithmetic, another in English, and a third in forms–relations. The psychologist intends to use the test scores as an advising aid. He also wants to study the differences between the student populations enrolled in each major, as well as the primary factors that lead a student to choose a particular program of study. Descriptive DA is used to identify the combination of scores on the three measured variables that best discriminates the four majors.

Example 2 (McGrath, 1960)

An automobile dealer has to make credit decisions on the basis of the personal history information provided by potential customers in their credit application forms. Initially, the dealer uses her own judgment, but as her business volume increases, she realizes that she needs a more consistent and expedient way of assessing the likely creditworthiness of each applicant. To this purpose, she decides to implement a numeric rating system, on the basis of her experience with previous credits. She also wants to develop an objective procedure for measuring the probability that a particular loan will default and how much money she can expect to lose because of bad loans. In this instance, predictive DA is used to categorize whether an applicant is a good or bad credit risk and for estimating probabilities of loan default and expected losses due to loan default.

Example 3 (Shubin, Afifi, Rand, & Weil, 1968)

A patient just admitted to the hospital is diagnosed with myocardial infarction. Systolic blood pressure, diastolic blood pressure, heart rate, stroke index, and mean arterial pressure are obtained. The medical staff wants to use this information in predictive DA to estimate the probability that the patient will survive.

Even though originating from very different situations, these three example problems have several features in common. In fact, as will become clear below, all three problems, as well as many other problems from completely different areas, can be analyzed using DA. Hence, it is of interest to identify characteristics common to all DA problems.

First, DA uses entities, such as students, subjects, businesses, or patients. The set of entities is categorized into distinct and well-defined groups or classes, different diagnostic categories, poor credit risks versus good credits, patients who die versus those who survive, or experimental and control groups. There must be at least two groups, and each entity belongs to one and only one group. Thus, the groups are *mutually exclusive* (each entity belongs to only one group) and *collectively exhaustive* (each entity belongs to some group). Each entity is described by a set of attributes, or variables, such as standardized test scores, items on the credit application form, blood pressure, or heart rate.

Second, each DA group should be *well-defined*, that is, the grouping should reflect true differences between the entities. In other words, the division of entities into groups should be done in a natural and objective way, rather than arbitrarily. The groups should differ *qualitatively*. Different diagnostic categories are one example of qualitatively different groups. For example, arbitrarily defining "high IQ" and "low IQ" by using a median split on an intelligence scale is not an appropriate way of establishing groups. One can find DA studies in which groups are formed by dividing some quantitative variable into classes. For example, one could classify families into classes of high, medium and low income, on the basis of a quantitative variable that measured family income. However, such a division of entities into categorical classes to create groups may be inappropriate for an analysis with DA techniques, because this division does not necessarily involve a naturally identified grouping of the entities. The creation of DA groups through the partitioning of quantitative variables is justifiable only if there exist natural and easily identifiable gaps at the points of division among the groups. For instance, in a drug study, three groups can be justifiably formed on the basis of different dosages of a medication (e.g., 10 mg, 20 mg, and 30 mg of valium). If there are more than two groups and these groups can be ordered, special-purpose methods should be used that explicitly take the group ordering into account, such as ordinal logit analysis, rather than traditional DA (see, e.g., Walker & Duncan, 1967). In this chapter, we will restrict ourselves to DA problems with groups that cannot be ordered.

Third, in DA, groups should be defined before collecting the data. In contrast, techniques in *cluster analysis* are used to define groups on the basis of the characteristics of the sample itself. Although the clustering problem is important, interesting, and related to DA, it is not the subject of this chapter.

Fourth, the set of attributes ideally should represent as complete and accurate a description of the entities as possible, for accurate discrimination between the groups. A particular attribute with characteristics that are similar for entities in all groups may describe the total set of entities very well but will not serve any discriminatory or classificatory purpose. For instance, all depressed and nondepressed patients can be assessed with respect to need for achievement, but this variable is unlikely to discriminate between the two groups.

An example of two attributes that describe the entities very well (for both groups, the data are increasing in both attributes) but fail to discriminate between Group A and Group B is depicted in Figure 1. Figure 2 shows a situation of two attributes that are reasonably effective in separating the two groups. The attribute values of the entities in Group C tend to have lower values on both attributes than the entities belonging to Group D. Note that there exists overlap between the groups, in which the classification and separation is not clear. Group overlap can be viewed as the complement of group separation. In most DA problems, there is a fair amount of group overlap.

In practice, it is often difficult to identify a priori which attributes are relevant for DA purposes. It is common practice to select a large set of potentially relevant attributes and identify the most important ones among this set in the course of the analysis. However, although a number of proven attribute-selection methods have been developed for descriptive DA, the corresponding theory for predictive DA, in which the use of an excessive number of unnecessary attributes can have serious negative consequences, is not as well-developed. The attribute (variable) selection problem is one of the most important and difficult problems in DA. To successfully address the variable selection problem, the analyst often needs

Figure 1

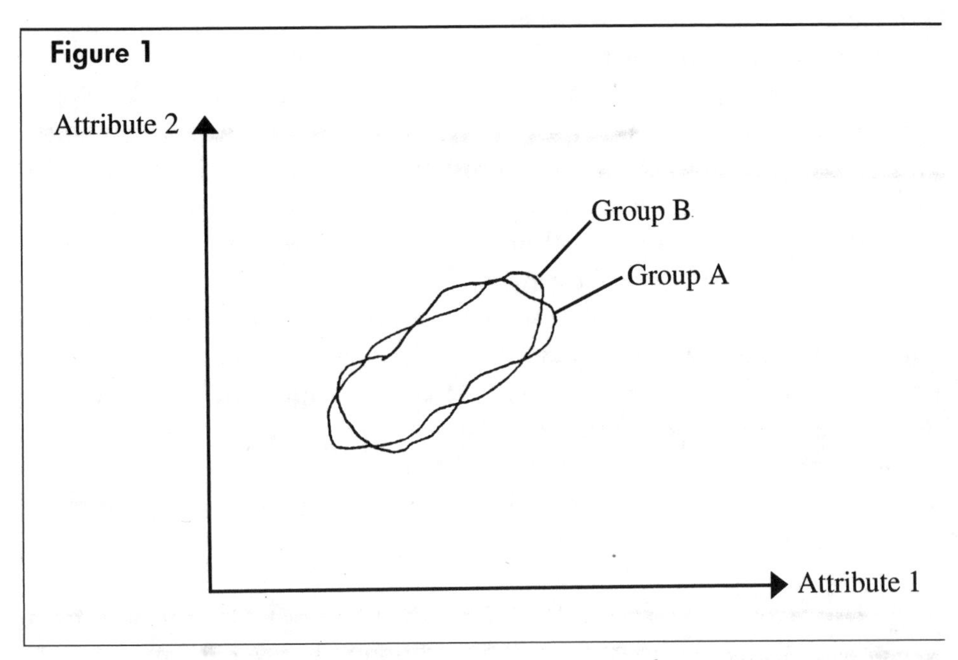

Two-group, two-attribute example with poor explanation of group separation.

Figure 2

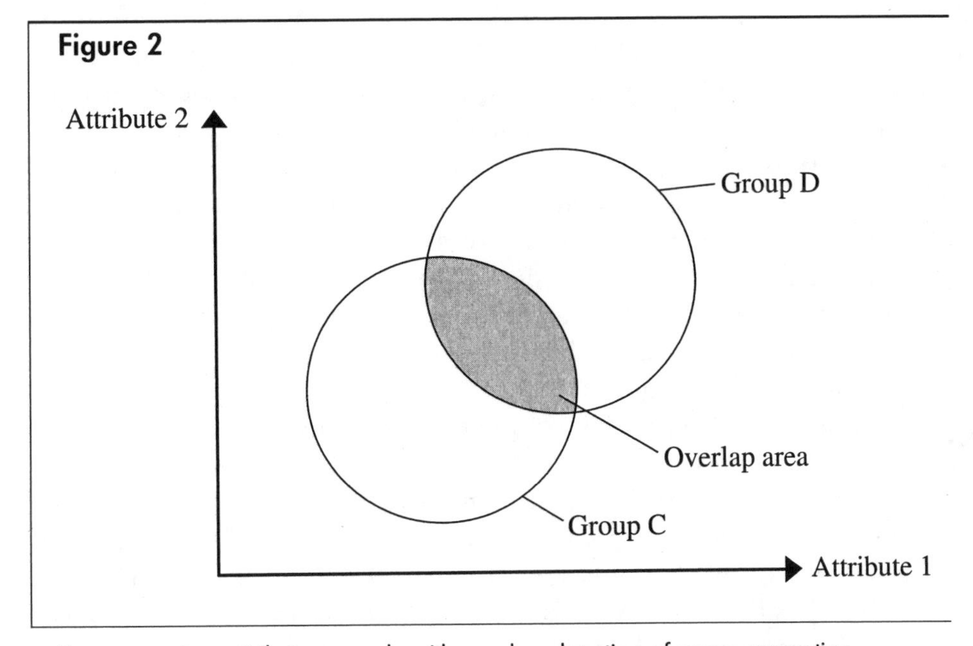

Two-group, two-attribute example with good explanation of group separation.

to combine substantive knowledge about the problem area with the relevant statistical techniques.

As mentioned above, DA is divided into two major areas: descriptive DA and predictive DA. Descriptive DA seeks to describe group differences on the basis of the attributes of the entities. Typical questions addressed in descriptive DA include the following: Which attributes contribute most to group separation, and which are irrelevant in explaining differences among the groups? Along how many dimensions (a dimension of group separation is usually associated with a *canonical variable*, which in turn is a linear combination of the attributes, also called a canonical variable) do the groups differ (i.e., which factors play an important role)? Can we determine a substantive interpretation of each dimension? Which subsets of the original groups can be separated by particular dimensions of group separation? Can we determine which attributes play an important role and find reasons for why certain groups differ with respect to a given attribute?

Predictive DA addresses the question of how to assign entities to groups, on the basis of the information contained in the attribute values. An entity is said to be misclassified if its predicted group membership is incorrect. The primary task in predictive DA is to use the attribute values to derive a classification rule that will be used to classify entities with unknown group membership. Usually it is important to consider misclassification probabilities and misclassification costs. Other issues of interest in predictive DA include the following: How many attributes should we use in the classification rule? How do we select the best set of attributes for prediction? What is the probability that a given entity will be classified correctly? What are the probabilities that this entity will belong to any of the remaining groups? What are the probabilities of misclassification and the expected misclassification costs for a classification rule?

Although descriptive and predictive DA have different objectives and use different methods of analysis, many studies combine techniques from both approaches, as well as from other methodologies such as multivariate analysis of variance (MANOVA). For example, in a typical study in which DA is used, the researcher begins the analysis by testing whether the set of attributes under consideration reveals statistically significant differences among groups. If such differences are indeed present, the analyst next explores and explains these differences in further detail by means of descriptive DA methods, after which predictive DA can be used to derive classification rules and estimate misclassification probabilities.

Of course, different problems may require different sequences of analysis. Sometimes, the focus of interest is purely descriptive, in which case a predictive DA analysis is unnecessary. In other cases, classification is the only purpose of the study, and descriptive methods are not needed. In some applications, it is possible to use predictive methods for descriptive purposes and, conversely, descriptive methods for prediction. For instance, after performing a traditional descriptive analysis, one may be interested in the degree of overlap among groups. In predictive DA, there are methods to estimate probabilities of misclassification and to measure group overlap. On the other hand, it is possible to use descriptive DA methods to select the set of attributes for use in predictive DA.

Descriptive Discriminant Analysis

As alluded to above, the principal objective of descriptive DA is to identify those factors that are able to separate or distinguish the predefined groups and to interpret these findings. The principal descriptive DA technique is *canonical* DA. Canonical DA successively identifies the linear combinations of attributes, known as *canonical discriminant functions*, which contribute maximally to group separation. For a problem with k different groups and m attributes, the number of canonical discriminant functions is either m or $k - 1$, whichever is smaller. Thus, canonical DA transforms the original set of m attributes into one or more new variables, known as *canonical variables*, ordered by their contribution to group separation (ordered according to their discriminatory power). As $k - 1$ is typically considerably smaller than m, the number of attributes usually exceeds the number of canonical discriminant functions. However, if $m \geq k - 1$, then the number of canonical discriminant functions will equal the number of attributes. Nevertheless, in most applications, a small subset of the canonical variables explains most of the differences among the groups. For this reason, canonical DA can be viewed as a dimension-reduction technique.

Each canonical variable corresponds with one dimension along which the groups differ. The identification of the appropriate dimensions is an important problem in descriptive DA. Other issues of interest include the interpretation of these dimensions, the assessment of their absolute and relative importance, to determine which groups can be separated using a particular dimension.

A substantive interpretation of the problem dimensions requires subjective judgment and specific knowledge of the problem domain and background. Statistical measures and methods useful in an interpretive analysis include the *standardized canonical coefficients* and a comparison of the correlation between the original attributes and the canonical variables. Canonical coefficients and canonical correlations are helpful in identifying which attributes are most strongly represented in the dimensions associated with each canonical variable. A measure of each canonical variable's contribution to group separation includes the canonical correlations, that is, the correlation between the canonical variables and the groups. Another useful canonical technique is to test the hypothesis that all remaining *canonical correlations* (i.e., associated with any variable not in the current model) are zero. The appropriate test statistic is a generalization of the Wilks's lambda statistic.

Wilks's lambda is the MANOVA equivalent of the F tests for the presence of effects in analysis of variance (ANOVA) models and tests for differences between the mean attribute vectors of the groups and for the presence of interaction effects in MANOVA models. The range of Wilks's lambda is from zero to unity. Lower values indicate larger mean differences, thus indicating stronger group separation. One can use transformations of Wilks's lambda statistic that follow an approximate chi-square distribution or an exact F-distribution, when the null hypothesis that all group means are equal holds. Huberty (1984) discussed partial Wilks's lambdas that test the contribution of additional variables to group separation.

The MANOVA test for equality of mean vectors across groups is equivalent to a test that none of the canonical variables of canonical DA contribute to group separation. However, in canonical DA, after testing the discriminant power of all the canonical variables, the analyst can proceed using a second Wilks's lambda to test if all the canonical variables, except the first, contribute anything to the discriminatory power already provided by the first canonical variable. The analysis will proceed in this fashion until all important dimensions of group separation (represented by the canonical variables with discriminatory power) have been identified. The canonical variables with large enough F values should be kept in the analysis; the other canonical variables do not significantly contribute to group separation and should be omitted.

Assumptions

Like many other statistical methodologies, descriptive DA methods are based on a series of assumptions about the true characteristics of the populations underlying the data. In canonical DA, these assumptions are as follows: (a) the attributes associated with different entities should be independent; (b) within each group, the attributes of the entities follow a multivariate normal distribution; (c) the variance–covariance structures of the entity attributes should be equal across all groups. A discussion of these assumptions follows.

The first assumption is common to most parametric methods and requires that the attribute values of one entity should not be affected by any of the other entities used in the study. If the observations are correlated, then the conclusions of canonical DA will no longer be valid.

The second assumption of canonical DA is that the attributes are *multivariate normally distributed*. The multivariate normal distribution is the generalization of the univariate normal distribution to more than one variable and assumes that the following conditions hold: (a) Each attribute is normally distributed, and (b) all regression models in which each attribute is in turn set to be the dependent variable, and the remaining attributes set to independent variables, should be linear.

The first step in checking the multivariate normality assumption is to analyze the univariate distributions of all variables involved. Although the univariate normality of all individual variables does not guarantee multivariate normality of their joint distribution, in practice, sets of random variables that are univariate normally distributed tend to follow the multivariate normal distribution. The univariate normality property can be verified visually by comparing histograms of the attribute distributions in each group with the normal curve. If the individual attributes are normally distributed, their histograms should be approximately symmetric and have a bell shape. More rigorous univariate normality tests are the Kolmogorov-Smirnov and the chi-square goodness-of-fit tests (see Neter, Wasserman, & Whitmore, 1988).

The second step in evaluating multivariate normality assumptions involves constructing two-dimensional scatter plots for all pairs of attributes, for each group individually. In the case of multivariate normality, these plots should reveal either no relation (in which case all points would be randomly scattered in the graph) or a linear relation among the attributes under consideration. If the plots indicate that some of the attributes are related in a nonlinear way, then the hypothesis of multivariate normality cannot possibly hold, even if each individual attribute follows a univariate normal distribution. An in-depth discussion of both informal and formal tests of multivariate normality can be found in Koziol (1986) and Baringhaus, Danschke, and Henze (1989).

The third assumption of canonical DA is that of equal variance–covariances across groups, requiring that the dispersion and dependence structure (i.e., the covariances) of the attribute distributions is the same for all groups. The assumption of equal dispersions (homogeneity of variances) across groups can be checked visually by comparing the groupwise histograms of the attribute distributions. The equality of the dependence structures can easily be verified by comparing two-dimensional scatter plots of paired attribute values across groups. The assumption of equal variance–covariances can be formally tested by the Bartlett test (Bartlett, 1937).

The extent to which deviations from the assumptions affect a useful interpretation of the results of a canonical DA depends on the nature of the analysis and the seriousness of the deviations. The canonical discriminant functions themselves can be derived regardless of whether the within-group attribute distributions are multivariate normal or not. However, if the variance–covariance structures are not equal across groups, some of the properties of the canonical variables will be lost. For instance, statistical tests used to evaluate the significance of the canonical discriminant functions require both equal variance–covariances and multivariate normality. If these assumptions are violated, the p values are invalid.

In terms of a useful interpretation of canonical DA, the most serious violation of the hypothesis of equal variance–covariances occurs when the correlations among the attributes structures are different across groups. The effects of not having the same variance–covariances across groups are relatively mild, as long as the correlations of attributes across groups are similar (Kendall, 1957). Even if the variance–covariances are different across groups, the groups can have similar correlation structures, provided that the variance–covariances are proportional across groups.

As is the case with other statistical methodologies, when testing the assumptions underlying DA methods, it is important to realize that these assumptions will rarely be satisfied perfectly. However, most of the DA methods discussed in this chapter are fairly robust and can give reasonable results, even in the presence of slight departures from the underlying assumptions. Note also that the statistics and associated p values used in hypothesis testing do not provide any information about the extent to which the assumptions are violated.

In addition, if the number of observations is small, it is only possible to detect substantial departures from the underlying assumptions. It is much easier to detect serious violations of the assumptions if the sample size is large, as large samples yield more precise estimates. In this case, sample information suggesting slight departures from the null hypothesis can provide strong evidence against the validity of the assumptions, for instance, in the form of very small p values. In this situation, it is good practice to use alternative methods, for example, an analysis of histograms and plots of the attribute values, to assess the extent to which the assumptions underlying the null hypothesis are violated. If the assumptions are grossly violated and the problem cannot be fixed by transforming the data, then one may use nonparametric DA.

Variable Selection and Ordering

Before starting the actual statistical analysis, the analyst will typically already have developed an explanatory theory about the problem and will have identified a collection of attributes that should contribute to explaining differences among the groups. Nevertheless, the relative importance of these attributes and the best way to combine information of several attributes (i.e., canonical variables) are usually not yet known. Therefore, in the data collection phase, information is usually gathered on a large set of variables, some of which may contribute to the explanation of group differences whereas others may not. The results of the statistical analysis are then used to decide which variables to keep in the model and which ones to drop from the model.

Thus, the problem of selecting a subset of variables from a larger set is important and frequently encountered in practice, and it is not surprising that the variable selection problem has received considerable attention in the DA literature. A related problem is that of ordering the variables according to their relative importance. In descriptive DA, *importance* is often defined as contribution to group separation. An importance ordering of the variables can be used to tentatively identify several best subsets of variables and can also be used for descriptive purposes.

The most widely used method for selecting and ordering variables in both descriptive and prescriptive DA is *stepwise* DA variable selection. *Forward* stepwise methods start the analysis without any variables and successively add and delete variables until a prespecified stopping criterion is satisfied. The stopping criterion used in most stepwise methods is based on the p value of an F statistic associated with a test that determines if including or deleting the next variable adds significantly to explaining group separation, beyond the contribution of those variables that are already present in the model. Conversely, stepwise *backward elimination* methods start the analysis with all variables included in the model and successively eliminate the least important variables from the model, until only significant variables remain. The *F* statistics used in stepwise DA are transformations of partial Wilks's lambdas, which are discussed in Huberty (1984).

In spite of their widespread use, stepwise methods for selecting and ordering variables in DA have some theoretical problems. The major problem is that only a subset of all possible variables is taken into consideration at each step of the analysis. However, this subset is not necessarily the best one or the one that will be used in the final analysis. Huberty (1984) pointed out that there are only two situations in which a stepwise analysis may be useful. One situation is when the initial set of variables is very large, in which case a stepwise procedure may serve as a useful preliminary tool for screening and discarding those variables that obviously contribute little or nothing to group separation, beyond the variables already present in the model selected by the stepwise procedure. Stepwise DA is also useful if the optimal ordering of the variables is known a priori, in which case the purpose of stepwise DA is to determine the appropriate number of attributes.

Several indicators of the relative importance of the attributes have been proposed in the literature. Among the most common are (a) the order of entry of the attributes in the stepwise program, (b) the correlation between each attribute and the first canonical discriminant function (the squared correlation of a given attribute and the first canonical variable is often used as an index of this attribute's contribution to group separation), (c) the standardized weight of the first discriminant function (the standardized weight of the *j*th variable of the first discriminant function equals the coefficient of X_j in the first discriminant function multiplied by the within-group standard deviation of X_j), and (d) the *F*-to-remove statistic for each attribute, given by a stepwise program when all variables are included in the analysis; this statistic measures the impact of removing one single attribute from the full set of attributes.

According to Huberty (1984), the *F*-to-remove statistic should be used, because it is the only one of these methods that considers all variables simultaneously; it is also appropriate as a single indicator across all dimensions of separation and interprets importance in a fairly clear and straightforward manner.

In descriptive DA, the goodness of discrimination afforded by a given subset of variables is often assessed by examining the Wilks's lambda statistic of the model (Morrison, 1990; Wilks, 1932). McKay and Campbell (1982a) noted that although Wilks's lambda is a reasonable index of total group separation, it fails to recognize the different dimensions in which the groups may differ. It may be the case that a subset of variables, with a low Wilks's lambda value, ignores one of the dimensions of group separation, whereas a subset with a slightly higher Wilks's lambda identifies all important dimensions of group separation. Thus, McKay and Campbell (1982a) suggested that in addition to Wilks's lambda, the eigenvalues associated with all of the important canonical variables should be considered as well. Subsets of variables for which the eigenvalue of some important canonical variable becomes too small should be avoided. In addition, theoretical considerations that are based on substantive knowledge of the problem domain can also play a useful role in the selection of variables.

Ideally, one would like to consider all possible subsets of variables and select a subset that includes most or all of the theoretically relevant variables, has a fairly low Wilks's lambda value, and identifies all important dimensions of group separation.

If the total number of variables exceeds 20, it would be very time-consuming or even prohibitive to consider all possible subsets of variables. In such cases, one might resort to an initial screening method to reduce the number of variables under consideration. The analysis involving all possible subsets would then be performed only for those variables not eliminated in the screening process. Potential criteria that can be used in the screening process include the p value associated with an F statistic to test univariate differences among group means and the order of entry of the variables provided by a stepwise discriminant method. If some of the variables are highly correlated, for instance, when several variables measure essentially the same phenomenon, the analyst may want to only use one in the analysis. Factor analysis may be an appropriate dimension-reducing procedure in such cases (see chapter 4).

Example 1

We exemplify the use of descriptive DA techniques using the example by Porebski (1966) introduced at the beginning of the chapter. The purpose of this analysis is to discover if, and the extent to which, four different types of college majors can be discriminated from one another on the basis of their test scores. The psychologist has collected the test results of 1,348 students enrolled at the college in previous years. The three test scores (Arithmetic, English, and Forms–Relations) are denoted by X_1, X_2, and X_3, respectively. The average test scores ($\bar{X}_1$, $\bar{X}_2$, and $\bar{X}_3$) and the

Table 1

Example 1, Mean Student Test Scores

Major	n	$\bar{X}_1$	$\bar{X}_2$	$\bar{X}_3$
E: Engineering	404	27.878	98.361	33.596
B: Building	400	20.650	85.425	31.513
A: Art	258	15.010	80.307	32.009
C: Commerce	286	24.378	94.941	26.686
Total	1,348	27.528	90.341	31.208

Note. From "Discriminatory and Canonical Analysis of Technical College Data," by O. R. Porebski, 1966, *British Journal of Mathematical and Statistical Psychology, 19,* p. 216. Copyright 1966 by the British Psychological Society. Reprinted with permission.

number of observations in each group for each major program of study (A = art, B = building, C = commerce, and E = engineering) are given in Table 1.

Initial Analysis

The validity and relevance of the conclusions drawn from the analysis will largely depend on the extent to which the assumptions underlying the methodologies used in the analysis are satisfied. The analysis begins by performing initial diagnostic checks to identify important general characteristics of the data set. The histograms of the sample distribution of the test scores appear to be approximately symmetric, with a shape resembling the bell form of the normal distribution. Furthermore, two-dimensional plots for all pairs of test scores (i.e., X_1 vs. X_2, X_1 vs. X_3, and X_2 vs. X_3) do not reveal any clear nonlinear relationship between the variables. For reasons of brevity, these histograms and plots are not shown in this chapter, nor are they typically displayed in journal articles. Thus, on the basis of the diagnostics, the psychologist concludes that there was no evidence that the assumption of multivariate normality was seriously violated.

Next, the researcher uses Bartlett's (1937) test to evaluate the hypothesis of equal variance–covariances across groups. The null hypothesis is rejected at the 1% significance level, thus suggesting that one of the basic assumptions of canonical DA was violated. At this point, the researcher has to make a choice between using a parametric or a nonparametric DA. An examination of the sample variance–covariance matrices (not shown) by group reveals that these matrices are roughly proportional. Therefore, this data set does not suffer from the most serious potential violation of the assumptions, that is, different correlations across groups.

As mentioned above, even small departures from the assumptions are easily detected if the sample size is large in relation to the number of attributes. With 1,348 observations, the data set is certainly large. Despite the differences in magnitude of the variance–covariances across groups, the researcher decides to continue the parametric analysis, because the correlation structures appear to be similar. However, the results should be interpreted with caution, because the p values associated with the parametric statistical tests (to be performed in the formal analysis) will not be strictly valid.

Formal Analysis

The researcher next conducts a MANOVA, with test scores serving as dependent variables and the student major serving as the grouping (class) variable. With a p value of less than .005, the resulting Wilks's lambda value of .629 is highly significant, indicating that the mean test scores among the four majors are statistically different at significance levels exceeding .005.

The one-way MANOVA analysis has already revealed that the mean test scores differ across the various majors, suggesting that some linear combinations of the attributes (dependent variables) discriminate the groups from one another. However, the MANOVA does not reveal which variables should be combined in what manner, to discriminate among the groups. A canonical DA would accomplish this task. Thus, the researcher is now interested in studying the nature of these differences in more detail, using a canonical DA. Since there are $m = 3$ attributes and $k = 4$ groups, there will be $k - 1 = 3$ canonical discriminant functions. Each discriminant function represents a unique linear combination of the attributes. The first discriminant function is the most important combination of variables to maximally discriminate among the groups. The third discriminant function is the least important, and the second discriminant function is intermediate in terms of its ability to discriminate among the groups. Each discriminant function defines a canonical variable.

In Table 2, the first discriminant function defines the canonical variable denoted by Z_1, the second discriminant function defines the canonical variable Z_2, and the third function defines Z_3. For example, the unstandardized first discriminant function (canonical variable) is $Z_1 = .117X_1 + .019X_2 - .027X_3$. Thus, Z_1 is a new variable, called a composite

Table 2

Canonical Analysis, Example 1

Canonical variable	Unstandardized discriminant coefficients						Standardized discriminant coefficients		
	X_1	X_2	X_3	Eigenvalue	% contribution	p	Arithmetic X_1	English X_2	Forms–Relations X_3
Z_1	0.117	0.019	−0.027	0.450	82.57	.000	0.869	0.364	−0.221
Z_2	0.011	−0.005	0.119	0.090	16.61	.000	0.083	−0.093	0.987
Z_3	−0.077	0.051	0.151	0.005	0.86	.003	−0.575	0.961	0.126

variable in chapter 1, which represents the best way to combine the attributes to discriminate among the groups. Note that the unstandardized discriminant coefficients do not measure the importance of each attribute in the discriminant functions, because they reflect different scales of measurement. To compare the relative importance of each attribute in each discriminant function, standardized discriminant coefficients, obtained by multiplying the unstandardized coefficients by the within-group standard deviation of the attributes, are provided as well.

The importance of the canonical variables (Z_1, Z_2, and Z_3) in terms of group separation is reflected by their corresponding eigenvalues. For a given canonical variable, the eigenvalue for that canonical variable, divided by the sum of all the eigenvalues, is a ratio that can be interpreted as the proportion of between-groups variation explained by that canonical variable. The larger the eigenvalue, the more important the canonical variable is for group separation. As seen in Table 2, the eigenvalue associated with Z_1 (.45) is larger than that for Z_2 (.09), which in turn is larger than the eigenvalue for Z_3 (.005). Furthermore, in terms of between-groups separation, Z_1 accounts for $.45/(.45 + .09 + .005) = .8257$, or 82.57%, of the between-groups variation. Note that Z_2 accounts for 16.61% of the between-groups variation and Z_3 explains less than 1% of the variation.

Finally, in the fourth column from the right of Table 2, for each canonical variable, the p value is provided for an F test (not shown) of the null hypothesis that the canonical variable (and all remaining canonical variables) is (are) zero. For example, for Z_1, $p = .000$, indicating that Z_1, and possibly Z_2 and Z_3, contribute to statistically significant levels of intergroup separation. If the F statistic is not significant (i.e., $p > .05$), then the corresponding canonical variable, and all remaining canonical variables, do not add significantly to between-groups discrimination.

In interpreting the p values in Table 2, keep in mind that these values are not strictly valid, due to the previously noted violation of the equal variance–covariance assumption. The (low) p values reveal that each of the canonical variables Z_1, Z_2, and Z_3 contribute significantly to explaining the differences among group means. Thus, the four student populations (the engineering, building, art, and commerce majors) appear to differ with respect to these three canonical variables. However, the first two canonical variables (Z_1 and Z_2) explain about 99% (82.57% + 16.61%) of these differences, whereas the third canonical variable (Z_3) contributes less than 1%.

The mean scores of each canonical variable for the four student

Table 3

Mean Values of Canonical Variables, Example 1

Major	Canonical variable means		
	Z_1	Z_2	Z_3
Engineering	4.240	3.813	3.333
Building	3.210	3.549	3.204
Art	2.436	3.570	3.388
Commerce	3.950	2.971	3.325

populations are given in Table 3. Because most of the group differences are explained by the first two canonical variables (99%), we study the differences by graphing the group means for each student population in the two-dimensional space of Z_1 and Z_2. From Figure 3, we see that Z_1 appears to clearly separate the four populations, whereas Z_2 accounts for most of the mean difference between Group C on the one hand and Groups A, B, and E on the other.

The next step in the analysis involves the substantive interpretation

Figure 3

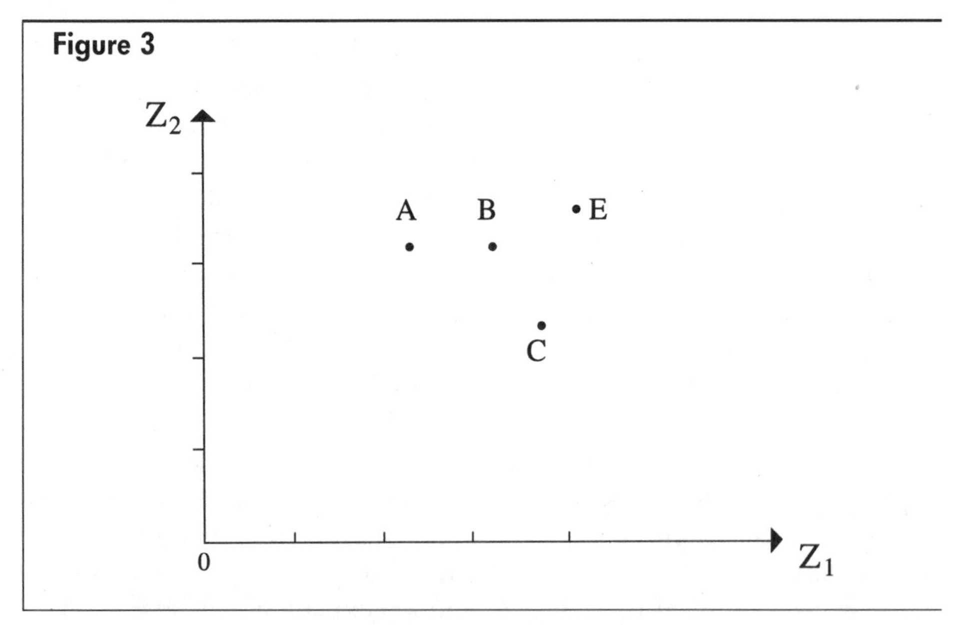

Mean scores of canonical variables Z_1 and Z_2 for each student population, Example 1. From "Discriminatory and Canonical Analysis of Technical College Data," by O. R. Porebski, 1966, *British Journal of Mathematical and Statistical Psychology, 19,* p. 216. Copyright 1966 by the British Psychological Society. Adapated with permission.

Table 4

Correlations, Example 1

	Test scores		
Canonical variables	Arithmetic X_1	English X_2	Forms–Relations X_3
Z_1	.915	.550	−.025
Z_2	.246	.017	.994
Z_3	−.322	.835	.106

of the dimensions reflected by each canonical variable. General knowledge about the nature of the particular abilities required in each major, combined with the relative position of each group mean in the canonical variable space (the plot of Z_1 vs. Z_2) can greatly facilitate the substantive interpretation of the canonical variables. In our example, one of the research questions the psychologist is interested in involves determining which abilities may be reflected in Z_1 and Z_2 and how these variables contribute to the separation among the majors. Thus, it is necessary to examine the relationship between the canonical variables and the original variables (test scores). The correlations between the canonical variables and the test scores are given in Table 4.

As seen from the standardized discriminant coefficients in Table 2, as well as from the canonical correlations in Table 4, the first canonical variable (Z_1) is strongly related to arithmetic (r = .915) and moderately related to verbal (English) ability (r = .550). Accordingly, the researcher interprets this canonical discriminant function as reflecting a measure of general learning ability. In contrast, Z_2 only correlates with forms–relations ability (r = .994), a measure of spatial ability. Finally, Z_3, which explains less than 1% of the between-groups variation, is associated only with verbal ability (r = .835). This interpretation of the canonical variables is accepted by the psychologist, because it is consistent with the correlations and group differences expected based on a priori knowledge of what the three tests should measure.

Summarizing, the researcher concludes that the four student populations differ in essentially two dimensions. The first dimension (Z_1) is a general learning ability with an emphasis on quantitative and, to a lesser extent, verbal skills. Along this dimension, engineering majors rank the highest, followed by students in commerce, building, and finally arts. The second dimension (Z_2) relates to spatial ability (see Table 3). Those students majoring in commerce tend to score lower on this dimension than the students in the other three programs, but no clear difference in spatial ability is found between students in art, building, and engineering. Finally, Z_3 is defined almost exclusively by English scores (r = .835). However, even though Z_3 is associated with significant between-groups discrimination, Z_3 explained less than 1% of the total between-groups variation. Thus, Z_3 does not contribute much beyond Z_1 and Z_2 in terms of an ability to separate or distinguish the different student groups.

Predictive Discriminant Analysis

The most important problem in predictive DA is that of *classifying* entities with unknown group membership, using some *classification rule*, which is based on the attribute values of the entity. A classification rule typically consists of a functional form of the attributes and a decision rule that is used to predict an entity's group membership based on the value of its classification function. An example of a simple (linear) classification rule for a two-group discriminant problem is to predict that an entity belongs to Group A if $2X_1 + 3X_2 - X_3 \geq 3$ and to Group B otherwise. In this function, X_1, X_2, and X_3 are attribute variables. An entity with $X_1 = 4$, $X_2 = -2$, and $X_3 = 1$, with a classification score of 1, which is less than 3, would be classified into Group B. Of course, there are various different types of classification rules, and these rules are not always as transparant as the one described above. Thus, most of the theory of predictive DA deals with the derivation and estimation of classification rules that are optimal in some sense, subject to certain assumptions (or no assumptions, depending on the method) about the underlying characteristics of the groups (populations).

Two distinct types of classification rules in predictive DA are *population-* and *sample-*based rules. A *population rule* optimizes some conceptual classification criterion, is mathematically derived, and requires exact knowledge of the probability distributions of the attributes. Of course, in practice, the true probability distributions of the attributes are usually unknown, so that a theoretically optimal population rule can rarely be determined. Consequently, population-based classification rules are mostly of theoretical interest.

In practice, it is necessary to estimate a classification rule based on a sample of entities with known group membership, replacing the (unknown) population parameters by their sample estimates. Such a rule is called a *sample-based classification rule*, and the sample used to build or estimate this rule is the *training sample*. The classification score of an entity is the value of the classification function for the entity's attribute values.

Common criteria used to derive optimal classification rules include the following: (a) minimizing the total probability of misclassification, (b) minimizing the expected total costs of misclassification, (c) minimizing the highest group-specific probability of misclassification, (d) minimizing the highest group-specific expected costs of misclassification, and (e) minimizing the total misclassification rate. As mentioned before, an entity is misclassified if the group to which it is predicted to belong is incorrect. The estimated misclassification rate is the proportion of misclassified observations in the sample. Naturally, the estimated misclassification rate equals 1 minus the estimated hit rate (the proportion of correctly classified observations in the sample).

Misclassification costs can vary across groups, reflecting that certain types of misclassification errors are more serious than others. For instance, when deciding on a loan application, the error of granting a loan that later defaults may be deemed more serious than the error of denying a loan to a customer who in fact would have paid up in full. Differential misclassification costs serve to penalize against the more serious types of errors. In a sense, the concept of differential misclassification costs is similar to that of the differential importance of the Type I and Type II errors in hypothesis testing.

To derive and estimate classification rules on the basis of one of the above criteria, it is necessary to define a priori (prior) probabilities of group membership. Moreover, one needs to prespecify the relative misclassification costs, if the criterion is based on expected misclassification costs. This information is typically based on previous knowledge of the problem.

If the training sample represents a random sample from the groups (populations), the prior probabilities of group membership can be estimated by the relative group proportions in the training sample. However, this procedure requires that the training sample is indeed a random sample, that is, the probability that a randomly selected entity from the training sample belongs to a particular group should equal the probability that a future observation will belong to this group.

Note that this requirement is often not satisfied in practice. For instance, consider the application in which one wishes to predict whether a firm will go into bankruptcy within the next year, implying two groups: those firms that do go bankrupt and those that do not. It will quite likely be much easier to collect data on the latter group.

In the bankruptcy problem, the group sizes will be highly unbalanced, because the population of nonbankrupt firms will be much larger than that of bankrupt ones. However, in many DA studies, the analyst artificially ensures that each group in the training sample has approximately the same number of entities, even though some populations may be much larger than others. This is a dangerous practice, because the sample may be strongly biased if the relative representation of both groups in the training sample does not correspond with the prior probabilities of group membership. Clearly, using such biased samples to estimate misclassification probabilities results in models that do not generalize well to independent samples, unless corrective action is taken in the interpretation of the classification results obtained by traditional methods. Thus, in this case, it is especially important to take special care that the population probabilities are correctly estimated and that the classification rule is based on these probabilities.

Classification Methods That Assume Multivariate Normal Attributes

The most widely used classification methods are those derived by minimizing the probability of misclassification, or expected misclassification cost, under the assumption that for each group the probability distribution of the attributes is *multivariate normal*. In this section, we focus on this class of methods, for which the theoretically optimal population rules are well-known. Therefore, we can use the population-based classification rules, replacing the unknown population parameters by their sample estimates.

If all variance–covariances are equal across the groups, the population classification rule is based on a set of k (one for each group) linear combinations of the attributes, known as linear classification rules (classifiers) or discriminant functions. Although widely used in practice, the term *discriminant function* is somewhat confusing in this context, because the canonical discriminant functions in descriptive DA are known as discriminant functions as well. Hence, we use the term *classification function*. Once the appropriate classification function has been established, each entity is assigned to the group for which it has the highest classification score. Unlike descriptive DA, in which the number of discriminant functions equals either the number of attributes or the number of groups minus one (whichever is smaller), in predictive DA the number of classification functions is equal to the number of groups. To predict the group membership status of a given case, enter the attribute values from that case into each of the classification functions. Finally, assign the case to the group for which the classification score is largest.

In the two-group case with variance–covariances that are equal across groups, the linear canonical discriminant function, known as *Fisher's linear discriminant function* (Fisher, 1936), can be used both for predictive (classification) and descriptive purposes (Fisher's research article from 1936 is credited with initiating the field of DA).

If the attributes follow multivariate normal distributions with different variance–covariances across groups, the optimal classification rule is to assign each entity to the group for which it has the highest classification score, using a *quadratic function* of the attributes (i.e., a combination of linear, squared, and cross-product terms of the attributes). Of course, nonlinear functions imply that the groups are separated by nonlinear surfaces. For example, for a problem with two attributes, X and Y, a quadratic model would have the terms X, X^2, Y, Y^2 and XY and the separating surfaces would be quadratic. In practice, the analyst does not need to worry about this point, as in the major statistical DA packages, the quadratic and cross-product terms are automatically included in the analysis if the variance–covariances are specified as different across groups. For two groups, the difference between the classification rules for the groups is known as *Smith's quadratic discriminant function* (Smith, 1947). Predictive DA based on quadratic classification rules is known as quadratic discriminant analysis.

The question of whether linear or quadratic classification functions are appropriate depends primarily on the nature of the variance–covariances across groups. However, other considerations may play a role as well. For example, the estimation of quadratic classification rules is more difficult (and less precise) than that of linear rules. When the variance–covariances differ across groups, there is a trade-off between using a classification rule of the correct form (quadratic) and using a rule whose parameters can be estimated more accurately (linear). Due to the quadratic and cross-product terms, the quadratic function requires the estimation of more parameters. If the size of the training sample is small, or if the number of attributes is large, the issue of accuracy of estimation becomes more important, and a linear classification rule is usually the preferred choice. This issue exemplifies the principle of *parsimony* (i.e., everything else being approximately equal, always select the simplest model), which applies to any statistical analysis. However, as the differences between the variance–covariances across groups become more pronounced, or as group separation decreases, the nonlinearity of the population rule becomes more accentuated, and a quadratic rule is increasingly preferred.

Marks and Dunn (1974) analyzed some of these trade-offs between linear and quadratic rules in a simulation study, for several different data conditions. For example, Marks and Dunn found that for training samples of 50 observations with two uncorrelated attributes and moderate group overlap, a linear sample-based classification rule would yield lower estimated misclassification rates than a quadratic rule, as long as the ratio between the variances in both groups was less than 2. A quadratic rule tends to perform better for ratios exceeding 2. For problems with 10 attributes, and the other factors the same as above, the use of quadratic models is preferred when the ratio of variances between groups is at least between 4 and 8. These results support the notion that rejection of the null hypothesis of equal variance–covariances across groups does not automatically imply that a sample-based quadratic rule will be preferred to a linear one. In case of doubt, it is good practice to evaluate the relative classification performance of both the linear and the quadratic models.

As mentioned above, classical linear or quadratic predictive DA is based on the assumption that the attributes follow a multivariate normal distribution. If this assumption is clearly not satisfied, alternative methods of analysis should be used. One frequently used approach is to apply some appropriate transformation to the original attributes, with the objective of reducing the degree of violation of the multivariate normality assumption. The parametric methods described above can then be applied to the transformed attributes. Several simulation studies have shown that the presence of asymmetrically distributed (skewed) attributes is one of the most serious violations of the normality assumption, in terms of the consequences for predictive DA (Clarke, Lachenbruch, & Broffit, 1979; Lachenbruch, Sneeringer, & Revo, 1973). The most widely used nonlinear data transformation to normalize skewed distributions is the natural logarithmic transformation. However, it is sometimes true that no transformation will produce symmetric distributions and homogeneous variance–covariances between groups. In such situations, a variety of nonparametric approaches, which do not make assumptions about variance–covariance structures, are appropriate alternative forms of DA. Of these alternative approaches, logistic regression analysis is the most frequently used (see chapter 7).

Evaluation and Interpretation of Classification Rules

Once a classification rule has been derived or estimated, it is important to assess its classificatory performance and to interpret its results. An analysis in which the classification performance is estimated on the basis of the same observations that were used to derive the classification rule is often called an *internal* classification analysis. An analysis that uses different observations to estimate the classification rules and to assess their classificatory performance is termed an *external* classification analysis.

The most commonly used performance measures for classification rules are the *hit rate* (1 minus the misclassification rate) and the *misclassification costs*. Counting estimates are particular estimates of hit rates. Hit rates are probabilities of correct classification, and misclassification rates are probabilities of misclassification. In addition to the overall misclassification rate, it is often useful to evaluate the estimated misclassification rates of the individual groups as well, because the misclassifications associated with the optimal classification rule that minimizes the overall estimated misclassification rate may be unevenly distributed among the different groups, rendering the overall rule undesirable.

Misclassification costs are simply misclassification probabilities weighted by their relative importance. Most frequently, the misclassification costs are assumed to be equal across groups, enabling a direct comparison and analysis of the estimated classification hit rates. Therefore, we focus on the estimation of hit rates.

Huberty, Wisenbaker, and Smith (1987) distinguished among three different types of hit rates: (a) the *optimal* hit rate $p^{(1)}$, (b) the *actual*, sample-based hit rate $p^{(2)}$, and (c) the *expected actual* hit rate $p^{(3)}$. The optimal hit rate, $p^{(1)}$, is the hit rate of the population classification rule and can be interpreted as a theoretical measure of group overlap. As such, $p^{(1)}$ is particularly useful in descriptive DA and can be used as a benchmark against which hit rates of sample-based classification rules can be compared. The actual hit rate $p^{(2)}$, also known as the *conditional* hit rate, is the hit rate for a given sample-based classification rule, that is, the estimated probability of correct classification when the estimated classification rule is applied to new observations. Because it is sample-based, and reflects the usual situation in practice, $p^{(2)}$ is the most important hit rate in applied predictive DA research. The expected actual hit rate $p^{(3)}$, also called the *unconditional* hit rate, is the expected value of $p^{(2)}$ when a particular method of deriving classification rules is applied to different training samples generated under the same sampling conditions.

There are parametric and nonparametric methods of estimating hit rates. The most important parametric hit rate estimators are based on the assumption that the attributes follow the multivariate normal distribution. For the two-group case, when the attributes are multivariate normal with the same variance–covariances, the optimal hit rate is a monotonic function of the *Mahalanobis distance* (1936) between group means. The Mahalanobis distance is a distance index between random vectors, which takes into consideration their variance–covariance structure. Distances between variables with low dispersion are weighted more heavily than distances between highly dispersed variables. One of the most common estimators of $p^{(1)}$ replaces the population Mahalanobis distance Δ^2 by the sample distance D^2 (i.e., uses the Mahalanobis distances based on sample, rather than population variance–covariances), in the formula for the true hit rates. This estimator (known as the *D method* estimator) is known to be optimistically biased (it tends to give estimates of $p^{(1)}$ higher than its true value). An almost unbiased estimator of $p^{(1)}$ replaces Δ^2 by the *corrected sample* Mahalanobis distance $\tilde{D}^2$, in the formula for the true hit rate. $\tilde{D}^2$ is a simple transformation of D^2 that corrects the bias of the D method estimator. A discussion of these and other parametric estimators of hit rates can be found in Sorum (1972), and in Huberty et al. (1987).

Many DA studies have reported sample-based counting estimates without specifying which particular hit rate is being estimated. Thus, it is not always clear which of the three hit rates ($p^{(1)}$, $p^{(2)}$ or $p^{(3)}$) the authors intend to estimate. In most applications, the hit rate of interest is $p^{(2)}$.

The proportion of correctly classified cases from an internal classification analysis is the most obvious of the counting estimators. Nonparametric estimators of the hit rate are usually based on a count of the correct classifications in the sample. This count is obtained by applying the classification rule under consideration to a sample of observations and counting the number of correctly classified observations. The estimated hit rate is the count of correctly classified observations divided by the total number of observations in the sample. Similarly, the estimated misclassification rate is the proportion of incorrectly classified observations. However, the estimated hit rate from an internal analysis yields an optimistically biased measure of any of the three rates: $p^{(1)}$, $p^{(2)}$, and $p^{(3)}$. The reason is that because the classification rule is estimated using the training sample data, applying this rule to these very same data will systematically overestimate the hit rate (and underestimate the misclassification rate). This bias can be avoided by assessing the classification performance by means of observations that were not used in estimating the classification rule, that is, through an external classification analysis.

One way to perform an external analysis is to split the available data in two separate data sets, one data set for estimating the classification rule (*training sample*), and the other data set for evaluating the classificatory performance of the estimated rule (*holdout* or *cross-validation sample*) . The estimated hit rate for the holdout sample is then taken as an estimate of $p^{(2)}$. Although it yields an unbiased estimate of the hit rate, the holdout method has several potential drawbacks. First, it requires a large sample size, which may not be available. Second, it is not clear how to divide the original data into training and holdout samples. If the training sample is small, then the classification rule itself may not be very accurate, because it is based on few observations. If the holdout sample is small, the hit rate estimate may not be very accurate, either.

A more useful estimator of $p^{(2)}$ is the *leave-one-out* (L-O-O) estimator, popularized by Lachenbruch (1967). In the L-O-O method, each observation is deleted in turn from the training sample of size n and is classified by means of the classification rule estimated on the basis of the remaining $n - 1$ observations. The proportion of observations left out from the training sample that are correctly classified is then taken as the hit rate estimate. The L-O-O estimator has been shown to be an almost unbiased estimator of $p^{(2)}$. However, it has a large sample variability, so it is not very precise.

A combination of the holdout and L-O-O methods is to hold out not one, but a randomly selected subset of n_1 observations; estimate a classification rule on the basis of the remaining $n - n_1$ observations; and evaluate the proportion of correctly classified holdout observations. Repeating this procedure a number of times, and calculating the average proportion of correctly classified holdout observations, yields an almost unbiased estimate of $p^{(2)}$.

Another estimator of $p^{(2)}$ that has recently gained popularity is the *maximum-posterior-probability* (M-P-P) estimator (Fukunaga & Kessell, 1973; Glick, 1978; Hora & Wilcox, 1982). This estimate is simply the average posterior group membership probability for all entities, for those groups to which the entities were assigned by the classification rule. This probability can be calculated by means of either an internal or an external analysis. The M-P-P estimator calculated by means of an internal analysis is optimistically biased. However, it is possible to combine the M-P-P and L-O-O methods, by calculating the average posterior probabilities of correct classification for a rule based on all observations except those being left out for classification. This estimator of $p^{(2)}$ is only slightly biased, has a small sample variability, and therefore is both relatively precise and accurate.

Variable Selection and Ordering

The issues of selecting the best set of variables and assessing the relative importance of each variable have not been studied in as much detail in the context of predictive DA as in descriptive DA. In fact, many of the variable selection and ordering methods and evaluation criteria originally developed for descriptive DA purposes are directly applicable to prescriptive DA. In particular, the descriptive stepwise variable selection procedures available in most statistical packages are based on F statistics that measure the importance of each variable or set of variables in terms of group separation and can easily be used in a predictive analysis.

However, several authors (Habbema & Hermans, 1977; Huberty, 1984; McKay & Campbell, 1982b) suggested that predictive DA should use a criterion directly related to the predictive power of the classification rule, for instance, an unbiased estimate of $p^{(2)}$, in the process of selecting and ordering variables. Habbema and Hermans (1977) proposed a forward selection procedure, in which the next variable to be included in the model would either maximize the value of the L-O-O estimate of $p^{(2)}$ or minimize the L-O-O estimate of misclassification costs. Their procedure sequentially adds variables to the model until the inclusion of new variables fails to improve the performance estimate of the classification rule (i.e., the value of either $p^{(2)}$ or the misclassification costs) by more than a prespecified threshold value. McKay and Campbell (1982b) note that although this procedure is a step in the right direction, it does not take into consideration the classification performance of the rule that includes all variables, which in their opinion should be used as a benchmark for the performance of any subset of variables.

Huberty (1984) recommended a forward selection process that used an estimate of $p^{(2)}$ based on an external analysis as the selection criterion to determine which variables should enter, until a stopping criterion was satisfied. However, Huberty did not specify which particular estimate of $p^{(2)}$ should be used. One candidate-stopping criterion would be to consider the best model for each possible subset size and to select the one among these subsets that yielded the highest value for $p^{(2)}$. Huberty also suggested that his proposed selection process be combined with a backward elimination procedure, but he provided no details on how this could be achieved.

Unfortunately, none of these methods have been implemented in the major statistical packages and may be computationally expensive. Thus, to date, the variable selection and ordering analysis remains an ad hoc effort, in spite of its importance in practice. We now present an example of a predictive DA.

Example 2

We exemplify the use of predictive DA with a study by Altman, Haldeman, and Narayanan (1977) to predict bankruptcy of nonfinancial firms. Broadly speaking, nonfinancial firms are companies whose primary purpose is not the management of money or credit. Models to predict bankruptcy are of considerable interest to the business community and abound in the finance literature. Investors can use such models to identify high-risk investments. Lenders can use them as an aid in evaluating the risk of loan default. Auditors can use them to help assess the relative financial health of a firm. Managers can use them as early warning signals, assisting them in identifying potential problems and initiating corrective action.

A decade before the Altman et al. (1977) study, Altman (1968) developed a predictive DA model for bankruptcy of nonfinancial firms. When applied to firms with financial problems, his model was able to predict firm bankruptcy 2 years in advance with an accuracy of close to 95% and predicted firm survival with 70% accuracy. In the early 1970s, several changes rendered Altman's (1968) model obsolete. During this period, the number of bankruptcies increased sharply, and the firms filing for bankruptcy tended to be of a larger size than in the 1960s. In addition, accounting and financial reporting standards changed, requiring an overhaul of the variables included in bankruptcy prediction models. In 1977, Altman et al. built a modified model for predicting bankruptcies, which incorporated these structural changes and overcame some methodological problems present in the earlier model developed by Altman (1968).

Here, a new model became necessary because of changes in the relationship among attributes and entities. This exemplifies a general caveat in the use of DA. For example, a predictive classification function may work well for one culture but not another, for females but not males, or for one time period but not 20 years later.

The sample used in the Altman et al. (1977) study consisted of 53 firms with assets of more than $20 million that filed for bankruptcy between 1962 and 1977 (50 of these 53 firms filed for bankruptcy after 1968), and 58 firms with similar business profiles that did not go bankrupt during the period considered. Initially, 58 firms that had filed for bankruptcy were selected, but 5 of them were later discarded due to unavailability of the data. The sample was obtained by a paired-sample method. That is, each of the 58 bankrupt firms was matched with a nonbankrupt firm of comparable size and in the same industry. The criterion of group division based on whether a firm had actually filed for bankruptcy differed from previous studies, in which bankruptcy or failure had been defined in broader terms, for example, including bond defaults and nonpayment of dividends as an indicator of failure. The group definition of Altman et al. facilitates an objective and natural separation of the two different groups.

Any classification rule in predictive DA depends on the prior probabilities of group membership and on the misclassification costs. Neither of these quantities was known with certainty in this study. Furthermore, bankruptcy probabilities change over time, and the correct misclassification costs depend on the context of the particular application of the predictive bankruptcy model. Thus, rather than deriving a single rule for fixed probabilities and costs, Altman et al. (1977) decided to analyze the performance of their model for various different cost levels and prior probabilities. Because prior probabilities and misclassification costs do not affect the classification function itself, but only the cutoff value used to divide the groups, such an analysis can easily be conducted (see, e.g., Soltysik & Yarnold, 1993).

Previous studies (see, e.g., Dun & Bradstreet, 1976) estimated that over the period of 1965 to 1975, the failure rate of nonfinancial firms in a given year was about 0.5%. However, Altman et al. (1977) decided to modify the scope of this previous study, for several reasons. First, Altman et al. designed their model for predicting bankruptcies several years in advance, not only for the current year. Obviously, a longer term model provides a much more effective warning against potential problems and a more useful tool for timely identification and correction of these problems. Second, many forms of business failures, such as judicial arrangements and takeovers by creditors, do not result in legal bankruptcies but have similar economic effects. These failures were not included in the estimate of 0.5% by Dun & Bradstreet (1976). Altman et al. concluded that prior probabilities of bankruptcy in the 1%–5% range were more realistic for their study.

Misclassification costs were estimated on the assumption that the model was to be used to evaluate loan requests in a commercial bank. In this case, the cost of misclassifying a firm that would go bankrupt equaled the principal loan balance that the bank would not be able to recover, plus administrative and legal costs. Using data from 1971 to 1975, Altman et al. (1977) estimated this cost to average between 60% and 80% of the loan amount. The cost of misclassifying a firm that did not go bankrupt equaled the opportunity cost of denying the loan application. This cost was estimated to be between 1% and 5% of the amount of the loan. For expository purposes, Altman et al. used some typical values within the above ranges to assess the efficiency of their model. However, they recommended revising the prior probabilities and misclassification costs according to the problem at hand in actual applications and changing the cutoff values accordingly.

Altman et al. (1977) used a total of 27 different attributes to measure various aspects of each firm's financial health. For those firms that went bankrupt, the data were collected for the year before bankruptcy. For the nonbankrupt firms, the data were collected for the year before the bankruptcy of the bankrupt firm they were matched with. Recall that the firms were matched on the basis of similar general characteristics. The attributes were divided into several major categories: profitability (6 variables), interest coverage and leverage (7 variables), liquidity (4 variables), capitalization ratios (5 variables), earnings variability (3 variables), and miscellaneous measures (2 variables).

An analysis of histograms (not shown here) was used to identify which variables appeared to have substantial deviations from normality. Variables with an approximately symmetric distribution were kept in their original form. A natural logarithmic transformation was applied to those variables with a distribution that was clearly skewed. Several different procedures were used to select and order the variables, including the following: F statistics to test univariate mean differences among groups, the relative contribution to the explanation of these mean differences, the F-to-remove statistic with all variables included in the analysis, the order of entry using a forward stepwise DA program, the order of elimination in a backward stepwise DA program, and an external analysis comparing misclassification cost estimates in holdout samples. These procedures consistently identified the following subset of variables, listed in the order of their importance (contribution to group separation): cumulative profitability (CP), stability of earnings (SOE), capitalization (CAP), size (SIZE), liquidity (LQD), debt service (DS), and return on assets (ROA).

Except for SIZE, none of these variables needed the logarithmic transformation. Altman et al. (1977) used a test derived by Box (1949) to evaluate the hypothesis that the attributes of the bankrupt and nonbankrupt firms have equal variance–covariances. This null hypothesis was rejected at any reasonable significance level. Thus, the variance–covariances were different, and a quadratic classification rule was applied. However, as the number of attributes kept in the analysis (seven) was relatively large for a data set with a total of $53 + 58 = 111$ observations, Altman et al. decided to estimate a linear rule as well, in anticipation of potential estimation difficulties.

The predictive model performance was evaluated in terms of expected accuracy in each group and expected misclassification cost. Although both the prior probabilities and the misclassification cost for the bankrupt and nonbankrupt groups are substantially different, they roughly balance each other, so that their product can be assumed to be approximately equal over the relevant range of values. This phenomenon of balancing factors frequently occurs in many other DA problems as well. The misclassification probabilities associated with predicting bankruptcy 1 year ahead were estimated by means of the L-O-O method (Lachenbruch, 1967). The model accuracy for predicting bankruptcy 2 to 5 years in advance was estimated using a holdout sample. The estimated hit rates are given, as percentages, in Table 5.

As seen from Table 5, the linear model yielded considerably more accurate estimates than the quadratic one, so that Altman et al. (1977) decided to proceed with the analysis using the linear model. Altman et al. also evaluated the efficiency of the model in terms of expected misclassification cost, for different combinations of prior probability of bankruptcy and misclassification costs (ZETA). The results of that analysis were compared with the expected costs associated with two naive approaches: (a) classify all firms into the nonbankrupt group (MAX), and

Table 5

Overall Classification Results, Example 2

Years prior bankruptcy	Bankrupt firms		Non-bankrupt firms	
	Linear	Quadratic	Linear	Quadratic
1	92.5	85.0	89.7	87.9
2	84.9	77.4	93.1	91.9
3	74.5	62.7	91.4	92.1
4	68.1	57.4	89.5	87.8
5	69.8	46.5	82.1	87.5

Note. Entries are given in percentages. From "Zeta Analysis: A New Model to Identify Bankruptcy Risk of Corporations," by E. I. Altman, R. G. Haldeman, and P. Narayanan, 1977, *Journal of Banking and Finance, 1*, p. 38. Copyright 1977 by Elsevier Science. Adapted with permission.

(b) randomly assign firms to the bankrupt and nonbankrupt groups, according to the prior probabilities of bankruptcy (PROB). The expected misclassification costs for these three strategies (ZETA, MAX, and PROB) are defined by the following expressions:

$$\begin{aligned} EC_{ZETA} &= q_1 e_{12} c_1 + q_2 e_{21} c_2 \\ EC_{MAX} &= q_1 c_1 \\ EC_{PROB} &= q_1 q_2 c_1 + q_2 q_1 c_2, \end{aligned}$$

where EC represents the expected misclassification costs, q_1 is the prior probability of bankruptcy, q_2 is the prior probability of nonbankruptcy (of course, $q_1 + q_2 = 1$), c_1 is the cost of erroneously predicting non-bankruptcy, c_2 is the cost of erroneously predicting bankruptcy, and e_{12}, e_{21} are the estimated misclassification probabilities for the bankrupt and nonbankrupt groups, respectively. To explain EC_{PROB}, the probability that a given firm will file for bankruptcy is q_1, whereas the probability that this firm will be assigned to the nonbankrupt group is q_2. Thus, the expected cost of failing to predict bankruptcies that in fact take place is $q_1 q_2 c_1$. Similarly, the expected cost of erroneously predicting bankruptcy is $q_2 q_1 c_2$.

The results of Altman et al.'s (1977) sensitivity analysis of the ZETA model for various levels of q_1, q_2, c_1 and c_2 (with e_{12} and e_{21} estimated by Lachenbruch's, 1967, L-O-O method) are given in Table 6.

The first section of Table 6, labeled Assumptions, provides five hypothetical business environments in which the model might be used. Of

Table 6

Model Efficiency, Example 2

Business environment	Assumptions				L-O-O estimate of misclassification probabilities			Expected misclassification costs (EC)		
	Prior probabilities		Misclassification costs							
	q_1	q_2	c_1	c_2	Cutoff score	Bankrupt e_{21}	Nonbankrupt e_{12}	ZETA	MAX	PROB
1	.02	.98	.70	.02	−0.33	.076	.070	.0024	.0140	.0141
2	.01	.99	.60	.05	−2.11	.226	.000	.0014	.0060	.0064
3	.01	.99	.80	.01	−0.21	.057	.070	.0011	.0080	.0080
4	.05	.95	.60	.05	−0.46	.076	.070	.0056	.0300	.0309
5	.05	.95	.80	.01	1.43	.000	.225	.0021	.0400	.0385

Note. From "Zeta Analysis: A New Model to Identify Bankruptcy Risk of Corporations," by E. I. Altman, R. G. Haldeman, and P. Narayanan, 1977, *Journal of Banking and Finance, 1*, p. 47. Copyright 1977 by Elsevier Science. Adapted with permission.

course, it is impossible to predict the precise nature of the future environment. Each hypothetical environment is represented as a row in Table 6 and is defined in terms of q_1, q_2, c_1, and c_2. For each hypothetical environment (each row), a classification function is computed. When the data for a given firm are entered into the classification function, for a given scenario, a classification score is obtained. Altman et al. (1977) referred to this score as a ZETA score and to the function as a ZETA classification function.

The middle section of Table 6, labeled L-O-O estimate of misclassification probabilities, includes the cutoff scores for each ZETA classification function and the estimated misclassification probabilities—estimated by means of the L-O-O procedure—for bankrupt (e_{21}) and nonbankrupt (e_{12}) firms. Note that for the first scenario, the cutoff score is -0.33. Thus, when the attribute values of a given firm are entered into the first ZETA classification function, if the resulting ZETA (classification) score is less than -0.33, this firm is predicted to remain solvent. The next row defines a different business environment, for which the cut-off value is -2.11. If a firm has a ZETA score that falls below -2.11, it would be predicted to remain solvent. Immediately to the right of the number -2.11 is the number .226. This number means that of all the firms that go bankrupt, 22.6% would have been erroneously predicted to remain solvent when using this classification function. Now examine the last row. Under the column labeled Nonbankrupt is the number .225. This means that given the business environment of this ZETA classification function, 22.5% of the companies that do not go bankrupt are erroneously predicted to go bankrupt.

Finally, the rightmost section of Table 6, labeled Expected misclassification costs (EC), contains the expected misclassification costs for each of the five scenarios, for three decision rules: the ZETA classification function (EC_{ZETA}), the strategy of always predicting solvency (EC_{MAX}), and the strategy of randomly predicting bankruptcy according to the prior probability of firms in the sample that go bankrupt (EC_{PROB}). Note that for the first scenario, EC_{ZETA} is lower than EC_{MAX} and EC_{PROB}. Moreover, we see that in all different scenarios, the ZETA model resulted in a significant improvement over the naive approaches, because the expected cost of the ZETA model tended to be about six times smaller than the expected cost of the MAX and PROB strategies.

Altman et al. (1977) concluded that their model was accurate in predicting bankruptcies 1 year in advance. At the same time, however, they did not recommend that the model be followed blindly. The major contribution of their model is that it provides an objective analytical tool for identifying firms with serious financial problems. The Altman et al. model calculates a firm's classification score (the ZETA score in Altman et al.'s terminology), which can be converted into a probability of future bankruptcy.

As discussed above, strictly speaking, the misclassification probabilities were not valid, because the variance–covariances across groups were different and not all variables in the model followed the normal distribution, even after transformation. However, Hamer (1983) found that in several studies in which the assumptions underlying linear DA were violated, bankruptcy probabilities computed by linear DA were similar to those computed by logistic regression. Thus, although not necessarily true probabilities of bankruptcy, these measures are still useful indicators of the seriousness of a firm's financial problems.

Firms with a low estimated probability of bankruptcy have similar financial characteristics as the training sample firms that did not file for bankruptcy, and any financial problems facing such a firm are likely to be short-term. Firms with a high estimated bankruptcy probability possess characteristics similar to those of training sample firms that filed for bankruptcy within a 1-year period. The extent to which a probability of bankruptcy should be interpreted as "high" depends on the ratio of the cost of failing to identify a firm with serious problems, and the cost of predicting bankruptcy for a firm that has a good chance of survival. Lenders and investors can use this information as an indicator of high-risk firms. Managers can use a high bankruptcy probability as an early warning signal of serious problems, and the DA results can help in identifying the causes of the problems and in taking corrective action.

By 1987, over three dozen financial institutions were using the commercial version of the Altman et al.'s (1977) model.

Summary

This chapter reviews the two major purposes of DA: description and prediction. Descriptive DA seeks to determine the factors that determine group separation, on the basis of a set of entities for which the group membership is already known, and to describe how the characteristics of the groups differ. Prescriptive DA focuses on the classification of future entities of which the group membership is unknown. Most of this chapter focuses on parametric methods for analyzing DA problems and is limited to a discussion of the most straightforward methods within this class. The intermediate and advanced analyst should probably consult some of the more advanced references noted in the Suggestions for Further Reading section. Although parametric methods are certainly the most important, it is also worthwhile to explore the Suggestions for Further Reading related to nonparametric methods, in particular if the data deviate considerably from multivariate normality.

Suggestions for Further Reading

Good nontechnical general references to DA include Flury and Riedwyl (1988), Altman, Avery, Eisenbeis, and Sinkey (1981), and Rulon, Tiedeman, Tatsuoka, and Langmuir (1967). At the intermediate level, good references (both in DA and related statistical topics) are Anderson (1984), Hand (1981), Johnson and Wichern (1988), Klecka (1980), Lachenbruch (1975), McLachlan (1992), Morrison (1990), Neter et al. (1989), Pindyck and Rubinfeld (1981), and Tatsuoka (1988). Ragsdale and Stam (1992) give a nontechnical introduction to concepts of DA targeted at master of business administration (MBA) students, including an extensive list of DA applications in business disciplines. They also indicate how Fisher's linear discriminant rule for two groups is equivalent to a regression analysis. Aldrich and Nelson's (1989) text is helpful with related topics, such as logistic regression and probit analysis. Moreover, the special issue of *Statistical Science* (1989) entirely devoted to the Panel on Discriminant Analysis, Classification and Clustering—a conference on DA—is a useful reference.

Recently, new estimators of hit rates have been proposed that are based on the bootstrap method of Efron (1979). A discussion of these methods can be found in Efron (1983), Konishi and Honda (1990), and McLachlan (1992, pp. 346–360).

Although the most widely used classification methods are based on the assumption of multivariate normally distributed attributes, we mention some alternative nonparametric methods that are of interest if the normality assumption is untenable. Some of these nonparametric methods, such as kernel methods and *k*-nearest neighbor methods (Hand, 1982; McLachlan, 1992) are well-established in the DA field.

A class of nonparametric methods that has recently attracted substantial interest in the literature is based on the fact that if the normality assumption is violated, the least squares estimation criterion is no longer optimal, and alternative criteria based on absolute distances or a direct count of the number of misclassified training sample entities may yield more accurate classification results. These methods may be particularly attractive if the data conditions are skewed, if the data have extreme observations (outliers), or if the data have categorical variables. An overview of various alternative nonparametric formulations is provided in Joachimsthaler and Stam (1990). Duarte Silva and Stam (1994) developed a number of SAS macros that implement many of these formulations. Soltysik and Yarnold (1993) developed several efficient procedures to minimize the number of misclassifications in the training sample. These procedures are available in their Optimal Discriminant Analysis (ODA) software package.

Glossary

CANONICAL DISCRIMINANT ANALYSIS Statistical method that identifies and studies the dimensions of group separation provided by a set of quantitative attributes.

CANONICAL DISCRIMINANT FUNCTIONS Linear combination of the attributes used in descriptive discriminant analysis to explain group separation.

CANONICAL VARIABLE Linear combination of attributes used in descriptive discriminant analysis to explain group separation.

CLASSIFICATION FUNCTION Function used in predictive discriminant analysis to predict group membership of an observation.

CONDITIONAL HIT RATE Hit rate of a given classification rule (same as p^2).

DESCRIPTIVE DISCRIMINANT ANALYSIS Analysis that aims to describe and explain the characteristics of and differences among the groups.

DEVELOPMENT SAMPLE Sample that is used to estimate the classification function (same as training sample).

DISCRIMINANT FUNCTION Linear combination of the attributes used in descriptive discriminant analysis to explain group separation; sometimes also used as classification function.

EXTERNAL CLASSIFICATION ANALYSIS Analysis of the classification accuracy of a holdout or validation sample that was not used in estimating the classification function.

GROUP CLASS Set of observations with homogeneous characteristics.

GROUP OVERLAP Extent of conflict in the attribute values of the observations in the different groups.

HIT RATE Probability of correct classification.

HOLDOUT (CROSS-VALIDATION) SAMPLE Nontraining sample, used to analyze the accuracy of a classification rule.

INTERNAL CLASSIFICATION ANALYSIS Analysis of the classification accuracy of the training sample that was used to estimate the classification rule.

MAHALANOBIS DISTANCE Index of distance between two multivariate populations, which takes into account the differences in dispersion across variables.

MISCLASSIFICATION COST Cost associated with assigning an observation to a group to which the observation does not belong.

$p^{(1)}$ Optimal hit rate.

$p^{(2)}$ Hit rate of a given classification rule (same as conditional hit rate).

$p^{(3)}$ Expected conditional hit rate (same as unconditional hit rate).

PREDICTIVE DISCRIMINANT ANALYSIS Statistical methodology that studies the assignment of observations into well-defined groups, based on a set of attributes.

STANDARDIZED CANONICAL COEFFICIENTS Coefficients of canonical discriminant functions, adjusted by the measurement scale of the original attributes.

STEPWISE VARIABLE SELECTION Procedure in which the variables are added to (in forward selection) or deleted from (in backward selection) the model, one at the time.

TRAINING SAMPLE Sample that is used to estimate the classification rule (same as development sample).

UNCONDITIONAL HIT RATE Expected conditional hit rate (same as $p^{(3)}$).

WILKS'S LAMBDA A criterion of general use in multivariate analysis for testing hypotheses concerning multivariate normal populations, especially hypotheses of homogeneity of means.

Z Canonical variable.

References

Aldrich, J. H., & Nelson, F. D. (1989). *Quantitatve applications in the social sciences: Vol. 45. Linear probability, logit, and probit models*. Beverly Hills, CA: Sage.

Altman, E. I. (1968). Financial ratios, discriminant analysis, and the prediction of corporate bankruptcy. *Journal of Finance, 23*, 589–609.

Altman, E. I., Avery, R. B., Eisenbeis, R. A., & Sinkey, J. F. (1981). *Application of classification techniques in business, banking and finance*. Greenwich, CT: JAI Press.

Altman, E. I., Haldeman, R. G., & Narayanan, P. (1977). Zeta analysis: A new model to identify bankruptcy risk of corporations. *Journal of Banking and Finance, 1*, 29–51.

Anderson, T. W. (1984). *An introduction to multivariate statistical analysis* (2nd ed.). New York: Wiley.

Baringhaus, L., Danschke, R., & Henze, N. (1989). Recent and classical tests for normality—A comparative study. *Communications of Statistics—Simulation, 18*, 363–379.

Bartlett, M. S. (1937). Properties of sufficiency and statistical tests. In *Proceedings of the Royal Society of London, A160*, 268–282.

Box, G. E. P. (1949). A general distribution theory for a class of likelihood criteria. *Biometrika, 36*, 317–346.

Clarke, W. R., Lachenbruch, P. A., & Broffit, B. (1979). How nonnormality affects the quadratic discrimination function. *Communications in Statistics—Theory and Methods, A8*, 1285–1301.

Duarte Silva, A. P., & Stam, A. (1994). *BestClass: A SAS-based software package of nonparametric methods for two-group classification* (Working Paper No. 94-396). Terry College of Business, University of Georgia, Athens.

Dun & Bradstreet. (1976). *The failure record*. New York: Author.

Efron, B. (1979). Bootstrap methods: Another look at the jacknife. *Annals of Statistics, 7*, 1–26.

Efron, B. (1983). Estimating the error rate of a prediction rule: Improvement on cross-validation. *Journal of the American Statistical Association, 78*, 316–331.

Fisher, R. A. (1936). The use of multiple measurements in taxonomic problems. *Annals of Eugenics, 7*, 179–188.

Flury, B., & Riedwyl, H. (1988). *Multivariate statistics: A practical approach*. New York: Chapman & Hall.

Fukunaga, K., & Kessell, D. L. (1973). Nonparametric Bayes error estimation using unclassified samples. *IEEE Transactions on Information Theory, IT-19*, 434–440.

Glick, N. (1978). Additive estimators for probabilities of correct classification. *Pattern Recognition, 10*, 211–222.

Habbema, J. D. F., & Hermans, J. (1977). Selection of variables in discriminant analysis by *F*-statistic and error rate. *Technometrics, 19*, 487–493.

Hamer, M. (1983). Failure prediction: Sensitivity of classification accuracy to alternative statistical methods and variable sets. *Journal of Accounting and Public Policy, 2*, 289–307.

Hand, D. J. (1981). *Discrimination and classification*, New York: Wiley.

Hand, D. J. (1982). *Kernel discriminant analysis*, New York: Research Studies Press.

Hora, S. C., & Wilcox, J. B. (1982). Estimation of error rates in several-population discriminant analysis. *Journal of Marketing Research, 19*, 57–61.

Huberty, C. J. (1984). Issues in the use and interpretation of discriminant analysis. *Psychological Bulletin, 95*, 156–171.

Huberty, C. J., Wisenbaker, J. M., & Smith, J. C. (1987). Assessing predictive accuracy in discriminant analysis. *Multivariate Behavioral Research, 22*, 307–329.

Joachimsthaler, E. A., & Stam, A. (1990). Mathematical programming approaches for the classification problem in two-group discriminant analysis. *Multivariate Behavioral Research, 25*, 427–454.

Johnson, R. A., & Wichern, D. W. (1988). *Applied multivariate statistical analysis* (2nd ed.). Englewood Cliffs, NJ: Prentice Hall.

Kendall, M. G. (1957). *A course in multivariate analysis*. London: Griffin.

Klecka, W. R. (1980). *Quantatitive applications in the social sciences: Vol. 19. Discriminant analysis*. Beverly Hills, CA: Sage.

Konishi, S., & Honda, M. (1990). Comparison of procedures for estimation of error rates in discriminant analysis under nonnormal populations. *Journal of Statistical Computing and Simulation, 36*, 105–115.

Koziol, J. A. (1986). Assessing multivariate normality: A compendium. *Communications in Statistics—Theory and Methods, 15*, 2763–2783.

Lachenbruch, P. A. (1967). An almost unbiased method of obtaining confidence intervals for the probability of misclassification in discriminant analysis. *Biometrics, 23*, 639–645.

Lachenbruch, P. A. (1975). *Discriminant analysis*. New York: Hafner.

Lachenbruch, P. A., Sneeringer, C. & Revo, L. T. (1973). Robustness of the linear and quadratic discriminant function to certain types of non-normality. *Communications in Statistics*, 1, 39–57.

Mahalanobis, P. C. (1936). On the generalized distance in statistics. *Proceedings of the National Institute of Science of India, 12*, 49–55.

Marks, S., & Dunn, O. J. (1974). Discriminant functions when covariance matrices are unequal. *Journal of the American Statistical Association, 69*, 555–559.

McGrath, J. J. (1960). Improving credit evaluation with a weighted application blank. *Journal of Applied Psychology, 44*, 325–328.

McKay, R. J., & Campbell, N. A. (1982a). Variable selection techniques in discriminant analysis: I. Description. *British Journal of Mathematical and Statistical Psychology, 35*, 1–29.

McKay, R. J., & Campbell, N. A. (1982b). Variable selection techniques in discriminant analysis: II. Allocation. *British Journal of Mathematical and Statistical Psychology, 35*, 30–41.

McLachlan, G. J. (1992). *Discriminant analysis and statistical pattern recognition*. New York: Wiley.

Morrison, D. F. (1990). *Multivariate statistical methods* (3rd ed.). New York: McGraw-Hill.

Neter, J., Wasserman, W., & Kutner, M. H. (1989). *Applied linear regression models* (2nd ed.). Homewood, IL: Irwin.

Neter, J., Wasserman, W., & Whitmore, G. A. (1988). *Applied statistics* (3rd ed.). Newton, MA: Allyn & Bacon.

Panel on Discriminant Analysis, Classification and Clustering. (1989). Discriminant analysis and clustering. *Statistical Science*, *4*, 34–69.

Pindyck, R. S., & Rubinfeld, D. L. (1981). *Econometric models and economic forecasts* (2nd ed.). New York: McGraw-Hill.

Porebski, O. R. (1966). Discriminatory and canonical analysis of technical college data. *British Journal of Mathematical and Statistical Psychology*, *19*, 215–236.

Ragsdale, C. T., & Stam, A. (1992). Introducing discriminant analysis to the business statistics curriculum. *Decision Sciences*, *23*, 724–745.

Rulon, P. J., Tiedeman, D. V., Tatsuoka, M. M., & Langmuir, C. R. (1967). *Multivariate statistics for personnel classification*. New York: Wiley.

Shubin, H., Afifi, A., Rand, W. M., & Weil, M. H. (1968). Objective index of haemodynamic status for quantitation of severity and prognosis of shock complicating myocardial infarction. *Cardiovascular Research*, *2*, 329.

Smith, C. A. B. (1947). Some examples of discrimination. *Annals of Eugenics*, *13*, 272–282.

Soltysik, R. C., & Yarnold, P. R. (1993). *ODA 1.0: Optimal data analysis for DOS*. Chicago: Optimal Data Analysis.

Sorum, M. J. (1972). Estimating the expected and the optimal probabilities of misclassification. *Technometrics*, *13*, 935–943.

Tatsuoka, M. M. (1988). *Multivariate analysis techniques for educational and psychological research* (2nd ed.). New York: Macmillan.

Walker, S. H., & Duncan, D. B. (1967). Estimation of the probability of an event as a function of several independent variables. *Biometrika*, *54*, 167–179.

Wilks, S. S. (1932). Certain generalizations in the analysis of variance. *Biometrika*, *24*, 471–494.

10 Understanding Meta-Analysis

Joseph A. Durlak

Do young children who have participated in Head Start programs display higher levels of academic achievement than children who have never enrolled in Head Start? Is a new medication to treat migraine headaches more effective than the most widely used current medication? Is group psychotherapy more effective in treating depression in adults than individual forms of treatment? These are the types of substantive research questions that can be investigated using a procedure called *meta-analysis*. Meta-analysis has become a popular research strategy within the behavioral and social sciences. By 1991, more than 1,000 reports involving meta-analyses were available (Durlak & Lipsey, 1991). Despite its popularity, meta-analysis is only now being routinely covered in statistical and research texts. As a result, many find the procedure and findings of a meta-analysis difficult to understand or interpret. Therefore, the primary intent of this chapter is to help the reader understand the basic goals and procedures of meta-analysis.

What Is Meta-Analysis?

Meta-analysis is a method used to review or survey research literature. It differs from other reviewing strategies in its focus on the statistical integration and analysis of research findings. In many ways, meta-analysis is analogous to individual experiments conducted in the behavioral and social sciences. In a typical experiment involving human subjects, for example, an experimenter uses a particular procedure to gather information on a group of participants. One or more hypotheses are often made regarding experimental outcomes. Steps are usually taken during the conduct of the research to control for bias or error that might distort the findings. The data collected from individuals are aggregated and then analyzed statistically to confirm or disconfirm initial hypotheses. Finally, the findings are interpreted; conclusions are drawn, and suggestions for future research are offered.

Meta-analysis follows this same pattern, but instead of collecting original data, a meta-analyst uses the information already collected by others. A meta-analyst identifies relevant research studies, collects copies of the actual research reports, examines them, and then draws data from each one. In other words, individual studies become the "subjects" in a meta-analysis. Data from these studies are pooled (aggregated), and just as in an individual experiment, the meta-analyst applies statistical tests to these pooled data and then interprets the results.

Meta-analysis quantifies the data from each study in two important ways: (a) Descriptive features of each study are coded using categorical or continuous coding schemes and, (b) the outcomes or results of each study are transformed into a common metric across studies called the effect size (ES).

Meta-analysis is also analogous to an individual experiment in terms of containing both independent and dependent variables. In individual studies, usually the number of independent variables is limited and there are only a few dependent variables. In meta-analysis the dependent variable is the ES drawn from each study (i.e., the results or outcome of the study); however, there are many potential independent variables. These possibilities include each feature or characteristic of the reviewed studies, such as characteristics of the subjects, interventions, outcome measures, and the like. Basically, meta-analysts test possible relationships between independent and dependent variables (i.e., between study characteristics and ESs) by assessing which of the former account for significant variation in the latter. Each study feature may make a difference, or none might make a difference. It is up to the meta-analyst to clarify this matter for the reader. In summary, meta-analysis is an effort to review the results of a research domain in quantitative terms with the intent of identifying what significant relationships exist between study features (the independent variables) and outcomes (expressed as an ES and representing the dependent variable).

The previous description refers to explanatory meta-analyses, which are only one type of a meta-analysis. For example, many meta-analyses are descriptive in purpose and their intent is to summarize findings from a research domain in broad terms without examining how or why such findings occur. Meta-analyses also vary in terms of the data that are collected and analyzed. As the name implies, correlational meta-analyses use product–moment correlations or their variants and are common in the areas of industrial–organizational psychology and for assessing test validity. Hunter and Schmidt (1990) discussed correlational meta-analyses in depth. This chapter focuses on explanatory meta-analyses, which use standardized group mean differences as the index of effect.

As the name implies, treatment effectivness meta-analyses assess the impact of some type of treatment, program, or intervention. In this type of meta-analysis, research questions are framed in relative terms. Is this type of program more effective than that type? Is this type of treatment for a particular problem better than no treatment at all? Do youth who have been exposed to an educational campaign regarding AIDS have more information on the risks of various sexual practices than youth who have not received specific education on the topic?

When properly conducted, meta-analyses offer several benefits. Meta-analyses are able to synthesize the findings from many studies succinctly, which aids in dissemination of knowledge, particularly to nonscientific communities. The data emanating from meta-analyses can also have public policy implications. Meta-analytic findings that illustrate the amount and relative impact of different programs on different criteria can assist policymakers in the decision making process. Finally, a meta-analysis can identify the most effective programs and highlight particular gaps or limitations in the literature that illustrate priorities for future research.

Major Steps in Meta-Analysis

There is no one standardized way to conduct a meta-analysis. As a result, meta-analysts might use different procedures and make different decisions depending on the purpose of their work. Although I cannot offer a how-to-do-it guide for every possible situation, in the following sections I describe the major steps in a meta-analysis and the decisions that each meta-analyst customarily faces.

A meta-analysis contains six major steps, and each must be successfully accomplished or the findings will be compromised. In this chapter I emphasize three of these steps, but I discuss the others briefly to place the statistical analyses into a general context. The six steps in meta-analysis are (a) formulating the research question; (b) searching the literature for relevant studies; (c) coding studies; (d) calculating the index of effect, including selecting an appropriate unit of analysis; (e) conducting the statistical analyses of effects; and (f) offering conclusions and interpretations.

Step 1: Formulating the Research Question

The first step in a meta-analysis is formulating the specific research question. There are several ways a meta-analyst attempts to "explain" data (Cook et al., 1992). This chapter focuses on explanatory meta-analyses that are devoted to specifying the factors or conditions that take into account the variability in effects that invariably occurs in any research domain.

These explanatory meta-analyses should begin with specific hypotheses that, in turn, rest on important theoretical, conceptual, or empirical aspects of prior research. To offer good hypotheses, the meta-analyst must be acquainted with prior findings in the relevant research area. Thus, a meta-analysis actually begins with a critical conceptual evaluation of prior research that identifies the major research issues that need clarification and that provide the basis or rationale for specific hypotheses.

Not all meta-analyses begin with formal hypotheses, but the importance of offering such hypotheses cannot be underestimated. Similar to other statistical procedures, findings in a meta-analysis that emerge in relation to specific a priori hypotheses are given greater credence than those obtained from multiple post hoc analyses that captitalize on chance. (This is so because if one conducts enough statistical analyses, something will emerge by chance as statistically significant.) Formulating the research question also involves defining the relevant research domain in clear operational terms. Two issues often arise in defining this domain.

Should Unpublished Studies Be Included?

Findings from several meta-analyses (Smith, 1980) have indicated that published studies (articles and books) yield significantly more positive findings than do unpublished studies (dissertations, convention papers, and technical reports). Such a result is referred to as publication bias and suggests that a meta-analysis focusing only on published research is probably overestimating the results of all studies (both published and unpublished research).

Publication bias is related to the so-called "file drawer problem"

(Rosenthal, 1979), which is the tendency for authors not to submit and journal editors not to accept for publication studies that fail to achieve statistically significant results. As a result, experiments that do not turn out as expected languish in investigators' file drawers, whereas the published literature contains a preponderance of positive findings. Therefore, care must be taken when interpreting the results of a meta-analysis evaluating only published work.

Methodological Quality

One disagreement among meta-analysts concerns the use of methodological criteria to include or exclude studies in a meta-analysis (Durlak & Lipsey, 1991). One side of the argument is that unless studies meet minimal design criteria they should be excluded, because findings and conclusions should not be offered using research that is of poor quality. The response to this concern is that methodologically poor studies may not yield different results than better controlled investigations and should not be excluded ipso facto. Moreover, there are no absolute standards for defining methodological quality; different features may be important in different areas.

Generally speaking, meta-analysts who do not exclude studies on the basis of methodological criteria attempt to assess how methodological features relate to ES. Before meta-analysts attach any substantive or theoretical meaning to results, it is important to rule out methodological factors as an alternative explanation. The degree of experimental rigor, for example, may be a more important factor affecting study findings than any other study characteristic.

Step 2: Literature Search

The ultimate goal of a literature search is to obtain a representative and nonbiased sample of relevant investigations, and it is up to the meta-analyst to convince the reader the search has been successful in this regard. Usually, three major techniques are combined to locate relevant studies: computer searches, manual searches, and examination of the reference lists of each identified study. Literature searches using each of these paths are essential because different studies are located with each strategy. Although computer searches are quick and can be conducted using databases targeted for both published (e.g., *PsycLIT* or MEDLARS) or unpublished (e.g., *Dissertation Abstracts*, ERIC) literature, they are unreliable in terms of identifying the relevant literature. Numerous irrelevant citations are frequently obtained and many relevant studies are omitted (Durlak & Lipsey, 1991).

Therefore, meta-analysts often hand-search the contents of outlets most likely to publish studies in the area and list the sources examined. That is, they examine journal contents study by study. This procedure is time consuming but often captures many relevant studies. Finally, the references of each relevant study are frequently examined to identify additional research.

It is impossible to be absolutely comprehensive in obtaining all relevant published and unpublished research, in the latter case particularly because of the file drawer problem described earlier. As a result, a "fail-safe *n*" is often calculated in a meta-analysis after the data are analyzed. This fail-safe *n* is a formula (Orwin, 1983) that estimates the number of additional studies that would have to be located in order to change obtained results. Unfortunately, there is no absolute standard for assessing what value of the fail-safe *n* is critical. In general, two factors to consider in gauging how likely it is that additional studies can be easily obtained are the extent of the effort expended in the initial literature search and the number of studies that are reviewed.

Step 3: Study Coding

Meta-analysts develop coding procedures to translate the features of each study into usable quantitative data. There is no single way to code studies. The aim is to develop a coding system that is general enough for all studies that are being reviewed but specific enough to capture unique study features. This is not easy to do. In order to illustrate the nature of study coding, consider the following brief narrative summaries of three hypothetical studies of child psychotherapy.

In Study 1, 25 hyperactive first-grade children from School A received eight individual sessions of cognitive–behavioral treatment that emphasized self-instructional training procedures designed to help the children manage their behavior more effectively. Twenty hyperactive children were identified at School B and served as controls. Children's cognitive problem-solving skills were evaluated, and their classroom behavior was also assessed using both teacher ratings and independent behavioral observations.

In Study 2, teachers identified 60 preschoolers who did not play with any other children and seldom spoke or interacted with others. This group received 20 group-play therapy sessions designed to assist them in social interaction, and this group was matched with 60 children from another preschool who were rated similarly by their teachers. Outcome measures consisted of teacher ratings of social behavior and a peer sociometric measure designed to assess whether the children were liked, disliked, or ignored by their peers.

In Study 3, screening of all junior high school students in one school district identifed 240 students with moderate to high levels of test anxiety. These students were randomly assigned to a systematic desensitization therapy program or to a waiting-list control condition. Students' self-reports of test anxiety were used to evaluate the impact of the treatment.

The first three rows in Table 1 depict one way in which the study characteristics can be coded. Codes generated from seven additional studies are also presented. Note how the types of therapists (column 4), control conditions (column 5), and treatments (column 6) are translated into categorical codes (e.g., behavioral treatment = 1; nonbehavioral treatment = 2). The number of treatment sessions (column 8) is entered as a continuous variable.

Many variables can be coded for each study, and the process is time consuming. When meta-analyses are published, coding procedures are summarized only so that the reader is usually unable to understand all of the procedures used for coding. However, footnotes usually indicate that coding sheets and "code books" are available to interested readers. The latter describes particularly troublesome coding problems and how the meta-analyst resolved them. This information is particularly important for those who wish to replicate this aspect of the meta-analysis. Meta-analysts also should report on the reliability of coding procedures to assure the reader that coding was a systematic procedure that can be replicated by others. Perfect reliability is 100%, but reliability figures that reach or exceed 80% are usually considered acceptable. Several sources (Hartmann, 1982; Stock et al., 1982) contain useful information on coding procedures and reliability.

A major issue in meta-analysis involves whether coding procedures effectively capture the essential features of original research. Note that in column 5 of Table 1, only the general type of treatment (behavioral or nonbehavioral) is coded. How important is it to code the specific treatment procedures (e.g., systematic desensitization or play therapy) that occur within these broad treatment categories? Such matters can become critical because there are various options for deriving ESs from study data that are often dependent on the characteristics of study coding. (This matter is discussed later.) In addition, important information is sometimes

Table 1

Selected Characteristics and Outcomes From a Hypothetical Sample of Child Psychotherapy Studies

Study	Subjects	General design[a]	Therapists[b]	Control group[c]	Treatment[d]	Presenting problem	Number of sessions	Method of administration
1	45 First graders	1	2	1	1, Self-instruction training	Hyperactivity	8	Individual
2	120 Preschoolers	2	6	1	2, Play therapy	Social isolation	20	Group
3	240 12–14-year-olds	3	2, 3, 4	2	1, Systematic desensitization	Test anxiety	—	Individual
4	40 Mothers of third graders	3	1	1	1, Reinforcement	Mixed	6	Group
5	65 8–10-year-olds	1	1, 4	1	2, Psychodynamic	Noncompliance	25	Individual
6	18 8-year-olds	2	5	2	1, Bell and pad	Bedwetting	4	Individual
7	Fourth and sixth graders	3	—	3	1, Transactional analysis	Noncompliance, learning problems	45	Group
8	10 10-year-olds	—	6	2	2, Adlerian	Low self-esteem counseling	—	Group
9	54; Age unknown	3	2	3	2, Rogerian counseling	Depression or anxiety	12	Group
10	30 Third graders	1	2	—	1, Modeling	Aggression	20	Group

[a]For general design: 1 = nonequivalent control group design; 2 = matched design; 3 = randomized true experiment. [b]For therapists: 1 = professionals; 2 = graduate students; 3 = parents; 4 = teachers; 5 = undergraduates; 6 = mix of 1–5. [c]For control condition: 1 = no treatment; 2 = waiting list; 3 = attention placebo. [d]For type of treatment: 1 = behavioral; 2 = nonbehavioral. Dashes indicate unascertainable for all codes.

missing and cannot be coded (the dashes in Table 1 indicate missing data). Furthermore, some features must be estimated from limited data (e.g., research with children often lists only grade levels instead of ages of the children). A good meta-analysis will acknowledge any limitations in coding and how they might affect subsequent findings.

Step 4: The Index of ES

In treatment effectiveness meta-analyses, the most important variable is the standardized difference between group means. This difference is called the ES or, alternately, *d* or *g*. The ES is a critical feature of meta-analysis, and it is important to understand how it is calculated and what this index reflects.

Calculating ESs

The ES is usually calculated by subtracting the mean of the control group at posttreatment from the mean of the treatment group at posttreatment and dividing by the pooled standard deviation of the two groups. A positive score thus reflects that the treated group outperformed the control group, and a negative score has the reverse meaning. If the group means and standard deviations are not available, ESs can be estimated from other statistical information contained in the study (Holmes, 1984, and Wolf, 1986, describe these procedures). Effects can also be calculated using the pre- and postscores of a single group in studies without a control condition, but such effects are often higher than those attained when control groups are used (see Posavac & Miller, 1990, for an example).

Conceptual Meaning of Effects

The ES is an index that transforms the unique data from each reviewed study into a common metric (i.e., it is based on standard deviation units, the denominator used when calculating ESs). Therefore, in experimental versus control group studies, an ES of 1.0 reflects that the experimental group changed 1 *SD* more than did the controls. ESs thus reflect the relative magnitude of effect in a term common across studies. By comparing the performance of the treatment and control groups, ESs indicate how relatively effective a treatment has been in each reviewed study. This is in marked contrast to the typical statistical information available in studies that reflects only whether the treated and control groups differ significantly at posttreatment. Because ESs transform all data to a common metric, they permit the data from different studies to be combined (aggregated) and compared. ESs from all reviewed studies can be averaged to determine an overall mean effect for the literature under review. Moreover, ESs of the same magnitude are equivalent across studies because they have been standardized. That is, an ES from Study A of 0.50 means exactly the same thing as an ES from Study B of 0.50 in the sense that the interventions evaluated in these two studies are equally effective or powerful.

Range of Effects

Generally speaking, ESs can be of any magnitude. Therefore, unlike a product–moment correlation coefficient, it is possible to have an ES below −1.0 and above +1.0, although the vast majority do fall within this range. A convention in the behavioral and social sciences has been to interpret ESs of around 0.20 to be "small" in magnitude, those around 0.50 to be "moderate" and those larger than 0.80 to be "high" in magnitude (Cohen, 1988). These are only conventions, however. In some research areas, mean effects of 0.50 may be considered large effects.

Variability in Effects

Although the average ES achieved across all reviewed studies is important, explanatory meta-analyses attempt to explain the variability in obtained effects across studies. Effects vary across studies; if they did not, there would be little purpose for the review in the first place, because it would be common knowledge that studies in a particular area consistently lead to the same result. Life is, of course, not that simple. Therefore, an intriguing question in explanatory meta-analyses is, Why do studies differ in their effects? In other words, a meta-analyst hopes to explain the variability that is obtained in effects by identifying those procedural variables that exert on influence on the dependent variables (i.e., the ESs).

The most popular measure of variability is the standard deviation, which is usually reported along with its corresponding mean and gives an indication of the variability of obtained ESs. Recalling the basic properties of a normal distribution that 68% of all cases fall within 1 *SD* (above and below) the mean and 98% are within 2 *SD*s of the mean, one can tell immediately how much variability is present in any group of effects. For example, a mean ES of 0.50 that has a standard deviation of 0.50 means that 98% of all the effects range between −0.50 and 2.50 (assuming the ESs are normally distributed). That is a wide range and suggests substantial variability in effects. Yet, it is not unusual to have standard deviations greater than or equal to the obtained mean.

Statistical Significance of a Mean Effect

Some meta-analysts conduct *t* tests to determine whether a mean effect that is obtained from a group of studies will differ significantly from zero. This is a *t* test with $N - 1$ degrees of freedom for a difference between a sample mean (obtained in the meta-analysis) and the population mean effect (assumed to be zero). If multiple *t* tests are conducted for several groups of studies, some control for Type I error should be made by using, for example, a Bonferroni correction procedure or by making the alpha level more stringent (see chapter 5 in this book).

It is more useful to calculate confidence intervals around a mean ES obtained from a group of studies (Hedges & Olkin, 1985). These intervals portray the range of effects that might exist in the true population given the presence of error and variation in the calculation of sample effects. A mean ES is interpreted as statistically significant from zero if its confidence interval does not include zero. However, whether a mean ES is statistically different from zero is usually not as important as its magnitude and whether effects drawn from different groups of studies differ from each other.

Variability in Study Features

It is necessary to digress a bit to discuss variability in study features. Everything said so far about the equivalency of ESs drawn from different studies is correct, with all other things being equal across studies. Invariably, however, most other things are not equal across studies. That is, not only do ESs differ across studies, but also studies differ in methodological and procedural features as well as sample characteristics.

The study features listed in Table 1 illustrate this issue. Note that although these 10 studies are obstensibly studying the same general phenomenon—child therapy—they vary on many variables, such as the type of treatment conducted, child characteristics, and various methodological features (sample size and type of control groups used). Deciding how these variables should be examined for their influence on outcome is discussed at length in a later section.

Practical Significance

What do ESs mean in practical terms? The answer is complicated and depends on many factors. The statistical magnitude of an ES is not related to its practical importance in any straightforward manner. Relatively small effects might have considerable practical significance, and the reverse can be true for large effects.

Therefore, if the results of a review indicate an average ES of 0.20 for a new medical treatment for AIDS, one could not determine the practical value of such findings without more information. For one thing, it would be important to know the characteristics of the outcome measure on which the ES is based. Is one talking about long-term survival rates from treatment, days of hospitalization, cost of treatment, or subjective self-reports of pain or discomfort? In summary, effects of any magnitude can have practical value or significance depending on the circumstances; at the same time, however, the meta-analyst should make the necessary translation for the reader, if possible, to demonstrate the social or practical significance of outcomes. Procedures for doing so are now available (Durlak, Fuhrman, & Lampman, 1991; Rosenthal & Rubin, 1982; Trull, Nietzel, & Main, 1988)

Unit of Analysis

Choosing an appropriate unit of analysis is a critical feature in a meta-analysis, and there are three major options that are illustrated in Table 2. Table 2 shows the ES information for the 10 studies coded in Table 1. Although ESs can be calculated for each dependent measure in a treatment effectiveness meta-analysis, they usually have to be combined or averaged on some basis. This is because studies can vary considerably in the number of dependent measures they contain, from only 1 to 20 or more. The studies in Table 2 contain one (Studies 3, 6, and 9) to six (Study 5) outcome measures.

The first option for a unit of analysis is simply to enter the ESs for each dependent measure, in each study into the analysis. Using this approach, the 10 studies in Table 2 would yield 26 total effects, with a mean ES of 0.49 (see the bottom of column 3 in Table 2), but Study 5, with its six effects, has six times more weight in the data analysis than Studies 3, 6, or 9, which can each contribute only one ES. For this reason and because of the statistical interdependence that results with effects drawn from the same study, each possible ES is rarely used as the unit of analysis.

Instead of including the ES for each variable, a second option is to use each study as the unit of analysis. This is done by averaging across all effect sizes within each study, as illustrated in the fourth column of Table 2. This strategy is popular but has a potential drawback. Different types of dependent measures may yield effects of different magnitudes, and averaging across all the measures would obscure these differences. Note that Studies 3 and 7 obtain the same average effect (0.79), but this finding is based on different types of outcomes. Results for Study 3 are

Table 2

Findings as a Function of Different Units of Analysis for Studies in Table 1

Study	Type of outcome measure[a]	Per measure	Unit of analysis per study	Per construct
1	1, 3, 6	1 = 1.24, 3 = 0.21, 6 = 0.54	0.66	1 = 1.24, 3 = 0.21, 6 = 0.54
2	3, 5	3 = 0.60, 5 = 0.00	0.30	3 = 0.60, 5 = 0.00
3	2	2 = 0.79	0.79	2 = 0.79
4	2, 2, 3, 6	2 = 0.09, 2 = 0.17, 3 = 0.25, 5 = 0.33	0.21	2 = 0.13, 3 = 0.25, 5 = 0.33
5	1, 1, 2, 3, 4, 5	1 = 0.30, 1 = 0.45, 2 = 0.00, 3 = 0.37, 4 = 0.12, 5 = 0.00	0.21	1 = 0.38, 2 = 0.00, 3 = 0.37, 4 = 0.12, 5 = 0.00
6	3	3 = 2.75	2.75	3 = 2.75
7	4, 6, 6	4 = 0.24, 6 = 0.89, 6 = 1.25	0.79	4 = 0.24, 6 = 1.07
8	1, 2, 4	1 = 0.82, 2 = 0.20, 4 = 0.10	0.37	1 = 0.82, 2 = 0.20, 4 = 0.10
9	2	2 = 0.14	0.14	2 = 0.14
10	5, 6	5 = 0.00, 6 = 0.90	0.45	5 = 0.00, 6 = 0.90
M		0.49	0.66	1 = 0.81, 2 = 0.25, 3 = 0.84, 4 = 0.15, 5 = 0.08, 6 = 0.84

[a]1 = Cognitive process measures; 2 = personality self-reports (anxiety, depression, self-concept); 3 = teacher or parent ratings of classroom or home behavior; 4 = academic achievement; 5 = peer sociometrics; 6 = behavior observational data.

based on data from a personality self-report measure, whereas those for Study 7 are derived from a measure of academic achievement and two different behavioral observation procedures. Therefore, the average effects for these studies are similar numerically, but not conceptually. In many cases, it is important to identify this conceptual dissimilarity. As another example, note how the average effect produced by Study 10 is 0.45, but this average results from a high effect for one dependent measure (behavioral observations), and a zero effect for the other measure (a sociometric measure). If such results were consistent across reviewed studies, this information should be presented. Using an average effect per study as the unit of analysis would not reflect such potentially important findings.

The third strategy is to calculate an effect for each distinct construct represented in studies and to keep these effects separate in the analyses. For example, the outcome measures listed in Table 2 are divided into six distinct categories, and effects are calculated for each category per study, although not every study has an effect for all categories.

Using each construct as the unit of analysis (the last column in Table 2) indicates that effects are highest when parent and teacher rating scales or behavioral observations are the outcome measures (mean ES = 0.84 for both categories), lowest when peer sociometric status is assessed (mean ES = 0.08), and intermediate in magnitude for the other types of outcome measures. This strategy yields specific information on how psychotherapy affects different aspects of children's adjustment. Therefore, meta-analysts recommend that effects representing different constructs should be kept separate in initial analyses (Bangert-Drowns, 1986; Durlak & Lipsey, 1991; Hedges & Olkin, 1985). As I demonstrate in a later section, the results of initial analyses can confirm whether keeping designated constructs separate in the analysis is important or whether some effects can be combined across categories.

Step 5: Statistical Analysis of ES Distributions

Meta-analytic data can be evaluated by either one of two related procedures: multiple regression (see chapter 2 in this book) or by analyzing group mean differences (see chapter 5 in this book). Because the former method is treated elsewhere in this book, I discuss it is only briefly here. Regardless of which general method is chosen, the general purpose is the same: to examine how study characteristics account for variability in ES.

Multiple Regression

Using multiple regression, study characteristics are entered as independent variables to predict ES, which serves as the criterion or dependent variable. It is important to emphasize that instead of entering all possible study characteristics into the regression simultaneously, the analysis should be guided by specific hypotheses. Therefore, hierarchical regression procedures in which variables are entered in order of their presumed importance are generally preferred. Hedges and Olkin (1985) provided examples of meta-analytic procedures involving regression.

There are no hard-and-fast rules regarding when multiple regression is used to analyze meta-analytic data. Regression is helpful when more than one significant predictor of effect size is identified and the meta-

Table 3

Hypothetical Results of Two Meta-Analyses Assessing the Effects of Surgical Treatment for Breast Cancer

Analysis and group	*n*	Effect size[a]	
		M	*SD*
Meta-Analysis 1			
Lumpectomy	50	0.35	0.30
Radical mastectomy	50	0.35	0.30
Meta-Analysis 2			
Positive family history			
Lumpectomy	25	0.10	0.09
Radical mastectomy	25	0.40	0.10
Negative family history			
Lumpectomy	25	0.35	0.10
Radical mastectomy	25	0.32	0.15

[a]Effect size is a measure of survival rates 5 years postsurgery.

analyst wishes to evaluate their relative importance (i.e., how much unique variance is predicted by each variable). Regression is often used when study characteristics are continuous in nature (e.g., number of treatment sessions). In many treatment effectiveness meta-analyses, however, the prime variables of interest are categorical (e.g., types of treatments or types of problems). Under these circumstances, researchers often analyze group mean differences.

Analysis of Group Mean Differences

To examine group mean differences, the meta-analyst can divide the total sample of studies into two or more subgroups that differ on certain variables believed to be important, such as type of treatment, type of subjects, research design, and so on. The mean effects of these subgroups can be statistically compared to determine whether the subgroups differ significantly (e.g., on the basis of an independent samples *t* test, or analysis of variance if there are more than two groups). If significant differences in mean ESs are obtained, the investigator would probably attribute the difference to the variable that was used to divide the studies.

There is a serious potential problem to this relatively straightforward approach. To illustrate this point, Table 3 shows the findings of two hypothetical meta-analyses investigating the same question. What is the relative effectivness of two surgical procedures for early detected breast cancer? In the first meta-analysis, the results of 50 studies that used a radical mastectomy procedure (removal of most of the breast tissue) were compared with 50 studies of lumpectomy (removal of the cancerous tumor and very little surrounding breast tissue). Data in the top half of Table 3 indicate that the mean ESs that are based on patient survival rates 5 years after the operation were identical for the two groups (0.35 vs. 0.35), leading to the conclusion that, because the two surgical procedures were equally effective in extending women's lives, the more limited and less disfiguring procedure (lumpectomy) was the treatment of choice, and radical mastectomy was unnecessary.

The importance of family history, however, was considered by a second meta-analysis, and the same 100 studies were evaluated. This time, women who had a positive family history for breast cancer (i.e., whose mother or grandmother had had breast cancer) and who received either type of surgery were considered separately, so that instead of two groups of studies there were four, as depicted in the bottom half of Table 3. Results revealed the importance of this four-category grouping. For those without any previous cancer in their family history, the surgical procedures were equally effective (compare the ESs in the last two rows of Table 3). For any woman with a family history of breast cancer, however, the radical mastectomy is clearly more effective; the ES is four times higher (0.40 vs 0.10). In other words, creating only two groups of studies in the first meta-analysis had obscured an important difference in effects occurring in some studies because an important moderator variable (family history) had been confounded in the analyses. As a result, an incorrect interpretation regarding the impact of surgical treatments was made. In considering family history, the conclusions of the second meta-analysis were that radical mastectomy was the treatment of choice for any woman with a positive family history for breast cancer and that lumpectomy was the surgery of choice for those with a negative family history.

The foregoing discussion highlights a central and controversial question in meta-analysis. How does one know that the manner in which studies have been grouped for analysis is appropriate and that the reviewer has (a) not mistakenly created overly broad categories for comparison purposes or (b) mistakenly attributed findings to the action of one variable when another (or more than one) variable is actually responsible for the obtained results?

Information from Tables 1 and 2 can be used to illustrate the potential interpretative problems that may arise. Whenever studies are subdivided on one variable of interest (e.g., treatment), they will also invariably differ on a host of other potentially important variables (e.g., research design and subjects). This confounding of variables would seem to preclude reaching any definite conclusions about the role of any one variable on which studies are compared. For example, in referring to Tables 1 and 2, if one were to create two subgroups of five studies each on the basis of the type of treatment administered (behavioral vs. nonbehavioral), these two subgroups would also differ in terms of length of treatment, the child's presenting problem, and type of outcome measure used to evaluate the clinical results. When the impact of behavioral and nonbehavioral treatments is compared, how could the meta-analyst be confident that it was not any of the latter variables that accounted for the obtained results? For example, the 5 behavioral studies treated children who had different types of problems (hyperactivity and test anxiety) than the children in the nonbehavioral studies (low self-esteem and social isolation); most of the nonbehavioral studies (4 of 5) used group interventions, but only 1 of the 5 behavioral treatments involved groups. Moreover, if one were to regroup the studies on the basis of whether a group or individual intervention was attempted, then the general types of treatments conducted (behavioral or nonbehavioral) would be confounded. The situation seems hopeless. Thus, the dilemma whenever studies are grouped for analysis becomes, What criteria should be used to group studies that would lend credence to the meta-analyst's interpretation of any obtained results?

Unfortunately, there is currently no universally accepted answer to this question. Procedures developed by Hedges and Olkin (1985) are becoming increasingly common, and it is likely the reader will encounter meta-analyses using their techniques. Therefore, in the following discussion, I describe their approach, although one should realize that disagreements exist over the appropriateness of this strategy under different circumstances (cf. Hedges & Olkin, 1985; Hunter & Schmidt, 1990). For example, the procedures described here are useful whenever the meta-analyst can assume that the variability in effects obtained in a sample of studies is attributable to systematic variation rather than merely to random fluctuation or error.

The Q Statistic and Model Testing

A central feature of Hedges and Olkin's (1985) approach is the use of the Q statistic (the homogeneity test), a test that assesses whether the effects produced by a group of studies vary primarily because of sampling error or represent systematic differences among the studies in addition to sampling error.

The Q statistic used to test for homogeneity in the distribution of ESs for a designated group of studies is distributed as a chi-square variable with $k - 1$ degrees of freedom, where k equals the number of studies. A statistical table is consulted to determine what critical value is needed for statistical signficance at a desired probability level and for the relevant degrees of freedom. A nonsignificant Q indicates homogeneity of effects, whereas a significant Q indicates heterogeneity of effects.

The finding from the Q statistic has important implications. If the effects produced by a group of studies are found to be homogenous, then the studies are considered to be from the same population and analysis of group mean effects is warranted; if, on the other hand, heterogeneity is obtained, then the group actually contains two or more distinct subpopulations of studies. These studies should be subdivided on the basis of one or more variables to identify these subpopulations and achieve homeogeneity within each group.

If the reader is confused at this point regarding tests of homogeneity and the Q statistic, it might be helpful to keep in mind that the Q statistic is useful for model testing. Basically, the Q statistic is a statistical way of confirming or disconfirming that the model chosen by the meta-analyst provides a good fit for the obtained data. By the term *model*, I mean the way the meta-analyst has grouped studies for analysis on the basis of one or more variables. In such groupings, the goal is to identify a single population of studies whose ESs vary only because of sampling error.

For example, if one applies the Q statistic to the two meta-analyses of cancer treatment depicted in Table 3, the results would be as follows. The Q values for the two groups of studies in Meta-Analysis 1 would each be statistically significant (at the .05 level or less), indicating that the distribution of effects for each group of 50 studies reflects both sampling error and systematic differences among studies. In other words, the data do not fit the model proposed in Meta-Analysis 1, which uses only one variable (type of surgical procedure) to explain variability in ESs. On the other hand, the Q values for the four groups of studies created in Meta-Analysis 2 would each be nonsignificant. This finding would confirm that the model in Meta-Analysis 2 using two variables (type of surgery and family history) is a good fit for the data (i.e., a better way to explain variability in ESs).

Establishing homogeneity always comes before analysis and interpretation of group means. It is not unusual to fail to obtain homogeneity in some or all study groupings (e.g., Suls & Wan, 1989). In such cases, the meta-analyst cannot make confident interpretations regarding which variables or factors are contributing to the obtained findings.

Study Artifacts

The term *artifact* is used here to refer generally to any type of methodological, statistical, or measurement error or bias that is present in studies. The use of an outcome measure that has low reliability is one example of a study artifact; in this case unreliability in measurement is a source of error that leads to an underestimation of the true ES.

Hunter and Schmidt (1990) listed 11 common study artifacts. Space limitations do not permit a detailed discussion of these artifacts except to emphasize their potential importance. For example, the meta-analyst may find that methological variables have a stronger effect on outcome than other study features such as the type of treatment conducted. This would, of course, have important implications for the types of conclusions eventually offered in a meta-analysis (see the next section). Therefore, the meta-analyst should assess the impact of study artifacts whenever possible.

Step 6: Offering Conclusions and Interpretations

In the sixth and final step of a meta-analysis, care must be taken to offer conclusions that are specific to the literature being evaluated and consistent with any limitations that exist in the database. For example, suppose a meta-analysis evaluating educational programs to teach math skills to young children offered the following conclusion: All programs are equally effective and longer programs do not achieve better results than shorter programs.

What if this conclusion were based on literature that involved only African American children from inner-city schools, could compare only two popular educational programs because few studies examined any other interventions, and involved only relatively brief interventions (i.e., those that varied from 1 to 4 weeks)? Obviously, this conclusion would be misleading.

Finally, meta-analysts should offer recommendations to improve future research. In particular, it is often useful to indicate how study artifacts in the reviewed studies might have affected current findings and how these problems can be corrected by subsequent researchers.

To illustrate the steps in a meta-analysis, the next section contains a detailed hypothetical example. The data were drawn from both published (Durlak et al., 1991) and unpublished (Durlak, Lampman, Wells,

& Carmody, 1990) work by the author and his colleagues, but they have been modified in several respects for illustrative purposes. Several sections of the example are greatly abbreviated compared with what would appear in a published report. The parenthetical material explains the text in key spots.

An Example of a Meta-Analysis

Purpose of Review and Hypotheses

The major purpose of this review was to determine which variables affect (moderate) the results of child psychotherapy. It was predicted that the three most important variables affecting therapy outcome would be, in order, the type of outcome measure used, the general type of treatment conducted, and the child's presenting problem. (The rationale for these hypotheses would be clearly explained in the text.)

All forms of child therapy were reviewed, but there was one important exception to the hypotheses. In a previous meta-analytic review (Durlak et al., 1991), it was reported that the developmental level of the children was the only variable that moderated outcomes for cognitive–behavioral therapy. Therefore, for cognitive–behavioral studies only, cognitive–developmental level was used as the only moderating variable in the analysis, because it was expected this prior finding would be replicated. (It is acceptable to use different variables in the analysis for different groups of studies as long as clear justifications for doing so are provided.)

Method

Studies Reviewed

Studies eligible for review consisted of reports appearing through the end of 1990 in which some form of psychotherapy for maladjusted children younger than 13 years of age was compared with a control group. (Specific criteria used to identify relevant studies would be presented, such as which types of psychotherapy were used, how maladjustment was defined, and what control groups were used.)

Search Procedures

Three procedures were used to search for studies. First, computer searches were conducted. (The databases should be identified.) Second, the contents of 15 journals in which child research is most likely to appear were examined study by study. (The journals would be identified.) Third, the reference lists from each included study were examined. In order to assess the possibility of publication bias, both a computer and manual search of *Dissertation Abstracts* were conducted. (These search procedures would be explained.)

Search procedures identified 367 reports evaluating child psychotherapy; 301 were from journals or books, and 66 were unpublished dissertations. These figures do not include 16 additional, relevant published studies that could not be evaluated because of insufficiently reported data. (It is helpful to know how many relevant studies could not be included and why.)

Coding of Studies

The outcome measures in the study sample fell into six major categories and were coded as behavioral observations, normed rating scales and checklists, nonnormed scales and checklists, various nonacademic performance measures, peer sociometrics, and measures of academic achievement. The treatment administered was coded as primarily behavioral, primarily nonbehavioral, or as cognitive–behavioral therapy. Finally, the child's presenting problem was classified as internalizing (shy, withdrawn), externalizing (acting-out, or aggressive), or mixed. (Examples of coded variables would be provided, and it is likely that other variables would be coded and explained in case they were needed for possible analysis.)

Reliability of Coding Procedures

Three research assistants were trained to code studies. One half of the studies were randomly selected, and kappa, a percentage agreement procedure corrected for chance (Cohen, 1960), was used to assess coding reliability. Coding reliabilities averaged 88% and ranged from 80% to 99%. (It is important to describe who did the coding, how reliability was estimated, and with what results.)

Unit of Analysis

Consistent with experimental hypotheses, separate ESs were calculated in each study for each of the six categories of outcome measures listed previously (behavioral observations, peer sociometrics, etc.). (In other words, the decision was to use each type of outcome measure as the unit of analysis to examine the hypothesis that type of outcome measure would influence ESs. Details on the mean number and range of outcomes per study would be provided.)

Calculation of Effects

ESs were computed using the pooled standard deviation of the treatment and control groups; positive scores indicated that the treated group was superior to the control group. Following Hedges and Olkin (1985), ESs were adjusted to correct for bias attributable to small sample size, and then weighting procedures were used to combine ESs from different studies to give greater weight to studies whose effects were more reliable (i.e., were based on larger sample sizes). (These adjustments and weighting procedures are important. In the former case, ESs that are based on small sample sizes [less than 30 subjects] in the original study yield overestimates of true effects and thus must be reduced accordingly. In the latter case, weighting procedures help adjust the data for sampling error; some form of sample weighting procedure should be used before any analyses are conducted.)

Results

Although it is customary to summarize characteristics of the reviewed studies, this information is not provided in this example. Such data would be important in judging, for instance, if the studies were unusual in any way and possessed appropriate data for the conduct of the analyses.

General Analytic Approach

Hedges and Olkin's (1985) categorical, fixed-effects analytic approach was followed. To help control for experimentwise error, a .01 probability level for statistical significance was used to evaluate whether the hypothesized variables (treatment, type of outcome measure, and child problem) moderated child therapy outcomes. Therefore, the first step was to test the homogeneity of ESs combined across all of the studies (N = 658). As expected, Q was large and significant (Q = 1,895.03, $p < .01$).

Table 4

Tests of Homogeneity Within Groups Defined by Type of Outcome Measure

Outcome measure	Q	n
Behavioral observations	474.68*	142
Peer sociometrics	70.18*	44
Normed rating scales	255.07*	116
Nonnormed rating scales	595.74*	187
Achievement tests	124.94*	67
Performance measures	284.06*	102
Q total	1,804.67*	658

*$p < .01$, indicating significant heterogeneity.

Results for the Outcome Measure

Because the Q statistic indicated heterogeneity for all reviewed studies combined and suggested the need to subdivide studies, my colleagues and I began to investigate our hypotheses. First, the ESs were divided into six groups according to the type of measure used to assess outcome. The results are presented in Table 4. The overall Q statistic ($Q = 1{,}404.67$, $p < .01$) remained significant, and within-group homogeneity was not found for any group of studies. The studies were then subdivided further to try to identify sources of heterogeneity. (Actually, the decision to continue to subdivide while retaining the type of outcome measure represents a judgment call in the analyses. It is supported by the finding that the overall Q statistic was reduced by approximately 20% when the type of outcome measure was considered. The Qs were 1,865 vs. 1,404, respectively. If there is no reduction in the variability of ESs when a variable is introduced into the analysis [i.e., the Q value is not reduced], the meta-analyst might elect to eliminate the variable from further consideration).

Results for Type of Treatment

The second classification variable entered into the model was the general type of treatment administered to the children (behavioral or nonbehavioral). Thus, a Q statistic was calculated for each of the 12 possible combinations of type of treatment crossed with type of outcome measure. (two treatments × six outcome measures). These data are presented in Table 5. This subdivision of studies was partially successful. Homogeneity was found in four of the cells, that is, the Q value was nonsignificant (rows 2 and 5). Furthermore, inspection of the homogeneity statistics and

Table 5

Tests of Homogeneity Within Groups Defined by Type of Outcome Measure and Type of Treatment

	Behavioral treatment		Nonbehavioral treatment	
Outcome measure	***Q***	***n***	***Q***	***n***
Behavioral observations	297.36*	106	85.42*	36
Peer sociometrics	24.80	16	41.41	28
Normed rating scales	137.51*	51	93.24*	65
Nonnormed rating scales	264.66*	94	191.85*	93
Achievement measures	57.24	39	35.78	28
Performance measures	222.36*	79	57.17*	23

Note. n = number of studies.
*$p < .01$, indicating significant heterogeneity.

their significance values indicate that this two variable model accounted for a large percentage of the variance in effects for the remaining cells. (In other words, the sizes of the Q statistics were close to those needed to reach the desired goal of nonsignificance; the data thus suggested that the combination of treatment and type of outcome measure in the analyses was making a substantial difference, but not producing the desired nonsignificance in all cells.)

Rather than introduce an additional moderator variable into the analysis at this point, it was decided to examine the data for potential outliers that might be skewing the results. To illustrate the distortion that might occur from outliers, return for a moment to examine the ES of Study 6 in Table 2. Using each study as the unit of analysis (column 4), the ES for Study 6 was 4 to almost 20 times larger than the ES for any other study. Retaining Study 6 in the sample yielded an overall mean ES of 0.66 with a standard deviation of 0.77. If Study 6 were removed as an outlier, however, the results would be changed dramatically; the mean ES for the remaining nine studies would be 0.43 (SD = 0.25). This is quite a difference.

Removing Outliers

ESs that were more than 2 *SD*s beyond the mean of their respective group (n = 35) were eliminated from the data set as outliers. (The criterion used to define outliers should be provided.) The homogeneity statistics were then recomputed for the cells where significant heterogeneity had occurred. Removal of the outliers made a major difference. The *Q* statistics presented in Table 6 are now all nonsignificant, indicating that the model of type of outcome measure by type of treatment created homogeneous populations of effects in all 12 cells.

Table 6 also shows the mean ESs and 99% confidence intervals for each group of studies. Each mean differs significantly from zero at the .01 level. (None of the confidence intervals contained zero.) However, the magnitude of effects varied as a function of both type of treatment and type of outcome measure. A discussion of how effects varied across treatment and outcome measure would be presented.

Inspection of the confidence intervals in Table 6 indicates that for three types of outcome measures (behavioral observations, nonnormed rating scales, and academic achievement measures), behavioral treatment produced significantly better results than did nonbehavioral treatment. (Once homogeneity within groups of studies is achieved, then comparisons between groups of studies are appropriate. Means differ significantly if their confidence intervals do not overlap. The corresponding treatment means in rows 2, 3, and 6 in Table 6 do overlap and thus do not differ significantly from each other.)

Findings for Cognitive-Behavioral Therapy

As noted earlier, the hypothesis for cognitive–behavioral therapy studies was that different outcomes would occur on the basis of the child's cognitive–developmental level. This finding would replicate prior research. There were 64 studies that used cognitive–behavioral therapy, and the findings are presented in Tables 7 and 8. (Note that each study is used as the unit of analysis, because the hypothesis was that cognitive–developmental level and not the type of outcome measure was going to influence clinical outcomes.) Results supported the hypothesis. When children are grouped according to cognitive–developmental level, using customary Piagetian age guidelines (preoperational level = under 7 years of age; concrete operational level = 7–11 years, and formal operations = 11–13 years old), each of the three subgroups of studies yielded homogeneous effects.

The mean posttreatment effect sizes in Table 8 are all significantly different from zero at the .05 level. (None of the confidence intervals contain zero.) Fail-safe *ns* were calculated in terms of the number of additional studies that would be needed to reduce each obtained mean effect to 0.20, a level that, by convention, represents a "small" effect in

Table 6

Tests of Homogeneity Within Groups Defined by Type of Outcome Measure and Type of Treatment and Mean Effects and Confidence Intervals for Each Group (Outliers Removed)

Outcome measure	Behavioral treatment	Nonbehavioral treatment
Behavioral observations		
Q	116.48	54.07
Mean ES	0.65_a	0.25_b
n	90	34
99% CI	0.54–0.75	0.16–0.33
Peer sociometrics[a]		
Q	24.80	41.44
Mean ES	0.43	0.25
n	16	28
99% CI	0.20–0.65	0.12–0.39
Normed rating scales		
Q	62.88	56.99
Mean ES	0.47	0.24
n	45	61
99% CI	0.31–0.62	0.15–0.32
Nonnormed rating scales		
Q	108.93	92.79
Mean ES	0.62_a	0.19_b
n	81	84
99% CI	0.52–0.73	0.14–0.24
Achievement measures[a]		
Q	57.24	35.78
Mean ES	0.45_a	0.18_b
n	39	28
99% CI	0.32–0.57	0.09–0.26
Performance measures		
Q	86.18	21.30
Mean ES	0.54	0.43
n	66	21
99% CI	0.43–0.66	0.25–0.60

Note. If the confidence interval (CI) does not contain zero, then the mean effect size (ES) is significant at the .01 level. Means in the same row that have different subscripts differ significantly at the .01 level.
[a]These cells were homogeneous without removing outliers.

Table 7

Summary of *Q* Statistics for Cognitive–Behavioral Therapy as a Function of the Child's Cognitive Developmental Level

Source	χ^2	*df*
Q total	101.48*	63
Between groups	34.58*	2
Within groups	66.90	61
Preoperational	8.83	8
Concrete operations	47.16	45
Formal operations	10.91	8

Note. The data from this table were taken from "Effectiveness of Cognitive–Behavior Therapy for Maladapting Children: A Meta-Analysis" by J. A. Durlak, T. Fuhrman, and C. Lampman, 1991, *Psychological Bulletin, 110,* pp. 208–209.

the social and behavioral sciences (Cohen, 1988). These fail-safe *n*s indicated that the most confidence can be placed in the findings for the concrete operational group. (In other words, it is probably unlikely that 80 additional studies involving 7–11-year-old children would be found in an extensive literature search to change the obtained results for this group.) Examination of between-groups differences indicated that only the mean ESs for the 11–13-year-olds and 7–11-year-olds differed significantly. (These confidence intervals do not overlap.)

Table 7 was prepared to illustrate how the *Q* statistical analysis is analogous to an ANOVA in the sense that *Q* total = *Q* between-groups

Table 8

Mean ESs, 95% Confidence Intervals, and Fail-Safe *n*s[a] for Groups

Source	*n*	*M*	95% confidence interval	Fail-safe *n*
All groups	64	0.56	0.46–0.66	115
Preoperational	9	0.57_{ab}	0.27–0.72	17
Concrete operations	46	0.55_{a}	0.44–0.60	80
Formal operations	9	0.92_{b}	0.61–1.30	32

Note. ESs = effect sizes. Means with different subscripts differ significantly at $p < .05$ in the Tukey honestly significant difference comparison. The data for this table were taken from "Effectiveness of Cognitive–Behavior Therapy for Maladapting Children: A Meta-Analysis" by J. A. Durlak, T. Fuhrman, and C. Lampman, 1991, *Psychological Bulletin, 110,* pp. 208–209.
[a]Number of additional studies needed to reduce the mean effect size to 0.20.

+ Q within-groups (see chapter 5 in this book). The logic of the analysis is as follows: Q total for all 64 studies indicates significant heterogeneity (i.e., there is a wide distribution of ESs, suggesting that there is more than one population of studies represented in the 64 studies). The Q statistics for studies involving children at different developmental levels were found to be nonsignificant in each case, indicating a single population of effects was represented in each of the three groups and confirming the hypothesis that developmental level influences therapeutic outcomes. Similarly, the value for Q between groups was significant, suggesting that differences existed among the three groups. Although children in the preoperational stage had virtually the same ES as those in concrete operations, the overlapping confidence intervals indicated that outcomes for the preoperational children and children in formal operations narrowly failed to reach statistical significance.

Examination of Outliers

Two patterns appeared among the outliers that are worth noting. In one cell involving nonbehavioral treatment, six of eight outliers involved relationship treatment for early detected school maladjustment. These six studies achieved a mean ES of 1.28, compared with the other nonbehavioral studies in this cell (row 4 of Table 6) that collectively obtained a mean effect of only 0.19. When behavioral treatment was evaluated using nonnormed measures, 10 outliers used the bell and pad conditioning procedure (e.g., DeLeon & Mandell, 1966) to treat enuretic children. These interventions were highly successful (mean ES = 2.52) compared with a mean ES of 0.62 for the other behavioral studies in this cell (row 4). (Although it may be necessary to identify outliers and then exclude them from the primary analyses because of their extreme values, outliers should subsequently be examined for their possible heuristic value. At issue is whether these outliers simply reflect aberrant results or whether these investigations might provide clues suggesting particularly effective or ineffective programs. In the current example, inspection suggested two unusually successful forms of treatment occurring among the reviewed treatments; the meta-analyst would discuss these studies and their possible implications.)

Before reading the next section, the reader is asked to compare the previous results with the previous explanatory material concerning meta-analysis. If a basic understanding of meta-analysis has been attained, the reader should be able to anticipate the major conclusions offered in the

Discussion. Moreover, there is one important omission in the Results section: What additional analysis should be reported?

Discussion of the Hypothetical Example

As predicted, type of outcome measure and type of treatment were significant moderators of treatment effects. These conclusions are based on the empirical demonstration that the use of these variables permitted an initially hetergeneous sample of 367 studies to be divided into 12 homogeneous subgroups in terms of outcomes. Furthermore, homogeneity of effects was not obtained until both variables were entered into the analyses. Therefore, the variables must be considered simultaneously when interpreting findings in much the same way that significant interactions are interpreted when they occur in ANOVA analyses. This qualification is important; the implication is that in order to understand the effects of child treatment, one must know what general type of treatment was administered and how outcome was assessed; both dimensions are important.

With respect to type of treatment, although behavioral therapies tended to produce larger effects than nonbehavioral interventions, significant between-groups differences emerged only in three of six possible cells (see rows 1, 4, and 5 in Table 6). It is important to note that specific types of behavioral and nonbehavioral treatments were not evaluated. The data do not indicate whether some specific forms of either treatment are more effective than others. This is an important limitation in practical terms in the sense that readers cannot discern exactly what the therapists did with the children.

Analyses failed to confirm the importance of the presenting clinical problem as an influence on therapy outcomes. In other words, the child's presenting problem never entered the analyses because it was not needed to obtain homogeneity. The implication is that children with externalizing, internalizing, and mixed problems respond similarly to each form of treatment and display similar changes regardless of which type of measure is used to assess outcome.

Finally, results replicated prior findings that the child's cognitive–developmental level affects the impact of cognitive–behavior therapy. Children in the formal operational period (aged 11–13 years old) appear to benefit almost twice as much from cognitive–behavioral therapy as do younger children (5–7 or 7–11 years old). Mean ESs were 0.92 versus 0.57 and 0.55, respectively.

What Is Missing From the Analysis?

No attempt was made to assess the influence of methodological quality or the possibility of publication bias. At the least, the influence of methodological quality on ESs should be examined. Design features of child therapy studies, such as those listed in Table 1 (random assignment and type of control group), could be entered into a regression analysis to see how much of the variance in effects is accounted for by methodological features. Alternatively, effects from studies involving random assignment could be compared with those not having this feature. Similarly, effects generated from published and unpublished studies should be compared. If unpublished studies yield lower effect sizes, then their distribution should be examined in each of the cells of the final analysis. Perhaps the cells for nonbehavioral studies contain more dissertations than those for behavioral treatment; this would account for the former's lower mean effects. The major goal of the previous analyses was to determine whether methodological features or publication status could serve as a possible alternative explanation for any obtained findings.

Chapter Summary

In this chapter I have summarized the main features of meta-analysis, which is a multivariate statistical method of integrating and analyzing research findings. Explanatory meta-analyses that analyze standardized group mean differences were discussed. In these situations, meta-analysts strive to identify what significant relationships exist between study characteristics (the independent variables) and outcomes (expressed as an ES and representing the dependent variable). Although there is no single, standardized approach to be taken in conducting a meta-analysis, several major issues and decisions confront the meta-analyst at each important step in the process. These steps include formulating clear research questions prior to the meta-analysis, searching the literature for relevant studies, coding of studies, calculating the index of effect, conducting the appropriate statistical analyses, and offering conclusions and interpretations of the data. Hopefully, this chapter has desmystified the complexities of meta-analysis and has been helpful for those who either wish to read further on the topic or who need assistance in understanding and interpreting the many meta-analyses that now appear in the medical, social, and behavioral sciences.

Suggestions for Further Readings

In addition to other references cited in this chapter, the following are highly recommended. The two best places to begin are Cook and Leviton (1980) and Light and Pillemer (1984). The former compares meta-analysis with traditional narrative research reviews and illustrates the potential strengths and limitations of each method; the latter is an excellent text that describes meta-analysis in nontechnical terms. Once familiarity with meta-analysis in general is attained, the following describe several features of meta-analysis in greater detail: Durlak and Lipsey (1991); Glass, McGaw, and Smith (1981); Hunter and Schmidt (1990); Light (1983); and Wolf (1986). In particular, Glass et al. (1981) present an extended discussion of coding procedures, Hunter and Schmidt (1990) discuss meta-analyses involving correlations, and Light (1983) presents meta-analyses conducted in several disciplines. Finally, the Russell Sage Foundation casebook (Cook et al., 1992) and some special issues of journals (Garfield, 1983; Michelson, 1985) contain a compendium of views about meta-analysis and its applications.

Glossary

ADJUSTED (OR UNBIASED) EFFECT Effects that have been reduced slightly in magnitude because they were based on small samples. In practical terms, adjustments are necessary only for sample sizes less than 30.

CODING Method by which study features (types of subjects, control group, etc.) are transformed into quantitative independent variables for the statistical analyses. Coding procedures usually yield a combination of continous variables representing some study features (e.g., the number of medical symptoms) and categorical variables representing other features (e.g., surgery only = 1; medication only = 2; surgery plus medication = 3; no treatment = 4).

CONFIDENCE INTERVAL (CI) The range of effects around the obtained mean that might occur at a particular probability level given the presence of error and variation in the calculation of effect sizes (ESs) from a sample of studies. If the CI does not contain zero (e.g., a mean ES of 0.45 has a CI ranging from 0.20 to 0.70), then the mean ES is considered to be statistically different from zero.

EFFECT SIZE (ES, *d* OR *g*) The quantitative dependent variable in meta-analyses that indicates the strength of association between study features and outcomes. Will appear as a standardized mean difference in group treatment (treatment effectiveness) meta-analyses.

FAIL-SAFE *n* Number of additional studies needed to reduce an obtained effect to a particular criterion level. Used to gauge the confidence one can place in current findings in the sense of indicating many more studies with low ESs must be found to reduce obtained findings to level of nonsignificance or to a very low ES. Author should specify the magnitude of the additional effects (usually zero) and the criterion level (usually statistical nonsignificance or an ES of 0.20) used in fail-safe formula. (The formula is available in Wolf, 1986, p. 39.)

FILE DRAWER PROBLEM The tendency for authors not to submit and journal editors not to accept for publication the results of experiments that fail to achieve statistically significant findings.

HOMOGENEITY–HETEROGENEITY The variability in the distribution of ESs occurring in a group of studies. Generally speaking, distributions of ESs whose variability is due only to sampling error are considered to be homogeneous, whereas distributions that reflect both sampling error and systematic differences among the studies are heterogenous.

MODERATOR VARIABLE A variable (study characteristic) that accounts for significant variability in effect sizes among reviewed studies; studies that differ on a presumed moderator variable thus should differ in the magnitude of their effect sizes.

n Unless otherwise indicated, *n* in a meta-analysis refers to the number of studies being reviewed, whereas *N* refers to the number of subjects in the reviewed studies.

OUTLIER An extreme value in a distribution of effects. Authors should specify how an outlier is defined. Even though outliers are excluded from eventual analyses so as not to distort the findings, they should be inspected for their heuristic value: Outliers might suggest particularly successful or unsuccessful programs.

PUBLICATION BIAS Bias refers to published studies yielding higher effects than unpublished reports (dissertations, technical reports, or convention papers). The possibility of bias cannot be estimated or corrected statistically; it should be determined empirically by comparing findings from published and unpublished investigations.

Q STATISTIC The *Q* statistic is a statistical test used to confirm whether the meta-analyst's chosen model provides a good fit for the data. *Q* is distributed as a chi-square variable with degrees of freedom equal to the number of studies minus 1. A nonsignificant *Q* statistic indicates homogeneity and suggests that the studies come from the same population.

STUDY ARTIFACTS Any type of methodological, statistical, or measurement error or bias that is present in studies.

WEIGHTED EFFECTS Weighting effects differentially based on the number of subjects on whom the effect is based. Effects from larger samples receive greater weight because they yield more accurate estimations of true population effects. Weighting helps adjust for sampling error and should occur prior to all statistical analyses.

References

Bangert-Drowns, R. L. (1986). Review of developments in meta-analytic method. *Psychological Bulletin, 101*, 213–232.

Cohen, J. (1960). A coefficient of agreement for nominal scales. *Educational and Psychological Measurement, 20*, 37–46.

Cohen, J. (1988). *Statistical power analysis for the behavioral sciences* (2nd ed.). Hillsdale, NJ: Erlbaum.

Cook, T. D., Cooper, H., Cordray, D. S., Hartman, H., Hedges, L. V., Light, R. J., Louis, T. A., & Mosteller, F. (1992). *Meta-analysis for explanation: A casebook*. New York: Russell Sage Foundation.

Cook, T. D., & Leviton, L. C. (1980). Reviewing the literature: A comparison of traditional methods with meta-analysis. *Journal of Personality, 48*, 449–472.

DeLeon, G., & Mandell, W. (1966). A comparison of conditioning and psychotherapy in the treatment of functional enuresis. *Journal of Clinical Psychology, 22*, 326–330.

Durlak, J. A., Fuhrman, T., & Lampman, C. (1991). Effectiveness of cognitive–behavior therapy for maladapting children: A meta-analysis. *Psychological Bulletin, 110*, 204–214.

Durlak, J. A., Lampman, C., Wells, A., & Carmody, J. (1990, May). *Effectiveness of child psychotherapy: A meta-analytic review*. Paper presented at the meeting of the Midwestern Psychological Association, Chicago.

Durlak, J. A., & Lipsey, M. W. (1991). A practitioner's guide to meta-analysis. *American Journal of Community Psychology, 19*, 291–332.

Garfield, S. L. (Ed.). (1983). Special section: Meta-analysis and psychotherapy. *Journal of Consulting and Clinical Psychology, 51*, 3–75.

Glass, G. V., McGaw, B., & Smith, M. L. (1981). *Meta analysis in social research*. Newbury Park, CA: Sage.

Hartmann, D. P. (Ed.). (1982). *Using observers to study behavior: New directions for methodology of social and behavioral sciences*. San Francisco: Jossey-Bass.

Hedges, L. V., & Olkin, I. (1985). *Statistical methods for meta-analysis*. San Diego, CA: Academic Press.

Holmes, C. T. (1984). Effect size estimation in meta-analysis. *Journal of Experimental Education, 52*, 106–109.

Hunter, J. E., & Schmidt, F. L. (1990). *Methods of meta-analysis: Correcting errors and bias in research findings*. Newbury Park, CA: Sage.

Light, R. J. (Ed.). (1983). *Evaluation Studies Review Annual: Vol. 8. Meta-analysis*. Beverly Hills, CA: Sage.

Light, R. J., & Pillemer, D. B. (1984). *Summing up: The science of reviewing research*. Cambridge, MA: Harvard University Press.

Michelson, L. (Ed.). (1985). Meta-analysis and clinical psychology [Special issue]. *Clinical Psychology Review, 5* (1).

Orwin, R. G. (1983). A fail-safe *N* for effect size in meta-analysis. *Journal of Educational Statistics, 8*, 157–159.

Posovac, E. J., & Miller, T. Q. (1990). Some problems caused by not having a conceptual foundation for health research: An illustration from studies of the psychological effects of abortion. *Psychology and Health, 5*, 13–23.

Rosenthal, R. (1979). The "file drawer problem" and tolerance for null results. *Psychological Bulletin, 86*, 638–641.

Rosenthal, R., & Rubin, D. B. (1982). A simple, general purpose display of magnitude of experimental effect. *Journal of Educational Psychology, 74*, 166–169.

Smith, M. L. (1980). Publication bias and meta-analysis. *Evaluation and Education, 4*, 22–24.

Stock, W. A., Okun, M. A., Haring, M. J., Miller, W., Kinney, C., & Seurvorst, R. W. (1982). Rigor in data synthesis: A case study of reliability in meta-analysis. *Educational Researcher, 11*, 10–14.

Suls, J., & Wan, C. K. (1989). Effects of sensory and procedural information on coping with stressful medical procedures and pain: A meta-analysis. *Journal of Consulting and Clinical Psychology, 57*, 372–379.

Trull, T. J., Nietzel, M. T., & Main, A. (1988). The use of meta-analysis to assess the clinical significance of behavior therapy for agoraphobia. *Behavior Therapy, 19*, 527–538.

Wolf, F. M. (1986). *Meta-analysis: Quantitative methods for research synthesis*. Newbury Park, CA: Sage.

Index

D

E

F

G

H

K

L

M

P

Q

R

S

T

U

V

W

X

Y

Z

About the Editors

Laurence G. Grimm received his PhD in clinical psychology from the University of Illinois at Champaign-Urbana. He is currently associate professor of psychology at the University of Illinois at Chicago, where he serves as the director of clinical training. He has published broadly in the area of clinical psychology and is the author of *Statistical Applications for the Behavioral Sciences*, published by Wiley.

Paul R. Yarnold received his PhD in academic social psychology from the University of Illinois at Chicago. He is currently research associate professor of medicine at Northwestern University Medical School, Division of General Internal Medicine. He is also adjunct associate professor of psychology at the University of Illinois at Chicago. He serves on the editorial board of *Educational and Psychological Measurement* and has authored over 80 articles in the areas of medicine, psychology, and statistics. He is the coauthor of the statistical software packages, *Optimal Data Analysis*, *GenReg*, and *Tree*.